Methods in Molecular Biology™

AF787765

Series Editor
John M. Walker
School of Life Sciences
University of Hertfordshire
Hatfield, Hertfordshire, AL10 9AB, UK

For further volumes:
http://www.springer.com/series/7651

Plant Metabolomics

Methods and Protocols

Edited by

Nigel W. Hardy

Department of Computer Sciences, Aberystwyth University, Aberystwyth, UK

Robert D. Hall

Plant Research International, Wageningen, The Netherlands;
Centre for BioSystems Genomics, Wageningen, The Netherlands;
Netherlands Metabolomics Centre, Leiden, The Netherlands

Humana Press

Editors
Nigel W. Hardy
Department of Computer Sciences
Aberystwyth University
Aberystwyth, UK

Robert D. Hall
Plant Research International
Wageningen, The Netherlands
and
Centre for BioSystems Genomics
Wageningen, The Netherlands
and
Netherlands Metabolomics Centre
Leiden, The Netherlands

ISSN 1064-3745 e-ISSN 1940-6029
ISBN 978-1-4939-5748-4 ISBN 978-1-61779-594-7 (eBook)
DOI 10.1007/978-1-61779-594-7
Springer New York Dordrecht Heidelberg London

© Springer Science+Business Media, LLC 2012
Softcover re-print of the Hardcover 1st edition 2012
All rights reserved. This work may not be translated or copied in whole or in part without the written permission of the publisher (Humana Press, c/o Springer Science+Business Media, LLC, 233 Spring Street, New York, NY 10013, USA), except for brief excerpts in connection with reviews or scholarly analysis. Use in connection with any form of information storage and retrieval, electronic adaptation, computer software, or by similar or dissimilar methodology now known or hereafter developed is forbidden.
The use in this publication of trade names, trademarks, service marks, and similar terms, even if they are not identified as such, is not to be taken as an expression of opinion as to whether or not they are subject to proprietary rights.

Printed on acid-free paper

Humana Press is part of Springer Science+Business Media (www.springer.com)

Preface

Estimation of the metabolite complement of plant material involves a wide range of techniques and technologies and that breadth continues to increase. The plant metabolome is both highly complex and highly dynamic and its measurement requires very careful control of "noise", since biological, experimental, and technical variability at all stages of the experimental workflow threaten to overwhelm the biological signals. The workflow must start with detailed and statistically justified experimental design leading to careful identification and preparation of study material followed by harvest and quenching of metabolism. Metabolomics research typically involves multiple sites for material preparation and analysis and most investigations are "high throughput", meaning that chemical analysis of sample sets are inevitably carried out over an extended period of time. These factors mean that well-validated procedures for shipping and storage of biological materials are required prior to application of one or more of the wide range of chemical analysis techniques which yield highly multivariate metabolomic data. A range of data analyses procedures must be applied to these data, starting with data cleaning and alignment (pre-processing), proceeding possibly to chemical identification and finally to statistical modelling designed to produce justifiable and biologically relevant results. Across all stages of this workflow, up to and including the statistical analysis, accurate and detailed collection of meta-data are also essential for good process management, to satisfy reporting requirements and to ensure wider interpretability and reuse (durability) of results. This volume therefore presents methods for all the stages of the plant metabolomics workflow.

Aberystwyth, UK — *Nigel W. Hardy*
Wageningen, The Netherlands — *Robert D. Hall*

Preface

Acknowledgements

The origins of this book lie within the activities of the EU project META-PHOR (www.meta-phor.eu) where a large number of European metabolomics technology partners have been collaborating on method development. Content for the majority of the chapters has therefore been derived from this project and the others have been provided by experts in complementary fields to provide full coverage. We would therefore like to thank the European Union and the 22 project partners for financially supporting the META-PHOR project (FOOD-CT-2006-036220) and for making this book possible. RDH also acknowledges financial support from the Centre for BioSystems Genomics and the Netherlands Metabolomics Centre, both initiatives under the jurisdiction of the Netherlands Genomics Initiative. NWH acknowledges support from Aberystwyth University. The editors would like to thank Helen Jenkins for her work in the preparation of the book.

SIXTH FRAMEWORK PROGRAMME

META-PHOR

CBSG | 2012
Centre for BioSystems Genomics

Netherlands Metabolomics Centre

PRIFYSGOL ABERYSTWYTH UNIVERSITY

Acknowledgements

The origins of the book are within a number of [illegible]

[illegible]

Contents

Contributors

ASAPH AHARONI • *Department of Plant Sciences, The Weizmann Institute of Science, Rehovot, Israel*
J. WILLIAM ALLWOOD • *IBERS – Institute of Biological, Environmental and Rural Sciences, Aberystwyth University, Aberystwyth, UK; School of Chemistry, Manchester Interdisciplinary Biocentre, The University of Manchester, Manchester, UK*
JOHN M. BAKER • *National Centre for Plant and Microbial Metabolomics, Plant Science Department, Rothamsted Research, Hertfordshire, UK*
AHMED HMAIDAN BANIMUSTAFA • *Department of Computer Science, Aberystwyth University, Aberystwyth, UK*
MICHAEL H. BEALE • *National Centre for Plant and Microbial Metabolomics, Plant Science Department, Rothamsted Research, Hertfordshire, UK*
MANFRED BECKMANN • *IBERS - Institute of Biological, Environmental and Rural Sciences, Aberystwyth University, Aberystwyth, UK*
STÉPHANE BERNILLON • *INRA, Centre INRA de Bordeaux, Villenave d'Ornon, France; Metabolome-Fluxome platform of Bordeaux Functional Genomics Centre, Centre INRA de Bordeaux, Villenave d'Ornon, France*
BENOÎT BIAIS • *INRA, Centre INRA de Bordeaux, Villenave d'Ornon, France*
JOSEPH BURGER • *Department of Vegetable Research, Agricultural Research Organization, Newe Ya'ar, Ramat Yishay, Israel*
CÉCILE CABASSON • *Université de Bordeaux, Centre INRA de Bordeaux, Villenave d'Ornon, France*
DELIA-IRINA COROL • *National Centre for Plant and Microbial Metabolomics, Plant Science Department, Rothamsted Research, Hertfordshire, UK*
CATHARINA CRONE • *Thermo Fisher Scientific, Bremen, Germany*
NICOLAIE EUGEN DAMOC • *Thermo Fisher Scientific, Bremen, Germany*
RIC C.H. DE VOS • *Plant Research International, Wageningen, The Netherlands; Centre for BioSystems Genomics, Wageningen, The Netherlands; Netherlands Metabolomics Centre, Leiden, The Netherlands*
CATHERINE DEBORDE • *INRA, Centre INRA de Bordeaux, Villenave d'Ornon, France; Metabolome-Fluxome platform of Bordeaux Functional Genomics Centre, Centre INRA de Bordeaux, Villenave d'Ornon, France*
PHUC THI DO • *Max-Planck-Institut für Molekulare Pflanzenphysiologie, Potsdam-Golm, Germany*
JOHN DRAPER • *IBERS - Institute of Biological, Environmental and Rural Sciences, Aberystwyth University, Aberystwyth, UK*
ALEXANDER ERBAN • *Max-Planck-Institut für Molekulare Pflanzenphysiologie, Potsdam-Golm, Germany*
ALISDAIR R. FERNIE • *Max-Planck-Institut für Molekulare Pflanzenphysiologie, Potsdam-Golm, Germany*
KARL FRASER • *AgResearch Ltd., Grasslands Research Centre, Palmerston North, New Zealand*
YVES GIBON • *INRA, Centre INRA de Bordeaux, Villenave d'Ornon, France*

ROYSTON GOODACRE • *School of Chemistry, Manchester Interdisciplinary Biocentre, The University of Manchester, Manchester, UK; Manchester Centre for Integrative Systems Biology, Manchester Interdisciplinary Biocentre, The University of Manchester, Manchester, UK*

ROBERT D. HALL • *Plant Research International, Wageningen, The Netherlands; Centre for BioSystems Genomics, Wageningen, The Netherlands; Netherlands Metabolomics Centre, Leiden, The Netherlands*

T.H. HANSEN • *Plant and Soil Science Laboratory, Department of Agriculture and Ecology, The University of Copenhagen, Copenhagen, Denmark*

NIGEL W. HARDY • *Department of Computer Science, Aberystwyth University, Aberystwyth, UK*

JIM HEALD • *IBERS – Institute of Biological, Environmental and Rural Sciences, Aberystwyth University, Aberystwyth, UK*

MARTIN HORNSHAW • *Thermo Fisher Scientific, Hemel Hempstead, UK*

S. HUSTED • *Plant and Soil Science Laboratory, Department of Agriculture and Ecology, The University of Copenhagen, Copenhagen, Denmark*

HARRY JONKER • *Plant Research International, Wageningen, The Netherlands; Centre for BioSystems Genomics, Wageningen, The Netherlands*

JOACHIM KOPKA • *Max-Planck-Institut für Molekulare Pflanzenphysiologie, Potsdam-Golm, Germany*

GEOFFREY A. LANE • *AgResearch Ltd., Grasslands Research Centre, Palmerston North, New Zealand*

K.H. LAURSEN • *Plant and Soil Science Laboratory, Department of Agriculture and Ecology, The University of Copenhagen, Copenhagen, Denmark*

JENNIE LEWIS • *National Centre for Plant and Microbial Metabolomics, Plant Science Department, Rothamsted Research, Hertfordshire, UK*

AIMEE M. LLEWELLYN • *National Centre for Plant and Microbial Metabolomics, Plant Science Department, Rothamsted Research, Hertfordshire, UK*

AMANDA J. LLOYD • *IBERS – Institute of Biological, Environmental and Rural Sciences, Aberystwyth University, Aberystwyth, UK*

ARJEN LOMMEN • *RIKILT–Institute of Food Safety, Wageningen, The Netherlands*

ALEXANDER LUEDEMANN • *Max-Planck-Institut für Molekulare Pflanzenphysiologie, Potsdam-Golm, Germany*

WADE MACE • *AgResearch Ltd., Grasslands Research Centre, Palmerston North, New Zealand*

SONIA J. MILLER • *National Centre for Plant and Microbial Metabolomics, Plant Science Department, Rothamsted Research, Hertfordshire, UK*

SOFIA MOCO • *Laboratory of Biochemistry, Wageningen University, Wageningen, The Netherlands*

THOMAS MOEHRING • *Thermo Fisher Scientific, Bremen, Germany*

ANNICK MOING • *INRA, Centre INRA de Bordeaux, Villenave d'Ornon, France; Metbolome-Fluxome platform of Bordeaux Functional Genomics Centre, Centre INRA de Bordeaux, Villenave d'Ornon, France*

HELMUT MUENSTER • *Thermo Fisher Scientific, Bremen, Germany*

LUIS A.J. MUR • *IBERS – Institute of Biological, Environmental and Rural Sciences, Aberystwyth University, Aberystwyth, UK*

MADALINA OPPERMANN • *Thermo Fisher Scientific, Kungens Kurva, Sweden*

SONIA OSORIO • *Max-Planck-Institut für Molekulare Pflanzenphysiologie, Potsdam-Golm, Germany*

DAVID PARKER • *IBERS - Institute of Biological, Environmental and Rural Sciences, Aberystwyth University, Aberystwyth, UK*
ANTHONY J. PARSONS • *AgResearch Ltd., Grasslands Research Centre, Palmerston North, New Zealand*
D.P. PERSSON • *Plant and Soil Science Laboratory, Department of Agriculture and Ecology, The University of Copenhagen, Copenhagen, Denmark*
SUSANNE RASMUSSEN • *AgResearch Ltd., Grasslands Research Centre, Palmerston North, New Zealand*
ILANA ROGACHEV • *Department of Plant Sciences, The Weizmann Institute of Science, Rehovot, Israel*
DOMINIQUE ROLIN • *Université de Bordeaux, Centre INRA de Bordeaux, Villenave d'Ornon, France*
ARTHUR A. SCHAFFER • *Department of Vegetable Research, Agricultural Research Organization, Volcani Center, Bet-Dagan, Israel*
BERT SCHIPPER • *Plant Research International, Wageningen, The Netherlands; Centre for BioSystems Genomics, Wageningen, The Netherlands; Netherlands Metabolomics Centre, Leiden, The Netherlands*
J.K. SCHJOERRING • *Plant and Soil Science Laboratory, Department of Agriculture and Ecology, The University of Copenhagen, Copenhagen, Denmark*
YAAKOV TADMOR • *Department of Vegetable Research, Agricultural Research Organization, Newe Ya'ar, Ramat Yishay, Israel*
HARRIE A. VERHOEVEN • *Plant Research International, Wageningen, The Netherlands*
JACQUES VERVOORT • *Laboratory of Biochemistry, Wageningen University, Wageningen, The Netherlands*
LUISE VON MALOTKY • *Max-Planck-Institut für Molekulare Pflanzenphysiologie, Potsdam-Golm, Germany*
JANE L. WARD • *National Centre for Plant and Microbial Metabolomics, Plant Science Department, Rothamsted Research, Hertfordshire, UK*
HONG XUE • *AgResearch Ltd., Grasslands Research Centre, Palmerston North, New Zealand*

Chapter 1

Practical Applications of Metabolomics in Plant Biology

Robert D. Hall and Nigel W. Hardy

Abstract

The technologies being developed for the large-scale, essentially unbiased analysis of the small molecules present in organic extracts made from plant materials are greatly changing our way of thinking about what is possible in plant biology. A range of different separation and detection techniques are being refined and expanded and their combination with advanced data management and data analysis approaches is already giving plant scientists far deeper insights into the complexity of plant metabolism and plant metabolic composition than was imaginable just a few years ago. This field of "metabolomics", while still in its infancy, has nevertheless already been welcomed with open arms by the plant science community, partly because of these said advantages but also because of the broad potential applicability of the approaches in both fundamental and applied science. The diversity in application already ranges from understanding the considerable complexity of primary metabolic networks in Arabidopsis, to the changes which occur in the biochemical composition of foods occurring, for example, during the Pasteurization of tomato purée for long-term storage or the boiling of Basmati rice for direct consumption. The insights being gained are revealing valuable information on the strict control yet flexible nature of plant metabolic networks in many different systems. This volume aims to give a comprehensive overview of the approaches available for the performance of a "typical" plant metabolomics experiment, the choice of analytical techniques and to offer warnings on the potential pitfalls in experimental design and execution.

Key words: Technologies, Challenges, Data generation, Data analysis, Applications, Sample preparation

1. Introduction

In modern daily life, the influence of new technologies, enhanced computational capacity, developments in information technology, etc. has had a phenomenal effect on how we now live and work. Things which were considered, just 20–30 years ago, as being in the realms of "science fiction", such as mobile phones, Internet, ready access to terabyte or petabyte computing capacity, etc. have now become reality and have revolutionized how we approach

Nigel W. Hardy and Robert D. Hall (eds.), *Plant Metabolomics: Methods and Protocols*, Methods in Molecular Biology, vol. 860, DOI 10.1007/978-1-61779-594-7_1, © Springer Science+Business Media, LLC 2012

many different tasks. The same is true in many areas of science where similar or equivalent improvements in technologies, equipment, infrastructure, computing capacity, and bioinformatics tools have opened up new opportunities and have even generated new fields of scientific research. One such technology is metabolomics, where the combined advances in hardware, required for reliable and accurate metabolite separation and detection, and its associated software for subsequent data storage, treatment, and analysis have been of great benefit to progress in the field of the biochemical analysis of biological materials. This more detailed knowledge of the biochemistry of living tissues is, in turn, giving us a much deeper insight into how biochemical pathways function, how they are interconnected in complex, interactive networks and how these are, in turn, strictly controlled yet in the context of a significant metabolic flexibility. Adoption and exploitation of the technology has been rapid both in plant science and beyond—in Fig. 1 an analysis is presented of the steady growth in publication frequency on the topic.

The concept: Metabolomics is the technology which has been, and is still being, developed to assist in the biochemical analysis of complex mixtures. The ultimate aim is to have a technology which permits essentially unbiased, quantitative biochemical analysis of all the components in an extract of a biological material. Key to this is having a functional combination of comprehensiveness, analytical precision, and sample throughput (for a glossary of terms, please see Table 1). While full metabolite quantification is perhaps a Holy Grail, in many cases, semi-quantification or even relative values of one sample to another may be sufficient (1). In plants, the challenge is particularly daunting, as plants are renowned for the diversity of the chemicals they can produce and the complexity of the individual molecules involved (2). The chemical composition of plant tissues is also highly dependent both on internal (genetic) factors as well as external (environmental) factors, all of which must also be placed within the concept of tissue differences, where even adjacent cell layers may contain highly contrasting biochemical profiles (3). The epidermis of a tomato fruit for example is biochemically very different from the parenchyma layers just beneath (4). The inner flesh of a melon fruit is significantly different in terms of acids and sugars, compared to the outer flesh (5). The composition of a seed coat is usually in total contrast to the embryo it encases and indeed the embryo itself comprises root, hypocotyl and cotyledonous tissues which are also biochemically distinct, often even to the naked eye (6). Metabolomics is helping us to bring these diverse biochemical differences better into view (7).

The applications: As already intimated above, the areas in which plant metabolomics has already taken a foothold are extensive and diverse. There are many publications detailing where metabolomics

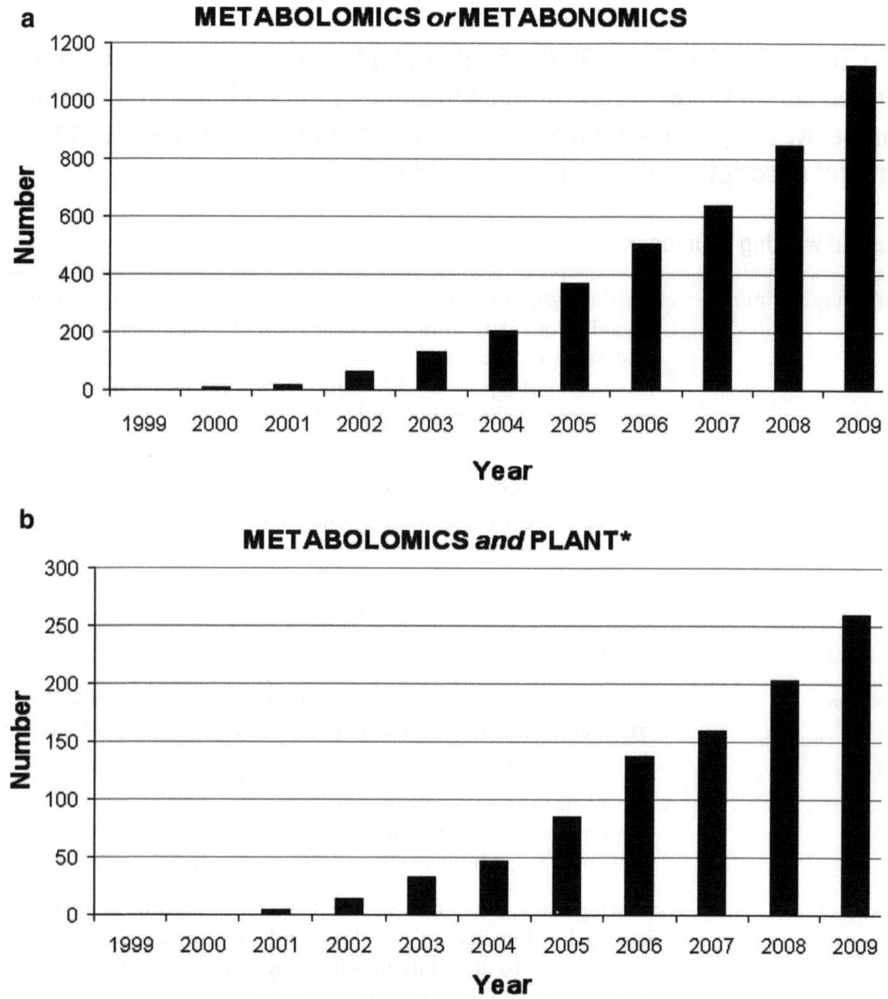

Fig. 1. A recent literature survey of the numbers of publications including (**a**) the terms "*metabolomics and/or metabonomics*" and (**b**) the terms "*metabolomics and plant**" per year since the paper of Oliver et al. (32).

approaches have been successfully applied in the field of fundamental plant biology and plant physiology. *Arabidopsis* has frequently have been used (8), but certainly not exclusively. Tomato is a regular subject both in terms of primary (9) and secondary metabolism related to both fruit development and the influence of genetics and environment on metabolite content (10). Other examples relate to for example stress physiology and investigating how plants cope with abiotic factors such as temperature, light, and salt stress (11), as well as biotic factors including fungal (12) and insect (13) pests. On the applied side, metabolomics has attracted much attention particularly in the field of food science and food quality. Changes during food preparation (14) and storage have been followed as have the effects of genotype and environment on fresh food quality (15).

Table 1
Readers becoming familiar with the field of plant metabolomics will be regularly confronted with a range of new terms which may at first glance appear rather similar. Below are given a number of the most common terms used together with a brief description of their meaning. (Modified from ref. 1)

Some useful working definitions

Metabolic fingerprinting	High-throughput qualitative screening of the metabolic composition of an organism or tissue with the primary aim of sample comparison and discrimination analysis. Generally no attempt is initially made to identify the metabolites present. All steps from sample preparation, separation, and detection should be rapid and as simple as is feasible. Often used as a forerunner to metabolic profiling.
Metabolic profiling	Identification and quantification of the metabolites present in an organism. For practical reasons this is generally only feasible for a limited number of components which are generally chosen on the basis of discriminant analysis or on molecular relationships based upon molecular pathways or networks.
Metabolome	The complete complement of small molecules present in an organism.
Metabolomics	The technology geared towards providing an essentially unbiased, comprehensive qualitative and quantitative overview of the metabolites present in an organism.
Metabonomics	A non-plant term generally used to define the technology used to measure quantitatively the metabolic composition of body fluids following a response to pathophysiological stimuli or genetic modification.
Targeted analysis	Following broad-scale metabolomics analysis, or based upon prior knowledge, biochemical profiling can be performed in greater detail on selected groups of metabolites by using optimized extraction and dedicated separation/detection techniques.

Applications in the food industry regarding for example tracing and tracking (16) and food adulteration (17) have also been described. However, perhaps the greatest area of application has been in the general field of plant breeding where different technologies are already being exploited to further our knowledge of how the composition of plant products are influenced by genotype and environment. The exact application depends on the crop, but many crops have already been subjected to a metabolomics assessment including major food crops such as rice (18), wheat (19), tomato (20), melon (21), *Brassica* (22), coffee (23), and potato (24). Such knowledge is highly complementary to that obtained from the more traditional and established methods and as such, metabolomics will likely provide additional tools to help advance plant breeding strategies and the speed of developing new varieties more suited to current demands.

2. Overview

In this volume, we have aimed to compile a series of chapters covering many of the key aspects related to designing, executing, and analyzing metabolomics experiments. Use can be made of a wide variety of analytical approaches, the choice of which is essentially determined by the biological question for which an answer is sought. There is no single approach, for extraction, separation, or detection which could be considered broadly optimal, and consequently, we have tried to cover most of the more standard and best developed approaches, using materials which are also generally available. The detailed, step-by-step methods provided are also generally ones which can readily be modified to suit other plant materials of interest. Inevitably, the infrastructure, instrumentation and other support facilities which are available, as well as the available expertise will also play a key role in the approach chosen. Many methodologies require not inconsiderable investment in hardware (and software) and this may also be of significant influence in deciding how to proceed. Consequently, the chapters provided cover a considerable variety of machines from different manufacturers and make use of software from a range of sources, much of which is now also available as "freeware".

Getting started. A metabolomics experiment will fail at the first hurdle if it has not been properly thought out and designed. While results will almost always be obtained, their reliability and biological relevance may be greatly in question if the approach has not been correct. Before beginning, the experiment has to be thought through thoroughly to the end and the inexperienced would be wise to already employ the services of a statistician at this stage to help assess the requirements concerning issues such as, sample size, sample pooling, replication, etc. which will ultimately determine the robustness of the dataset and its suitability for subsequent (multivariate) data analysis. In addition, the manner of sample provision and preparation, their handling and storage, as well as the actual performance of the extraction and analysis itself, are also of critical importance. Consequently, a number of chapters have been dedicated to these specific issues.

The technologies. For metabolomics and micronutrients analyses quite a number of choices are available, both for compound or element separation and detection. Most of the generally used approaches have been covered either separately or in various combinations. For a full list of the abbreviations of the most widely used technologies the reader is referred to Table 2. The range of methods presented for detection based on either mass Spectrometry (MS) or Nuclear Magnetic Resonance (NMR) reflects the main approaches used in the field and an upcoming, so-called, hyphenated approach, combining the two is also described. Separation

Table 2
The whole field of plant metabolomics is strewn with many abbreviations, often in hyphenated multiple combinations (e.g. FI-ESI-FT-ICR-MS or LC-PDA-SPE-NMR- MS!). This can be very daunting to the inexperienced reader. Consequently, in this table a list of the most common abbreviations and those regularly used in the various chapters to follow are given

AMDIS	Automated Mass Spectral Deconvolution and Identification System (26)
APCI	Atmospheric Pressure Chemical Ionization
APPI	Atmospheric Pressure Photo-ionization
CAS	Chemical Abstracts Service (27)
CE	Capillary electrophoresis
CID	Collision Induced Dissociation
DI (FI)-MS	Direct Infusion (or Flow Injection)-MS
ESI	Electrospray Ionization
FIA/DIA	Flow Injection Analysis/Direct Infusion Analysis
FT-ICR-MS	Fourier Transform—Ion Cyclotron Resonance—MS (or FTMS)
HCD	High-energy Collision-induced Dissociation
HPLC	High Performance (Pressure) Liquid Chromatography
HTP	High Throughput
ICP-MS	Inductively Coupled Plasma MS
LC/GC	Liquid/Gas Chromatography
LDA	Linear Discriminant Analysis
LTQ	Linear Trap Quadrupole
MALDI-MS	Matrix Assisted Laser Desorption Ionization–MS
MS	Mass Spectrometry
MS/MS; MS^n	Double (MS/MS) or multiple levels (MS^n)of molecular fragmentation/ re-fragmentation with MS detection
MSI	Metabolomics Standards Initiative (28, 29)
m/z	Mass/charge
netCDF	Network Common Data Form (30)
NIST	National Institute of Standards and Technology (metabolite database) (31)
NMR	Nuclear Magnetic Resonance
PCA	Principal Components Analysis
PDA (DAD)	Photodiode Array Detection (Diode Array Detection)
PI	Photo-ionization

(continued)

Table 2 (continued)

RF	Random Forest
SEC	Size Exclusion Chromatography
SPE/SPME	Solid Phase Extraction or Solid Phase Micro-Extraction
TOF	Time of Flight
UPLC	Ultra Performance Liquid Chromatography

methods based upon both GC and LC are also given and for non-volatile compounds the possibility to exclude full scale separation via Direct Injection is also touched upon.

Data pre-processing. Analytical equipment does not produce clean and comparable lists of metabolites in the samples, and raw data must be processed in a variety of ways to produce metabolically significant signals on which meaningful analysis of treatment differences can be based. This is known variously and confusingly as post-processing (after the chemical analysis) or pre-processing (before the data analysis). Principled removal of noise is a commonly required step. Chromatography-based techniques typically rely on *peak picking* methods to detect metabolite-based features in the data and this is necessarily followed by alignment of results from multiple runs to compensate for time and matrix-based variations such that they may be compared on a peak to peak basis. Comparable relative intensities may be calculated from chromatographic peaks. Pre-processed data may be understood to relate to distinct (possibly unidentified) metabolites, when it is typically know as profile data, or it may represent a metabolic "fingerprint" where the data values are a reflection of the chemical species present but are not associated one for one. Use of software packages supplied with instruments is covered in a number of chapters and application of instrument-independent general purpose packages is described in two chapters.

Metabolite identification. While profile data may relate to "unknown" (but repeatably detected) metabolites, identification of metabolites which are significant between experimental treatments is clearly important for biological understanding. This is typically achieved by comparison of signals with library data of common chemical species for the analytical technique. This is considered for a range of techniques. Accurate mass determination, for deriving empirical formulae, and analysis of multiple MS fragmentation patterns are additional indicative techniques which are covered.

Data analysis. Fingerprint or profile data may be analyzed in pursuit of the experimental objectives or for speculative investigation. By the nature of metabolomics and in contrast to traditional biochemical approaches, the data sets are multivariate. For each biological sample or replicate as appropriate, many intensity values are available. While univariate techniques may be used (with appropriate caution) multivariate analysis is more normally appropriate. The typical range of data analysis techniques are covered here in association with the various chemical analytical technologies and principled choice of alternative appropriate data mining and analysis techniques is considered.

Data reporting. Throughout the chapters, the need for accurate and comprehensive recording of all details of both materials and procedures is emphasized (the so-called "meta-data"). The origin of materials, their growth, harvest, storage and laboratory processing as well as details of the chemical and data analysis techniques applied are crucial to the execution of effective metabolomics and essential for the long-term usefulness of the data. The differences—signals—sought in this field are typically small and easily overwhelmed by the noise which can be introduced by the techniques used. Publishers and regulators are rightly expecting comprehensive datasets to substantiate findings. Consideration is therefore given to the collection and management of these large data sets.

3. Future Challenges

Plant metabolomics is a field of science which is still in a dynamic phase of development. Perhaps the achievements already booked in terms of analytical capacity, precision, and throughput raise even more new questions than have answered old ones. Nevertheless, the potential has clearly been demonstrated and examples of good practice are presented here. Techniques and equipment for both chemical and data analysis improve constantly, but robust procedures for their application will clearly always be required.

Acknowledgements

This work has been carried out under the auspices of the EU FPVI project META-PHOR—project number: FOOD-CT-2006-036220; (25). RDH acknowledges additional funding from the Centre for BioSystems Genomics (CBSG) and the Netherlands Metabolomics Centre (NMC), both part of the Netherlands Genomics Initiative (NGI). NWH acknowledges the support of Aberystwyth University.

References

1. Hall, R. D. (2006) Plant metabolomics: from holistic hope, to hype, to hot topic. *New Phytologist* **169**, 453–468.
2. Tohge, T. and Fernie, A. R. (2009) Web-based resources for mass-spectrometry-based metabolomics: A user's guide. *Phytochemistry* **70**, 450–456.
3. Saito, K., Dixon, R. A. and Willmitzer, L., eds. (2006) *Plant Metabolomics*. Biotechnology in Agriculture and Forestry, Vol. 57. T. Nagata, ed. Springer-Verlag: Berlin.
4. Bovy, A., Schijlen, E. and Hall, R. D. (2007) Metabolic engineering of flavonoids in tomato (Solanum lycopersicum): the potential for metabolomics. *Metabolomics* **3**, 399–412.
5. Biais, B., Allwood, J. W., Deborde, C., Xu, Y., Maucourt, M., Beauvoit, B., et al. (2009) H-1 NMR, GC-EI-TOFMS, and Data Set Correlation for Fruit Metabolomics: Application to Spatial Metabolite Analysis in Melon. *Analytical Chemistry* **81**, 2884–2894.
6. Fait, A., Hanhineva, K., Beleggia, R., Dai, N., Rogachev, I., Nikiforova, V. J., et al. (2008) Reconfiguration of the achene and receptacle metabolic networks during strawberry fruit development. *Plant Physiology* **148**, 730–750.
7. Hall, R. D., ed. (2011) *Biology of Plant Metabolomics*. Wiley-Blackwell, Oxford.
8. Beale, M. H. and Sussman, M. R. (2011) Metabolomics of Arabidopsis thaliana, in *The Biology of Plant Metabolomics* pp. 157–180 (R.D. Hall, ed.), Wiley-Blackwell.
9. Schauer, N., Zamir, D. and Fernie, A. R. (2005) Metabolic profiling of leaves and fruit of wild species tomato: a survey of the Solanum lycopersicum complex. *J Expt Bot* **56**, 297–307.
10. Bovy, A. G., Gomez-Roldan, V. and Hall, R. D. (2010) Strategies to optimize the flavonoid content of tomato fruit, in *The Handbook of Polyphenols: Recent Advances in Polyphenol Research* (C. Santos-Buelga, M.-T. Escribano-Bailon, and V. Lattanzio, eds.) pp. 138–162.
11. Ahuja, I., de Vos, C. H. R., Bones, A. and Hall, R. D. (2010) Plant molecular stress responses face climate change. *Trends in Plant Science* **15**, 664–674.
12. Allwood, J. W., Ellis, D. I. and Goodacre, R. (2008) Metabolomic technologies and their application to the study of plants and plant-host interactions. *Physiologia Plantarum* **132**, 117–135.
13. Jansen, J. J., Allwood, J. W., Marsden-Edwards, E., van der Putten, W. H., Goodacre, R. and van Dam, N. M. (2009) Metabolomic analysis of the interaction between plants and herbivores. *Metabolomics* **5**, 150–161.
14. Capanoglu, E., Beekwilder, J., Boyacioglu, D., de Vos, C. H. R. and Hall, R. D. (2010) The effect of industrial food processing on potentially health-beneficial tomato antioxidants. *Crit Rev Food Chem* **50**, 919–930.
15. Fernie, A. R. and Schauer, N. (2009) Metabolomics-assisted breeding: a viable option for crop improvement? *Trends in Genetics* **25**, 39–48.
16. Goodacre, R., York, E. V., Heald, J. K. and Scott, I. M. (2003) Chemometric discrimination of unfractionated plant extracts analyzed by electrospray mass spectrometry. *Phytochemistry* **62**, 859–863.
17. Steward, D., Shepherd, L. V. T., Hall, R. D. and Fraser, P. D. (2011) Crops and tasty, nutritious food – how can metabolomics help? in *The Biology of Plant Metabolomics* (R.D. Hall, ed.), Wiley-Blackwell pp. 181–218.
18. Hall, R. D., Brouwer, I. D. and Fitzgerald, M. A. (2008) Plant metabolomics and its potential application for human nutrition. *Physiologia Plantarum* **132**, 162–175.
19. Graham, S. F., Amigues, E., Migaud, M. and Browne, R. A. (2009) Application of NMR based metabolomics for mapping metabolite variation in European wheat. *Metabolomics* **5**, 302–306.
20. Moco, S., Bino, R. J., Vorst, O., Verhoeven, H. A., de Groot, J., van Beek, T. A., et al. (2006) A liquid chromatography-mass spectrometry-based metabolome database for tomato. *Plant Physiology* **141**, 1205–1218.
21. Moing, A., Aharoni, A., Biais, B., Rogachev, I., Meir, S., Brodsky, L., et al. (2011) Spatial and temporal metabolic profiling using multiple analytical platforms highlights the crosstalk between primary and secondary metabolites and mineral elements in melon fruit. *New Phytologist* **190**, 683–696.
22. Jahangir, M., Kim, H. K., Choi, Y. H. and Verpoorte, R. (2009) Health-Affecting Compounds in Brassicaceae. *Compr Rev Food & Sci Food Safety* **8**, 31–43.
23. Lindinger, C., Pollien, P., de Vos, R. C. H., Tikunov, Y., Hageman, J. A., Lambot, C., et al. (2009) Identification of Ethyl Formate as a Quality Marker of the Fermented Off-note in Coffee by a Nontargeted Chemometric Approach. *J Agric & Food Chem* **57**, 9972–9978.
24. Beckmann, M., Enot, D. P., Overy, D. P. and Draper, J. (2007) Representation, comparison,

and interpretation of metabolome fingerprint data for total composition analysis and quality trait investigation in potato cultivars. *J Agric & Food Chem* **55**, 3444–3451.

25. *META-PHOR: Metabolomics for plants, health and outreach.* (Accessed May 2010). Available at: http://www.meta-phor.eu/.
26. *AMDIS: The Automated Mass Spectral Deconvolution and Identification System.* (Accessed May 2010). Available at: http://chemdata.nist.gov/mass-spc/amdis/.
27. *CAS: Chemical Abstracts Service.* American Chemical Society. (Accessed May 2010). Available at: http://www.cas.org/.
28. Fiehn, O., Robertson, D., Griffin, J., van der Werf, M., Nikolau, B., Morrison, N., et al. (2007) The metabolomics standards initiative (MSI). *Metabolomics* **3**, 175–178.
29. *MSI: The Metabolomics Standards Initiative.* Metabolomics Society. (Accessed May 2010). Available at: http://msi-workgroups.sourceforge.net/.
30. *NetCDF (network Common Data Form).* Unidata. (Accessed May 2010). Available at: http://www.unidata.ucar.edu/software/netcdf/.
31. *NIST Scientific and Technical Databases.* National Institute of Standards and Technology. (Accessed May 2010). Available at: http://www.nist.gov/srd/analy.htm.
32. Oliver, S.G., Winston, M.K. and Kell, D.B. (1998) Systematic functional analysis of the yeast genome. *Trends Biotech* **16**, 373–378.

Part I

Material Preparation

Chapter 2

Aspects of Experimental Design for Plant Metabolomics Experiments and Guidelines for Growth of Plant Material

Yves Gibon and Dominique Rolin

Abstract

Experiments involve the deliberate variation of one or more factors in order to provoke responses, the identification of which then provides the first step towards functional knowledge. Because environmental, biological, and/or technical noise is unavoidable, biological experiments usually need to be designed. Thus, once the major sources of experimental noise have been identified, individual samples can be grouped, randomised, and/or pooled. Like other 'omics approaches, metabolomics is characterised by the numbers of analytes largely exceeding sample number. While this unprecedented singularity in biology dramatically increases false discovery, experimental error can nevertheless be decreased in plant metabolomics experiments. For this, each step from plant cultivation to data acquisition needs to be evaluated in order to identify the major sources of error and then an appropriate design can be produced, as with any other experimental approach. The choice of technology, the time at which tissues are harvested, and the way metabolism is quenched also need to be taken into consideration, as they decide which metabolites can be studied. A further recommendation is to document data and metadata in a machine readable way. The latter should also describe every aspect of the experiment. This should provide valuable hints for future experimental design and ultimately give metabolomic data a second life. To facilitate the identification of critical steps, a list of items to be considered before embarking on time-consuming and costly metabolomic experiments is proposed.

Key words: Biological error, Technical error, Experimental noise, Blocking, Pooling, Replication, Quenching of metabolism, Metadata

1. Introduction

The ultimate goal of biology is to understand living systems in sufficient detail to enable accurate quantitative predictions about their behaviour (1). In the second part of the twentieth century, progress in biological research was mainly driven by the revolutionary concepts and technologies of molecular biology, which links

Nigel W. Hardy and Robert D. Hall (eds.), *Plant Metabolomics: Methods and Protocols*, Methods in Molecular Biology, vol. 860, DOI 10.1007/978-1-61779-594-7_2, © Springer Science+Business Media, LLC 2012

information about genetic traits to physical entities such as DNA or proteins. Strikingly, this led biologists to think "molecular", eventually promoting reductionist approaches, which resulted in the attribution of biological phenomena to the actions of one or a few genes. Although reductionism is powerful in building logically simple hypotheses that are rather easy to test, it is very difficult to reconstitute a model for a whole biological system by combining the pieces of information it generates. Thus, from "functional genomics" biologists are moving to "systems biology", in order to identify and integrate at the functional level all gene products present in a given biological system (2, 3). This move, which is characterised by the development of multiparallel technologies, called 'omics, that produce massive data sets, now brings biologists to consider living systems as a whole again (4). However, this abundance of multiplexed information also presents many hurdles, starting with the major challenge of setting up the right experimental design. In particular, we may ask ourselves whether the unfocused nature of metabolomics conciliates the concept of *planning*. Have we set up the right experimental design? Have we the right number of samples for statistical analysis knowing that the number of metabolite peaks per sample usually exceeds the number of data points from an experiment? Do we need statistics to design valid experiments? Do we need pilot experiments before planning full-scale metabolomics analyses?

In this article, we briefly introduce the notions of experiment and experimental design; we then discuss some issues in designing 'omics experiments, before embarking on a checklist for the design of experiments in the field of plant metabolomics.

2. What Is Experimental Design?

In 400 BC, the philosophers Socrates, Plato, and Aristotle investigated the meaning of knowledge and the methods to obtain it, using a rational-deductive process. Later, scientists Ptolemy and Copernicus developed empirical-inductive methods that focused on precise observations and explanation of the stars. These early scientists were not experimenters. It is only when later scientists began to investigate earthly objects rather than the heavens, that they uncovered a new paradigm for increasing knowledge. In the early 1600s, Francis Bacon introduced the term "experiment" (5). The basis of this new paradigm called *experimentation* was a simple question, "If I do this, what will happen?" The key to understanding experimentation, and the characteristic that separates experimentation from all other research methods, is manipulating factors to see what happens. Explanations involve identifying the causes of what has been described and this involves finding out what factors influence the variables. The scientific aspect of experimentation is the manipulation of variables under controlled conditions while taking precise measurements. Today, especially for metabolomic approaches, the key feature is still the deliberate variation of something so as to discover what happens to something else, and later to uncover the effects of presumed causes. But the real hurdle is that biological systems are complex, and many possible variables could be implicated, from genotype variability to fluctuations in growth conditions.

The design of experiments can be defined as a procedure aimed at planning experiments in the most efficient way to obtain data that describe the relationship between the different factors/variables affecting a process and its outputs (6). Traditionally, plant biology experiments have been performed by changing variables one by one, but it became evident that it is difficult to exactly reproduce measured results (7). In the 1920s, Ronald A. Fisher, a renowned mathematician and geneticist, developed the concept of "factorial design", a powerful approach to deal with experimental error (8). An experiment with a factorial design can be defined as an experiment in which the effects of at least two factors are studied by testing all possible combinations. Environmental variables such as light intensity, temperature, moisture, or the availability of nutrients may vary across a field, a greenhouse, or even a growth chamber. In order to cope with such unavoidable but identifiable sources of variation, the arrangement of experimental units into groups (*blocking*) can be used. Then, in order to get an estimate of the errors that cannot be eliminated, multiple measurements (*replication*) can be performed randomly (*randomisation*) within each block. Factorial experiment design has proven efficient to evaluate the effects and possible interactions of several factors (9). This approach was built on the foundation of the *analysis of variance*, a collection of models in which the observed variance is partitioned into components due to different factors which are tested.

When a full factorial design requires too many samples to be processed, an "optimal design" can be used instead, given that combinations of various factor levels that are relevant have been identified. This is best achieved with the help of an algorithm (e.g. http://www.optimal-design.org). Optimal design is being used widely in all science and technology domains. For example, biotechnology companies always need to optimise production systems based on cell cultures, thus dealing with sometimes more than 50 variables (various nutrients, temperature, pH, speed of agitation, etc.). Running experiments one-factor-at-a-time would be extremely expensive, and would require months or even years. Recently, a complex cell-culture medium has been optimised by testing eight factors at five concentrations with 192 runs and within 8 days. By contrast, a full 5-level factorial experimental design would have required 390,625 runs (http://www.statease.com/pubs/invitrogen.pdf).

3. The Challenge of Designing 'Omics Experiments

The advent of the "'omics revolution" has forced us to re-evaluate our ability to acquire, measure, and handle data sets. In particular, many of us have had to realise that advanced statistics were inescapable.

3.1. Throughput

'Omics technologies provide unprecedentedly rich information about DNA, messenger RNA, proteins, and metabolites from complex biological systems. This is enabled by the development of a large variety of analytical platforms (e.g. DNA sequencers, microarrays, mass and nuclear magnetic resonance spectrometers), and by conceptual efforts in the areas of data management, biostatistics, data integration, computational modelling, and knowledge assembly protocols. However, 'omics face technical difficulties, high costs, and time-consuming data analysis, which dramatically limit the number of samples that can be processed within experiments. Such difficulties actually favour poor experimental design and there is a widespread idea that the large number of measurements obtained in gene expression array, protein or metabolite identifications would somehow make up for small sample sizes (10). This idea is reinforced by some confusion about the meaning of high throughput in the literature. Originally used in biology in the context of screens, this expression has drifted to qualify technologies capturing large numbers of analytes per sample. But unless the behaviour of groups of analytes is studied, as can be done for example in PageMan (11), an analyte is definitely not a replicate. We would actually tend to consider 'omics experiments as being rather low throughput, the first consequence of this being low replication, which appeals for great care during the design and analysis of experiments.

3.2. False Discovery

The very notion that measuring every possible output variable is desirable has been seen as a big delusion surrounding the 'omics, as system-wide measurements may violate statistical norms and have little precedent with respect to feasibility in analytical chemistry literature (12). 'Omics experiments typically involve comparing a group of control samples with one or more groups of treated samples, with data often being expressed in a "semi-quantitative" way, which means that "fold-changes" are evaluated by calculating a ratio between the data obtained in treated and control samples. Replication (typically around five replicates) then allows checking whether the fold-changes are significant, generally by performing a *t*-test. However, methods based on *t*-tests depend on strong parametric assumptions (e.g. normality, homogeneity of variance, and independent errors), which are often invalidated by the restricted number of replicates (13).

A further striking problem is that the larger the number of analytes being measured, the easier it is to find rare events and therefore the easier it is to make the mistake of thinking that there is an effect when there is none. This is intimately bound to the multiple testing nature of 'omics approaches and is called false discovery and requires multiplicity control (14). A range of methods and tools dedicated to the reduction of false discovery rate have been developed (e.g. a number of dedicated R-scripts can be found at http://strimmerlab.org/notes/fdr.html).

3.3. Significance

With respect to experimental design, we are tempted to put side by side 'omics and experimentation on animals. Indeed, both suffer from low replication, the one because of technological issues, the other for obvious ethical reasons. An interesting article published in the journal *Laboratory Animals* reports a survey of three experiments performed with dogs or mice, which reveals that better experimental design could have resulted in the use of fewer animals (15). Furthermore, it demonstrates that factorial experimental design would have resulted in better precision. The same reasoning is valid for 'omics experiments, as depicted below with a simple example.

Studies of metabolism usually face a large number of potential sources of variation. They can be biological (e.g. environmental, positional, temporal) or technical (e.g. experimenter, batch effect), some of them being unavoidable. To a certain extent, such interfering covariates can nevertheless be included in the analysis to adjust for their influences. For example, consider an experiment (see Table 1) in which two genotypes submitted to two treatments were grown in blocks corresponding to two shelves in a growth chamber (each shelf was characterised by slightly different growth conditions). A first option would be to perform a Student's *t*-test by grouping replicates from different blocks. Because, Student's *t*-test can only compare two treatments, it would also be necessary to transform the data into fold-changes. We chose to calculate treatment versus control ratios, by dividing each "treated" datum by averaged "control" data. However, such transformations imply the loss of two levels of information, eventually increasing the number of false positives or negatives. Indeed, we obtain a *p*-value of 0.16 (in Excel), which suggest that the response to the treatment was not significantly different between the two genotypes, or that sample size was too small. A more powerful option would be to perform a multifactorial analysis of variance (see Table 2). This time, we obtain a *p*-value of 6.52E-03, which indicates that there actually is a significant difference. A further interesting point is that a significant interaction is also found between treatment and shelf (*p*-value = 0.01), reinforcing the idea that the investigation of multiple factors at the same time can be more efficient and effective than a series of experiments aimed at each factor alone.

Table 1
Fake experiment, in which two genotypes were grown under two treatments, on two different shelves, and in which one variable was measured

Genotype	Treatment	Shelf	Variable
1	1	1	50
1	1	1	49
1	1	2	52
1	1	2	54
1	2	1	38
1	2	1	35
1	2	2	21
1	2	2	23
2	1	1	90
2	1	1	65
2	1	2	78
2	1	2	95
2	2	1	45
2	2	1	41
2	2	2	23
2	2	2	15

Table 2
Analysis of variance performed on the fake experiment shown in Table 1 using the functions "factor", "lm", and "anova" in R (http://www.r-project.org)

	p-Value
Genotype	3.50E-03**
Treatment	1.60E-05***
Shelf	0.14
Genotype × treatment	6.52E-03**
Genotype × shelf	0.81
Treatment × shelf	0.01*
Genotype × treatment × shelf	0.37

Only *p*-values are shown. Significance codes: "***", <0.001; "**", <0.01; "*", <0.05

4. A Checklist for the Design of Plant Metabolomics Experiments

Plants cannot escape their environment, but they have evolved a wide range of mechanisms to face sometimes highly fluctuating growth conditions. They make a variety of organs (leaves, roots, stems, tubers, etc.) composed of multiple specialised cell types (epidermis, guard cells, parenchyma, glandular hairs, etc.), each of them having a dedicated metabolism. In addition, abiotic (light, UV, water) and biotic (herbivore, parasitism, and pathogen attack) stress factors continually have to be dealt with and, for this, plants have developed a complex metabolic arsenal of compounds. Some of them are common, but many are restricted to one genus or perhaps even to one species. In addition to its high diversity, plant metabolism is also characterised by considerable robustness (e.g. metabolism operates under a wide range of temperatures), elasticity (e.g. metabolic fluxes can drop and recover within seconds when light fluctuates), and plasticity (e.g. plants are able to reprogram metabolism in response to many developmental, biotic, or abiotic challenges). The purpose of plant metabolomics is to capture instant pictures of this diversity, and integrate them into functional information (16). A major challenge is that such estimates must represent the amounts of the metabolites that were actually present in the harvested tissues when these were metabolising under the specified growth conditions (17).

One initial goal for metabolomics was to avoid exclusion of any metabolite by using well conceived sample preparation procedures and analytical techniques, thus allowing a comprehensive analysis of biological systems (18). However, unlike transcriptomics, and to a certain extent proteomics, technologies that are available to metabolomics are far from such comprehensiveness. Considerable progress has been achieved recently (19), but there is still no unique solution to extract and then determine every single metabolite simultaneously. This is further complicated by plant metabolomes being extremely diverse and complex (16). These limitations need to be taken into account so that the experimental design can be tuned to the biological question of interest (20).

4.1. Choose the Methodology

Globally, there are two types of data that can be generated in a metabolomics experiment, fingerprints and profiles. Fingerprinting, which is typically performed using FT-IR or ^{1}H-NMR, ignores time-consuming signal assignment and can thus be used to rapidly compare or classify samples in an unbiased way. It has been used to study the impact of environmental factors (21), cadmium toxicity (22), herbicide treatments (23) as well as to compare wild-type and transgenic plants (24, 25). Profiling, which is usually performed with MS- or NMR-based technology (see ref. 26 for review), provides quantitative data. Its success strongly depends on the "quality" of the biological material, sample preparation, and sample extraction. Gullberg et al. (27) have detailed a useful report on appropriate strategies for the design of metabolomic experiments with GC-MS technology. Several classes of metabolites are nevertheless unsuited to GC-MS or NMR metabolomics, thus requiring dedicated experimental strategies. In particular, most intermediates of primary metabolism or coenzymes are present at low concentration and are usually very unstable. Their analysis may necessitate specific extraction protocols and instrumentation, e.g. LC–MS/MS (28), in addition to careful sampling procedures (*see* below). The analysis of several classes of secondary metabolites (e.g. phenylpropanoids, terpenoids) also requires dedicated LC–MS methodologies. In brief, experimenters should, at first, be aware of the possibilities offered by available or future equipment, as there is no unique and universal methodology so far.

4.2. Evaluate Experimental Error

Experimental error results from both biological and technical variability. Evaluating them can make a major contribution to the experimental design, by giving hints to reduce experimental error and/or by helping in the choice of replication strategy.

As discussed above, 'omics approaches generalised the problem of making multiple hypotheses in a limited number of samples, eventually leading to new statistical concepts, or to the rediscovery of old ones (29), and tools dedicated to the optimisation of sample size were developed (30–32). False discovery has been considered as less challenging in metabolomics than in transcriptomics or proteomics because measureable metabolites are currently by far less numerous than measurable transcripts or proteins, and because variations in the metabolome are expected to be of much larger amplitude (33). However, this is counterbalanced by the high chemical diversity of metabolites which may cause experimental error. Indeed, nucleic acids consist of polymers of four nucleotides and share identical physico-chemical properties, and proteins are essentially made out of 22 primary amino acids, resulting in much lower chemical complexity than a metabolome (16). Accordingly, the chemical diversity of metabolites leads to unequal stability, matrix effects, differences in detection limits and linearity ranges. This is further complicated by the untargeted nature of metabolomics (18) because technical error cannot be defined for unexpected analytes.

To assess *technical variability*, pure chemicals can be used alone, but this is likely to be invalid due to matrix effects resulting from the high complexity of plant extracts. Recovery experiments, in which known metabolites of interest are mixed with plant extracts, are recommended instead (20). Then, the use of a range of concentrations of standards can be very helpful in determining the detection limit, which can be defined as the lowest level of a spiked metabolite that is statistically significant (34). Furthermore, spiking different amounts provides an idea about the linearity of the dose–response curve for a given metabolite. Another useful procedure involves spiking extracts with suitable isotopomers (28). However, spiking is usually restricted to a few known metabolites. A simple and useful additional strategy is to use reference biological material, which would be similar to the material under investigation, and ideally "isotopomerised" (35, 36). Such material could also be obtained by mixing samples obtained under various conditions (environment, phenology, and/or genotype), in order to maximise the diversity of the resulting mixed metabolome. Dilution gradients can be made to evaluate linearity for each analyte under study. Then, recovery experiments can be performed by mixing samples of interest and reference material. Finally, this material can be used as an additional reference to the usual "control" samples that are grown in each experiment, offering the possibility to integrate data from many experiments (37).

Assessing *biological variability* is much more difficult, as it can vary depending on the genotype, the developmental stage, and/or the growth conditions. It is, however, useful to evaluate it within a large number of biological replicates obtained under standard conditions. Furthermore, biological variability should be documented and made accessible.

4.3. Handle the Experimental Error

What counts is whether the differences between conditions are larger than can be explained by experimental variability, and determining this requires statistically valid analyses. The precision of an experiment depends critically on the size of the experiment and the homogeneity of the experimental material. Even quite a small reduction in the within replicates standard deviation can lead to a dramatic increase in precision (15). Apart from working carefully, there are ways to decrease experimental error.

First, given technical error and/or biological variability have been evaluated, the most *adequate number of replicates* can be defined in relation to the aim of the experiment; it will have a major influence on reliability and reproducibility. A range of methods allowing the estimation of sample size have been developed for microarray experiments comparing two or more conditions (32). We assume that such methods should prove useful to metabolomics, which to a certain extent, face the same problem of having the number of analytes greatly exceeding the number of samples (33). A further point to consider is that, whenever possible, biological replication should be preferred to technical replication. In fact, technical variability, which is generated alternatively by experimenters, techniques, and/or equipment, is usually small (less than 10%) in comparison to biological variability. When considering that the number of samples that can be processed is limited by the technology, biological samples should always be preferred.

Pooling samples corresponding to different individuals grown under identical conditions can prove very useful in decreasing biological variability given that costs per sample are relatively low and costs per analysis high, which is generally the case in metabolomics. As shown by Rocke (10), pooling samples can dramatically reduce costs: A study that would cost 33,600 US$ in which 56 samples are planned, using a simple experimental design with a standard deviation of 0.3 for biological replicates and 0.1 for analytical replicates, can be carried out for less than 10,000 US$ by pooling seven samples per array and using only eight arrays for the whole study. In general, the largest gains from pooling are obtained when the cost per sample is low and the cost per analysis high. However, pooling assumes that information will be lost, as variations in metabolite levels always have a biological significance (16). A further disadvantage is that a single bad or unusual sample can ruin a pool or even an entire experiment.

When experimental noise cannot be avoided, *blocking* should be used, as discussed above. For example, if an experiment requires the use of two growth chambers with theoretically identical growth conditions, replicates (e.g. corresponding to genotypes) should be equally distributed between the two chambers because growth conditions will be practically different. Blocking should, however, be restricted to major sources of noise as a too complex experimental design would require too many samples and result in difficult statistical analysis.

Randomisation is a further way to cope with experimental error. It can be used throughout the experimental pipeline. Fields, greenhouses, and even controlled growth chambers never deliver uniform growth conditions; there are always gradients and border effects, even within blocks. When various genotypes are compared under one theoretical growth condition, they must be randomised and when time-course experiments are performed, plants grown under theoretically identical conditions should be harvested randomly. However, randomisation can lead to fastidious harvesting procedures. For example, when randomly planted individuals are pooled, extra time will be required to move from one individual to the other. Randomisation should also be used throughout the analytical procedure, from sample storage to detection. An article of Scholz and colleagues (38) illustrates very nicely how useful such precautions can be.

4.4. Strike the Balance Between Sample Number and Throughput

The rather low throughput offered by metabolomic technologies still limits the number of samples that can be processed and thus the scale of experiments. Based on literature and on our own experience, we estimate that in academic research, current plant metabolomic experiments do not usually represent more than several hundreds of samples. Although such size is considerable, it is merely adequate for experiments in which a limited number of growth conditions are compared, but it is not suited for experiments in which effects of a number of potentially interacting environmental variables would be studied. For example, an experiment as simple as a time course with 10 harvest points, 6 replicates per harvest point, 2 growth scenarios and repeated once would already represent 240 samples.

Next, throughput of hundreds of samples is adequate for quantitative trait loci- (39–41) or association mapping studies (42), as they typically involve populations of hundreds of individuals. But again, this assumes that only one or two growth scenarios or harvest points would be used. Furthermore, the emergence of huge mapping populations such as the recently developed Maize nested association mapping population of 5,000 genotypes (43) poses an interesting challenge for 'omics approaches.

Nevertheless, the balance between sample number and throughput should always be optimised to decrease the risk of performing costly but badly designed experiments. In addition to literature and database searches, it can be very useful to perform preliminary experiments, in which visual or biochemical diagnostic markers are evaluated to identify the most appropriate set-up to reach the desired physiological and/or developmental state (44). For example, given a detailed time-course experiment has been performed, it might be easier to define the time of harvest that would be the most relevant to the biological question of interest. As an example, glucose-6-phosphate, which is relatively easy and cheap to measure in high throughput (34), has been suggested as a marker for carbon status in leaves of *Arabidopsis thaliana*. In response to carbon starvation, this metabolite drops and recovers partially within hours, indicating a time window that is also characterised by a dramatic response at the level of the transcriptome (45).

4.5. How to Grow Plants

There are no recommendations about how to grow plants that would be specific for metabolomics. However, because plant metabolomes reflect short- to long-term interactions between genotypes and their environments, variables that are the most likely to affect metabolism should be controlled in a reproducible way and/or monitored. As already mentioned, randomisation within an experiment is essential to cope with unavoidable "local" effects associated with gradients in essential variables such as light intensity, temperature, and air humidity that are typical for greenhouses but also frequent in controlled growth chambers.

Possible interactions between environmental variables should also be foreseen. For example, under high light intensities, plants will tend to grow faster, thus consuming more water and nutrients (46). If these variables are not controlled, plants growing the fastest might deplete resources quicker under high light, eventually running into limitations in water or nutrients. It is obvious that this would lead to erroneous results in situations where a mutant with stunted growth is studied.

Optimised labelling of individuals also belongs to a good experimental design. Bar codes are increasingly used in plant research, dramatically facilitating documentation and traceability of experiments.

4.6. When to Harvest

The metabolic composition of plants or plant organs varies throughout their lifecycle. However, *age* is a rather imprecise criterion to define maturity of plants, as dynamics of traits such as phenology and sex expression (47) or metabolic composition (48) can vary dramatically in response to the environment. Furthermore, such responses may vary depending on the genotype (mutant, transformant, ecotype, or cultivar), eventually leading to apparent metabolic phenotypes that would be indirect, and thus very difficult to explain at the functional level. It might therefore be very useful to search for diagnostic markers that are specific for the desired developmental or physiological stage. Ideally, such markers would be visual and/or very easy and cheap to determine.

The next issue is to define the most appropriate *time of the day* to harvest plant tissues, as many metabolites can show strong diurnal variations. A widespread habit involves taking samples in the middle of the day, assuming that everything is "on" or at steady-state. While this might be true for fluxes and levels of intermediates through pathways connected to photosynthesis, it is wrong for many metabolite levels. It is indeed well known that in leaves a range of metabolites such as major carbohydrates including starch (44, 49), amino acids (50), fatty acids (37), or organic acids (51) accumulate during the day in leaves. Although less marked, diurnal fluctuations in metabolite contents have also been reported in developing fruits (52).

Harvest should be as quick as possible, posing again the problem of the size of the experiment (the more samples the longer it takes to harvest them). It might be useful to estimate how much time one sample would require and thus predict harvest duration. For example, if one sample requires 1 minute, an experiment with 300 samples would require 5 h, which would be likely to introduce considerable variation into the experiment, unless logistics have been adequately tuned.

4.7. How to Harvest

Specific harvest and extraction protocols are available for plant metabolomics (see Chapter 4 for more details). However, some issues need to be taken into account at the level of the experimental design, mainly in terms of feasibility.

A major issue is that many metabolites are unstable due to particular chemical properties, or simply when the inactivation of particular enzymes proves ineffective during harvest and/or extraction. The latter is particularly problematic for metabolites with high in vivo turnover rates, such as many intermediates in the primary metabolism as well as coenzymes. It has for example been reported that ATP pools turn over within seconds in leaves (53). For such metabolites, harvested tissues have to be inactivated as quickly as possible and under the growth conditions of interest, e.g. under ambient light if photosynthetic tissues are to be harvested during the photoperiod. Freeze-clamping is probably the most appropriate method for such purpose (54), but it is rather time-consuming, thus limiting the number of samples than can be taken.

Plants produce a large and diverse panel of naturally volatile compounds, which are also technically challenging to study, especially when they are collected from the headspace of leaves, flowers, or fruits, thus requiring sophisticated experimental set-ups (55–57).

4.8. Giving Metabolomic Data a Second Life

Because they are rather expensive and slow, 'omics faces the paradox of measuring too many things in too few samples. When experiments have been thoroughly designed and described, it nevertheless becomes possible to perform meta-analyses with very large data sets. The implementation of public repositories further increases the amount of data that can be accessed to extract new information without needing to perform additional experiments, and to support and extend the interpretation of new data sets. The use of standardised conceptualisations with explicit specifications to report data and metadata (i.e. data about data) will be decisive. MIAME (Minimum Information About a Microarray Experiment) was the first initiative to impose the use of a controlled system to describe 'omics experiments (58). Quickly, major scientific journals then started to require publications describing microarray experiments to comply with the MIAME guidelines, thus greatly improving accessibility to transcriptomics data. This initiative inspired the emergence of a range of minimum information checklists for reporting diverse biological experiments (http://www.mibbi.org/index.php/MIBBI_portal). Thus, standardisation efforts aiming at obtaining metabolomic data that support evaluation, repetition, and/or extension of experiments and ultimately enable data mining are ongoing (59, 60), resulting in guidelines that cover almost every aspect of the experimentation, ranging from growth conditions to technical details of the analysis. Importantly, minimal information checklists standardise the data content, as they impose what terms have to be described. However, they do not necessarily constrain the format, as terms can usually be described using free text. This is probably better achieved using ontologies (61) dedicated to specific aspects of the biological experimentation (e.g. genotypes, phenology, or abiotic growth conditions), each of them providing structured terminologies that are implemented with precise specifications about the terms and their use (e.g. http://www.gramene.org; http://www.codeplex.com/XeO). The use of ontologies to generate metadata about experiments should ultimately prove more advantageous for data mining approaches, as they better constrain the format. Furthermore, given an experiment has been described using dedicated ontologies, the generation of the relevant minimum information checklist should be facilitated. Indeed, generating checklists can be time-consuming and frustrating, favouring mistakes as pointed out recently (62). Fortunately, tools dedicated to the planning of experiments and the generation of checklists and/or ontology-based descriptions have already been developed (45, 63), paving the way for a more efficient sharing of data and metadata. Recently, Xeml Lab, a platform that helps plant biologists to plan experiments, from setting environmental history to defining sampling strategy, and concomitantly generate machine-readable and ontology-based metadata files has been proposed (45). Thus, while filling in checklists is likely to become unavoidable for publication, it is worth considering the use of such tools at the very beginning of an experiment.

5. Conclusions

In the last decade, great efforts and energy have been invested in advancing technologies as well as dedicated bioinformatics (see Chapters in Parts 2 and 3 in this book for more details). However, without care we will continue to generate data while running the risk of losing sight of the primary goal of the production of knowledge. We need to identify and understand the limitations of the methods we are using at each step of the experimentation, and then formulate the most appropriate experimental design.

We propose a list of items to be considered for experimental design in the field of plant metabolomics (see Fig. 1). Typically, once the biological question has been clearly formulated, one needs to choose the most appropriate biological resource and analytical technology. Which factor(s) can be varied to reveal a metabolic response, and how can this response be monitored? There are usually a number of valid options at this stage, but there are probably even more non-valid ones. Then, in order to cope with biological error, it is important to decide how to grow plant material and when and how to harvest samples. There are no recommendations about how to grow plants, but there is a real and urgent need to document the history of growth conditions for each sample, in

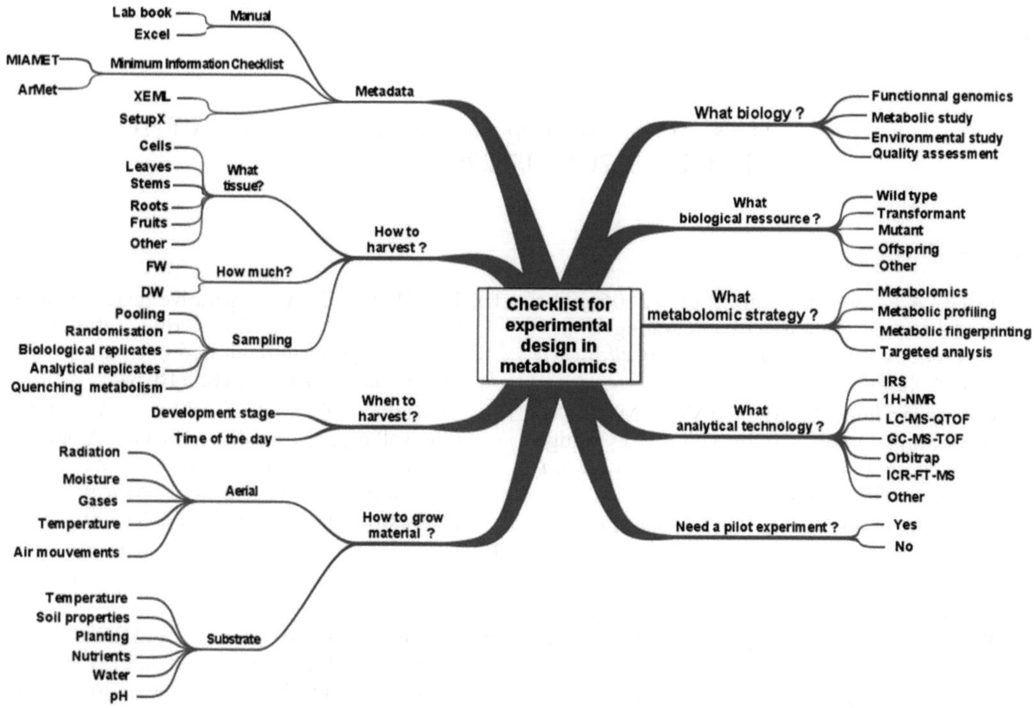

Fig. 1. List of items to be considered before starting a plant metabolomic experiment.

order to be able to compute metadata and analytical data in the near future. For this, minimum information checklists, ontologies, and tools such as Xeml Lab (45) are already available. Optimal reproducibility might be obtained by considering the following. First, it is essential to know how many samples are necessary to perform the analysis. Second, it is important to control both developmental stage and the time of the day at which samples are harvested. Third, the most adequate sampling procedure needs to be decided, mostly depending on how stable the metabolites of interest are. Experimental error needs to be evaluated and documented so that decisions about blocking and randomisation, pooling samples, and biological and technical replication may be made. It is also recommended that as many plants as possible are grown and that plant samples are as large as possible. Before embarking on time-consuming and costly experiments, it might be useful to perform a preliminary experiment verifying that all operational parameters are appropriate. Another good option would be to use tools dedicated to the planning of experiments and the generation of checklists and/or ontology-based descriptions. These could contribute to the success of plant metabolomics by enabling us to link environmental metadata with analytical data and to perform reliable and comparable analyses.

Acknowledgements

This work was supported by the EU META-PHOR Project (FOOD-CT-2006-036220).

References

1. Joyce, A.R. and Palsson, B.O. (2006) The model organism as a system: integrating 'omics' data sets. *Nature Review Molecular Cell Biology* 7, 198–210.
2. Ge, H., Walhout, A.J.M., and Vidal, M. (2003) Integrating 'omic' information: a bridge between genomics and systems biology. *Trends in Genetics* **19**, 551–60.
3. Van Dien, S. and Schilling, C.H. (2006) Bringing metabolomics data into the forefront of systems biology. *Molecular Systems Biology* **2**, 1–2.
4. Liu, E.T. (2005) Systems Biology, Integrative Biology, Predictive Biology. *Cell* **121**, 505–6.
5. Bacon, F. (1620) The new organon or true directions concerning the interpretation of nature, in *The Works Vol. VIII* (Spedding J., Ellis R.L., and D.D. Heath, eds.): Taggard and Thompson, Boston, USA; 1863.
6. Anderson, M.J. and Whitcomb, P.J. (2007) DOE simplified practical tools for effective experimentation. 2nd edition Productivity Press (New York).
7. Fernandez, L., Romieu, C., Moing, A., Bouquet, A., Maucourt, M., Thomas, M.R., and Torregrosa, L. (2006) The Grapevine fleshless berry mutation. A unique genotype to investigate differences between fleshy and non fleshy fruits. *Plant Physiology* **140**, 537–47.
8. Fisher, R. (1926) The arrangement of field experiments. *Journal of the Ministry of Agriculture of Great Britain* **33**, 503–13.
9. Peric-Concha, N. and Long, P.F. (2003) Mining the microbial metabolome: a new frontier for natural product lead discovery. *Drug Discovery Today* **8**, 1078–84.
10. Rocke, D.M. (2004) Design and analysis of experiments with high throughput biological assay data. *Seminars in Cell & Developmental Biology* **15**, 703–13.
11. Usadel, B., Nagel, A., Steinhauser, D., Gibon, Y., Bläsing, O.E., Redestig, H., et al. (2006) PageMan: An interactive ontology tool to generate, display, and annotate overview graphs for profiling experiments. *BMC Bioinformatics* 7, 535.
12. Lay, J.O., Liyanagea, R., Borgmannb, S., and Wilkins, C.L. (2006) Problems with the "omics". *Trends in Analytical Chemistry* **25**, 1046–56.
13. Pan, W. (2002) A comparative review of statistical methods for discovering differentially expressed genes in replicated microarray experiments. *Bioinformatics* **18**, 546–54.
14. Benjamini, Y. and Hochberg, Y. (1995) Controlling the false discovery rate: A practical and powerful approach to multiple testing. *Journal of the Royal Statistical Society: Series B (Statistical Methodology)* **57**, 289–300.
15. Festing, M. (1994) Reduction of animal experimental design and quality of experiments. *Laboratory Animals* **28**, 212–21.
16. Sumner, L.W., Mendes, P., and Dixon, R.A. (2003) Plant metabolomics: large-scale phytochemistry in the functional genomics era. *Phytochemistry* **62**, 817–36.
17. Ap Rees, T. and Hill, S.A. (1994) Metabolic control analysis of plant metabolism. *Plant, Cell and Environment* **17**, 587–99.
18. Fiehn, O. (2002) Metabolomics: the link between genotypes and phenotypes. *Plant Molecular Biology* **48**, 155–71.
19. Giavalisco, P., Hummel, J., Lisec, J., Inostroza, A., C, Catchpole, G., and Willmitzer, L. (2008) High-Resolution Direct Infusion-Based Mass Spectrometry in Combination with Whole C-13 Metabolome Isotope Labeling Allows Unambiguous Assignment of Chemical Sum Formulas. *Analytical Chemistry* **80**, 9417–25.
20. Kopka, J., Fernie, A.R., Weckwerth, W., Gibon, Y., and Stitt, M. (2004) Metabolite profiling in plant biology: platforms and destinations. *Genome Biology* **5**, 109.
21. Lommen, A., Weseman, J.M., Smith, G.O., and Noteborn, H.P.J.M. (1998) On the detection of environmental effects on complex matrices combining off-line liquid chromatography and ^{1}H-NMR. *Biodegradation* **9**, 513–25.
22. Bailey, N.J.C., Oven, M., Holmes, E., Nicholson, J.K., and Zenk, M.H. (2003) Metabolomic analysis of the consequences of cadmium exposure in *Silene cucubalus* cell cultures via ^{1}H NMR spectroscopy and chemometrics. *Phytochemistry* **62**, 851–8.
23. Ott, K.-H., AranÌbar, N., Singh, B., and Stockton, G.W. (2003) Metabonomics classifies pathways affected by bioactive compounds. Artificial neural network classification of NMR

spectra of plant extracts. *Phytochemistry* **62**, 971–85.

24. Noteborn, H.P.J.M., Lommen, A., van der Jagt, R.C., and Weseman, J.M. (2000) Chemical fingerprinting for the evaluation of unintended secondary metabolic changes in transgenic food crops. *Journal of Biotechnology* 77, 103–14.
25. Le Gall, G., DuPont, M.S., Mellon, F.A., Davis, A.L., Collins, G.J., Verhoeyen, M.E., and Colquhoun, I.J. (2003) Characterization and Content of Flavonoid Glycosides in Genetically Modified Tomato (*Lycopersicon esculentum*) Fruits. *Journal of Agricultural and Food Chemistry* **51**, 2438–46.
26. Saito, K., Dixon, R.A., and Willmitzer, L. (2006) Plant Metabolomics. Springer (Berlin Heidelberg).
27. Gullberg, J., Jonsson, P., Nordstrom, A., Sjostrom, M., and Moritz, T. (2004) Design of experiments: an efficient strategy to identify factors influencing extraction and derivatization of Arabidopsis thaliana samples in metabolomic studies with gas chromatography/mass spectrometry. *Analytical Biochemistry* **331**, 283–95.
28. Lunn, J.E., Feil, R, Hendriks, J.H.M., Gibon, Y., Morcuende, R., Osuna, D., et al. (2006) Sugar-induced increases in trehalose 6-phosphate are correlated with redox activation of ADPglucose pyrophosphorylase and higher rates of starch synthesis in Arabidopsis thaliana. *Biochemical Journal* **397**, 139–48.
29. Bonferroni, C.E. (1935) Il calcolo delle assicurazioni su gruppi di teste, in *Studi in Onore del Professore Salvatore Ortu Carboni*. Rome Italy; pp. 13–60.
30. Yang, M.C.K., Yang, J.J., McIndoe, R.A., and She, J.X. (2003) Microarray experimental design: power and sample size considerations. *Physiological Genomics* **16**, 24–8.
31. Pawitan, Y., Michiels, S., Koscielny, S., Gusnanto, A., and Ploner, A. (2005) False discovery rate, sensitivity and sample size for microarray studies. *Bioinformatics* **21**, 3017–24.
32. Jørstad, T.S., Langaas, M., and Bones, A.M. (2007) Understanding sample size: what determines the required number of microarrays for an experiment? *Trends in Plant Science* **12**, 46–50.
33. Broadhurst, D.I. and Kell, D.B. (2006) Statistical strategies for avoiding false discoveries in metabolomics and related experiments. Metabolomics **2**, 171–96.
34. Gibon, Y., Vigeolas, H., Tiessen, A., Geigenberger, P., and Stitt, M. (2002) Sensitive and high throughput metabolite assays for inorganic pyrophosphate, ADPGlc, nucleotide phosphates, and glycolytic intermediates based on a novel enzymic cycling system. *Plant Journal* **30**, 221–35.
35. Mashego M.R., Wu L., Van Dam J.C., Ras C., Vinke J.L., Van Winden W.A., et al. (2004) MIRACLE: mass isotopomer ratio analysis of U-C-13-labeled extracts. A new method for accurate quantification of changes in concentrations of intracellular metabolites. *Biotechnology and Bioengineering* **85**, 620–8.
36. Huang, X. and Regnier, F.E. (2008) Differential Metabolomics Using Stable Isotope Labeling and Two-Dimensional Gas Chromatography with Time-of-Flight Mass Spectrometry. *Analytical Chemistry* **80**, 107–14.
37. Gibon, Y., Usadel, B., Blaesing, O.E., Kamlage, B., Hoehne, M., Trethewey, R., and Stitt, M. (2006) Integration of metabolite with transcript and enzyme activity profiling during diurnal cycles in Arabidopsis rosettes. *Genome Biolology* 7, R76.
38. Scholz, M., Gatzek, S., Sterling, A., Fiehn, O., and Selbig, J. (2004) Metabolite fingerprinting: detecting biological features by independent component analysis. *Bioinformatics* **20**, 2447–54.
39. Schauer, N., Semel, Y., Roessner, U., Gur, A., Balbo, I., Carrari, F., et al. (2006) Comprehensive metabolic profiling and phenotyping of interspecific introgression lines for tomato improvement. *Nature Biotechnology* **24**, 447–54.
40. Keurentjes, J.J., Fu, J., de Vos, C.H., Lommen, A., Hall, R.D., Bino, R.J., et al. (2006) The genetics of plant metabolism. *Nature Genetics* **38**, 842–9.
41. Rowe, H.C., Hansen, B.G., Halkier, B.A., and Kliebenstein, D.J. (2008) Biochemical Networks and Epistasis Shape the *Arabidopsis thaliana* Metabolome. *The Plant Cell* **20**, 1199–216.
42. Fernie, A.R. and Schauer, N. (2009) Metabolomics-assisted breeding: a viable option for crop improvement? *Trends in Genetics* **25**, 39–48.
43. Yu, J., Holland, J.B., McMullen, M.D., and Buckler, E.S. (2008) Genetic Design and Statistical Power of Nested Association Mapping in Maize. *Genetics* **178**, 539–51.
44. Thimm, O., Bläsing, O.E., Usadel, B., and Gibon, Y. (2006) Evaluation of the transcriptome and genome to inform the study of metabolic control, in *Control of Primary Metabolism in Plants*. (Plaxton B, McManus M, eds.) Blackwell Publishing Oxford (UK). pp. 1–23.
45. Stitt, M., Gibon, Y., Lunn, J.E., and Piques, M. (2006) Multilevel genomics analysis of carbon

signalling during low carbon availability: coordinating the supply and utilisation of carbon in a fluctuating environment. *Functional Plant Biology* **34**, 526–49.

46. Hannemann, J., Poorter, H., Usadel, B., Bläsing, O.E., Finck, A., Tardieu, F., et al. (2009) Xeml Lab: a software suite for a standardised description of the growth environment of plants. *Plant, Cell and Environment* **32**, 1185–200.
47. Sultan, S.E. (2000) Phenotypic plasticity for plant development, function and life history. *Trends in Plant Science* **5**, 537–42.
48. Allan, W.L. and Shelp, B.J. (2006) Fluctuations of gamma-aminobutyrate, gamma-hydroxybutyrate and related amino acids in Arabidopsis leaves as a function of the light–dark cycle, leaf age, and N stress. *Canadian Journal of Botany* **84**, 1339–46.
49. Geiger, D.R. and Servaites, J.C. (1994) Diurnal regulation of photosynthetic carbon metabolism in C3 plants. *Annual Review of Plant Physiology* **45**, 235–56.
50. Winter, H., Lohaus, G., and Heldt, H.W. (1992) Phloem transport of amino-acids in relation to their cytosolic levels in barley leaves. *Plant Physiology* **99**, 996–1004.
51. Fahnenstich, H., Saigo, M., Niessen, M., Drincovich, M., F, Flügge, U.-I., and Maurino, V.G. (2008) Malate and fumarate emerge as key players in primary metabolism: Arabidopsis thaliana overexpressing C_4-NADP-ME offer a way to manipulate the levels of malate and to analyse the physiological consequences, in *Photosynthesis. Energy from the Sun* (J.F. Allen, E. Gantt, J.H. Golbeck and B. Osmond eds.) Springer-Verlag, Heidelberg, Germany pp. 971–5.
52. Ma, F. and Cheng, L. (2003) The sun-exposed peel of apple fruit has higher xanthophyll cycle dependent thermal dissipation and antioxidants of the ascorbate/glutathione pathway than the shaded peel. *Plant Science* **165**, 819–27.
53. Sharkey, T.D., Stitt, M., Heineke, D., Gerhardt, R., Raschke, K., and Heldt, H.W. (1986) Limitation of Photosynthesis by Carbon Metabolism: II. O_2-Insensitive CO_2 Uptake Results from Limitation Of Triose Phosphate Utilization. *Plant Physiology* **81**, 1123–9.
54. Ap Rees, T., Fuller, W.A., and Wright, B.W. (1977) Measurements of glycolytic intermediates during the onset of thermogenesis in the spadix of *Arum maculatum*. *Biochimica Biophysica Acta* **461**, 274–82.
55. Verdonk, J.C., de Vos, C.H.R., Verhoeven, H.A., Haring, M.A., van Tunen, A.J., and Schuurink, R.C. (2003) Regulation of floral scent production in petunia revealed by targeted metabolomics. *Phytochemistry* **62**, 997–1008.
56. Tikunov Y.M., Verstappen F.W., and Hall R.D. (2007) Metabolomic profiling of natural volatiles: headspace trapping: GC-MS. *Methods in Molecular Biology* **358,** 39–53.
57. Tikunov, Y., Lommen, A., de Vos, C.H., Verhoeven, H.A., Bino, R.J., Hall, R.D., and Bovy, A.G. (2005) A Novel Approach for Nontargeted Data Analysis for Metabolomics. Large-Scale Profiling of Tomato Fruit Volatiles. *Plant Physiology* **139**, 1125–37.
58. Brazma, A., Hingamp, P., Quackenbush, J., Sherlock, G., Spellman, P., Stoeckert, C., et al. (2001) Minimum information about a microarray experiment (MIAME) – toward standards for microarray data. *Nature Genetics* **29**, 365–71.
59. Jenkins, H., Hardy, N., Beckmann, M., Draper, J., Smith, A.R., Taylor, J., et al. (2004) A proposed framework for the description of plant metabolomics experiments and their results. *Nature Biotechnology* **22**, 1601–6.
60. Fiehn, O., Wohlgemuth, G., Scholz, M., Kind, T., Lee Do, Y., Lu, Y., Moon, S., and Nikolau, B. (2008) Quality control for plant metabolomics: reporting MSI-compliant studies. *The Plant Journal* **53,** 691–704.
61. Gruber, T.R. (1995) Toward principles for the design of ontologies used for knowledge sharing? *International Journal of Human-Computer Studies* **43**, 907–28.
62. Larsson, O. and Sandberg, R. (2006) Lack of correct data format and comparability limits future integrative microarray research. *Nature Biotechnology* **24**, 1322–3.
63. Scholz, M. and Fiehn, O. (2007) Setup X – A public study design database for metabolomic projects. *Pacific Symposium on Biocomputing* **12**, 169–80.

Chapter 3

Separating the Inseparable: The Metabolomic Analysis of Plant–Pathogen Interactions

J. William Allwood, Jim Heald, Amanda J. Lloyd, Royston Goodacre, and Luis A.J. Mur

Abstract

Plant–microbe interactions—whether pathogenic or symbiotic—exert major influences on plant physiology and productivity. Analysis of such interactions represents a particular challenge to metabolomic approaches due to the intimate association between the interacting partners coupled with a general commonality of metabolites. We here describe an approach based on co-cultivation of *Arabidopsis* cell cultures and bacterial plant pathogens to assess the metabolomes of both interacting partners, which we refer to as dual metabolomics.

Key words: Plant–microbe interactions, Pathogen, *Arabidopsis*, Plant suspension cultures, Co-cultivation, Dual metabolomics

1. Introduction

Plant interactions with microbes play a major role in defining physiology and development. Plants are continually under attack from pathogens of various species and the deployment of diverse defences represents a substantial cost to the host (1). Equally, pathogens can act as selective agents driving the selection of resistant germplasm (2). From an anthropogenic point of view, pathogen attack represents a considerable source of crop loss (3). However, plant–microbe interactions are not solely pathogenic; many symbiotic relationships exist which improve nutrient assimilation by the plant and hence improve productivity. The interaction between nitrogen-fixing bacteria and legume plants is especially well characterised (4, 5), but due to their wider host-species range, interactions with mycorrhizal fungi are possibly more important. Mycorrhizal fungi establish hyphal connections with cells in the root and extend into the surrounding soil, thereby

Nigel W. Hardy and Robert D. Hall (eds.), *Plant Metabolomics: Methods and Protocols*, Methods in Molecular Biology, vol. 860,
DOI 10.1007/978-1-61779-594-7_3, © Springer Science+Business Media, LLC 2012

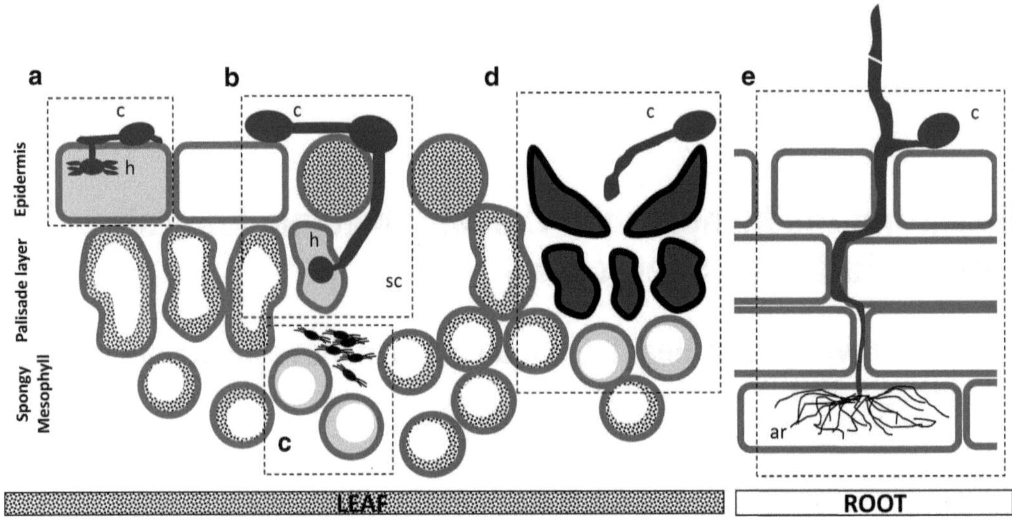

Fig. 1. Tissue heterogeneity as a result of various plant pathogen interactions. A schematic transverse section through a plant leaf and root illustrating interactions with a range of microbes. Green and healthy plant cells are filled with *dots*, and those which ware exhibiting disease symptoms are shown in *light grey*, whilst those which are dead are in *dark grey*. (**a**) The germinated condium (c) ultimately forms a digitate feeding structure—the haustorium (h)—which does not penetrate beyond the epidermal layer but supplies nutrients from the host to ectopic fungal development. (**b**) The infection structure of Rust fungal pathogens which target open stomata, penetrating into the substomatal cavity (sc). Within this area, the fungus forms haustoria-like feeding structures and elaborates *in planta* hyphal development until sporulation, where the rust-clusters of conidiophores burst through the epidermal surface (not shown). (**c**) Biotrophic bacterial pathogens (i.e. those which live off living plant tissue for extended periods) tend to infect via stomata or opportunistically at wound sites. They multiply within the apoplastic space surrounding the cells. The amphitrichous flagellate *Pseudomonas syringae* is shown. (**d**) A pathogenic interaction involving a necrotrophic fungus is shown. Host death arises through toxin production and/or enzymatic attack originating from the pathogen. Note that no obvious infection structure is observed with necrotrophic fungal pathogens. (**e**) A symbiotic interaction with an arbuscular mycorrhiza (plural mycorrhizae or mycorrhizas) is shown where the fungus (Phylum Glomeromycota) penetrates to the cortical cells of the roots of a vascular plant. This interaction is characterised by the formation of arbuscules (ar) and significant fungal growth from the root into the surrounding soil (indicated by a *broken hypha* in the diagram).

improving nitrogen and phosphate uptake (4) and encourage association with plant-growth-promoting rhizobacteria (6, 7).

A key feature of all such plant–microbe interactions is an intimate association between both partners (Fig. 1). It is important to note that reciprocal responses in both host and microbe results in altered molecular and physiological status so that it is distinct to their situations when considered in isolation. Further, when investigating these interactions it is often difficult as they frequently involve only a few participating plant cells, which themselves can show considerable spatial heterogeneity in their responses. Sampling therefore often includes large numbers of non-, or differently, responding cells so that any localised responses may be difficult to discern. Equally, it is often difficult to gather sufficient material from the microbial partner to make analysis possible. Therefore, there are considerable technical problems in investigating plant–microbe interactions.

If the aim is only to consider the host response and the focus is on gene transcript or proteomic changes, it is valid to simply ignore

Fig. 2. Approaches to assess changes in plant microbe interactions. (**a**) A widely used approach to inoculate *Arabidopsis thaliana* with bacterial pathogens involves the infiltration of the intracellular spaces of leaves with bacterial suspensions in 10 mM $MgCl_2$ (~10^6 cell/mL). Typically, the bacterial suspensions are infiltrated using a syringe via the stomata of the lower epidermal surface. Alternative approaches can involve dipping or spraying *Arabidopsis* with high titres of bacterial suspensions. Infiltration of leaf spaces has the advantage of producing a large area of synchronously responding plant tissue which reflects the nature of the interaction. (**b**) Inoculation with *Pseudomonas syringae* pv. *tomato* strain DC3000 (*Pst*) *avrRpm1* rapidly elicits cell death (a Hypersensitive Response (HR)) within the inoculated area (encompassed by the dashed lines and arrowed) (bar = 1 cm). Disease and elicitation of the HR is dependent on the delivery of bacterial protein effectors into the host, the nature of the response being dependent on the plant genotype. The bacterial effectors may be cloned and fused to an inducible promoter and introduced into *Arabidopsis* plants to generate transgenic lines. Two examples are given. (**c**) The *HopAB2* bacterial effector gene is fused to the glucocorticoid responsive promoter. This, along with the mammalian glucocorticoid receptor/transcriptional activator protein gene, was introduced into *Arabidopsis.* Application of glucocorticoid to *HopAB2* transgenic plants resulted in the elaboration of symptoms (*arrowed*) analogous to disease symptoms. Details of the inducible system can be found in ref. (65). (**d**) The *avrPpiA1* avirulence gene which elicits a HR in *RPM1* encoding *Arabidopsis* Col-0. The *avrPpiA1* gene was fused to the *Aspergillus nidulans niger* alcohol dehydrogenase (*alcA)* promoter. This, along with the alcohol responsive transcriptional activator (*AlcR*) protein gene, was introduced into *Arabidopsis* (66). Application of alcohol to *avrPpiA1* transgenic plants resulted in the rapid elicitation of cell death (*arrowed*)—which was reminiscent of the HR. (**e**) Plant–pathogen interactions can also be investigated in plant suspension cell cultures inoculated with bacterial pathogens. Illustrated is an *Arabidopsis* cell cluster from a suspension cell culture. (Bar = 200 μm).

the interacting microbe, since RNA transcript or protein sequence allows changes to be targeted to a specific partner. However, metabolites tend not to exhibit such specificity and therefore this poses a significant challenge to metabolomic analyses. This problem is further compounded if interactions of bacterial pathogens with, for example, *Arabidopsis thaliana* are being examined. In such cases, a commonly used approach is to infiltrate the intracellular species of leaves with high titres of bacterial suspension (Fig. 2a). This offers an excellent source of large areas of synchronously responding tissue, thereby mitigating the problem of tissue limitation referred to above. Depending on host and pathogen genotypes, inoculated areas can exhibit disease symptoms or a form of plant cell death—the Hypersensitive Response (HR; Fig. 2b)—which is localised to the point of attempted infection and is often linked to resistance (8). However, the high bacterial population makes a considerable contribution to the metabolome of a sample from plant–pathogen interactions so that the origin of a given metabolite may defy attribution.

A number of approaches exist that may possibly circumvent this problem. It may be assumed that the biomass of the interacting microbe is so low that its contribution to the metabolome will be insignificant and will therefore represent only that of the plant host. If such a strategy is followed, it would be necessary to confirm that no significant metabolite contribution to the samples has occurred. This necessitates screening for microbe specific metabolite biomarkers. In the case of fungal microbes, this may involve assaying for the steroid ergosterol (9) or the phospholipid acyl chain, arachidonic acid (C20:4; (10)), both of which are absent from plants. Alternatively, the metabolome of one or indeed both interacting partners can be isotopically labelled (11–13). Thus, comparison of labelled and non-labelled interactions will allow the relative contribution of the interacting partners to the metabolome to be defined.

Another approach is to avoid any plant-interacting metabolome at all, for example, by focusing on plant viruses. For example, metabolomic changes associated with tobacco infected with tobacco mosaic virus have been described (14). There is, however, no need to limit oneself to viral pathogens. The use of pathogen-derived elicitors to investigate plant defence responses is a very well-established approach (15). There are a range of elicitors available from pathogens from various kingdoms which can be used to examine resistance mechanisms associated with or without limited host cell death (Table 1). It is also possible to focus on mutants in model plant species which show the constitutive activation of plant defences. For example, *lesions simulating disease 1* (*lsd1*), exhibits the spontaneous exhibition of necrotic lesions, which is frequently equated with a HR (16, 17). There are many examples of such cell death mutants (18). Other mutants show constitutive exhibition of defence-associated signalling events linked to, for example, the hormones salicylic acid (SA), jasmonates (JA), or ethylene (19).

Yet another approach is suggested from the pathogenesis mechanisms of bacterial pathogens, particularly those of *Pseudomonas syringae* and *Xanthomonas* spp. In these species, bacteria deliver large numbers of effector proteins into plant cells to cause disease, but in certain plant genotypes an effector is recognised by a resistance gene to elicit the HR (20). In the latter situation, the recognised effectors are referred to as avirulence gene products. Bacterial effectors can be fused to an inducible promoter

Table 1
Some microbial elicitors of plants defence or symbiotic responses

Elicitor	Origin	Action	References
Chitin oligosaccharide	Fungal cell walls	General defence initiation. No cell death	(55)
13-Pep	*Phytophthora sojae*	General defence initiation. No cell death	(56)
Flagellin/flg22	Bacterial pathogens	Initiator of basal defences	(57)
Lipopolysaccharide (LPS)	Bacterial	Initiator of basal defences	(58)
Harpin	Bacterial pathogens	Initiator of cell death	(59)
Cryptogein	*Phytophthora cryptogea*	Initiator of defences including cell death	(60)
Victorin	*Cochliobolus*	Initiator of cell in *Vb* genotypes of oat	(61)
NIP1	*Rhynchosporium secalis*	Initiator of cell in *Rrs1* *genotypes of barley*	(62)
Nod factors (*lipochito-Oligosaccharides*)	*Rhizobium* spp.	Root deformation *in legumes*	(63)

Plant responses to pathogens may be crudely designated as cell death associated (often equated with the Hypersensitive Response, HR) or defences without initiation of macroscopic cell death. The table lists some of the elicitors of each form of defence that are available. Elicitors of plant defence leading to the induction of defence gene expression, but not cell death, include chitin and 13-Pep. Other elicitors initiate a subset of plant defences, which are linked to basal/innate resistance, which are displayed against any microbes and not only plant pathogens. Well-characterised examples of basal resistance are elicited by flagellin—flg22—or the bacterial LPS. Defences associated with cell death may be investigated by the addition of NIP1 (Necrosis inducing protein 1) which is secreted by the fungus *Rhynchosporium secalis* (the causal agent of scald disease) on *Rrs1* genotypes of barley or the general cell death elicitor harpin, which is an ionophore isolated from Pseudomonad bacterial pathogens. Other cell death elicitors are produced by pathogens which actively kill as part of their pathogenesis mechanism, but these are taken as parallels of the HR. These include cryptogein, a 98-amino acid proteinaceous elicitor from *Phytophthora cryptogea*—which elicits cell death on many plant species and victorin, a host-selective toxin produced by the fungus *Cochliobolus victoriae*, which acts only on *Vb* oat (*Avena sativa*) genotypes. Considering symbiotic interaction, although nitrogen fixation in legume-*Rhizobium* bacteria interactions takes place within root, some early aspects can be induced by pathogen-secreted Nod factors. Nod factors are lipochitooligosaccharides consisting of an acylated chitin oligomeric backbone with differing moieties conferring different host specificities (63, 64)

and introduced into plants to generate transgenic lines (Fig. 2c, d). This offers a substantial source of responsive tissue that can be linked to the action of particular bacterial cell effectors and be used to examine metabolomic changes.

However, such approaches are limited as the metabolome of the interacting microbe is absent. One way of assessing the complicated metabolomic changes associated with plant microbial interactions is to exploit the possibilities offered by in situ imaging of metabolites and thereby assigning key changes to one interacting partner or the other. For example, there have been recent advances in imaging metabolites based on matrix-assisted laser desorption ionisation

(MALDI) imaging techniques. In this approach, the MALDI matrix is applied to thin sections of sample and the laser desorbed proteins, peptides, or small molecules are assessed by Mass Spectrometry (MS). The spatial patterns of metabolites can be mapped using a range of suitable imaging processing software and related to an optical image of the sample. A particularly attractive aspect of this technique is the ability to describe the distribution of tentatively identified metabolites. MALDI-MS imaging has been used to map the distribution of a modified form of the CLAVATA3 (CLV3) secreted peptide hormone in *Arabidopsis* callus slices (21), the movement of pesticides on the leaves and within the stems of soya plants (22), and the distribution of carbohydrate in wheat stems (23). Most pertinent to this chapter, MALDI was used in conjunction with Fourier Transform-Mass Spectrometry (FT-MS) to compare the spectra of leaves from a healthy *Prunus persica* (peach) tree to those infected with the fungus *Taphrina deformans*. This approach identified differences in the abundance of various phospholipids (24).

In an allied approach, microspectroscopic techniques are now being applied to visualise biochemical changes in plant systems, although identification of actual metabolites is difficult. Infrared (IR) mapping with a synchrotron source using a focal plane array (FPA) allowed the analysis of Eucalyptus (25), aleurone cell walls in wheat grain (26), and maize seeds (27). Raman spectroscopy relies on inelastic scattering in monochromatic light (28) and has proven to be particularly useful for imaging lignin structure (29).

Although we have exploited FT-IR microspectroscopic approaches (30), here we describe an alternative approach based on the co-culture of *Arabidopsis* suspension cell clusters (Fig. 2f) and bacteria pathogens. This approach offers a ready source of plant material that is often used as a model for plant responses (12, 31–33). Further, as phytopathogenic bacteria are not phagocytosed, as occasionally occurs with disease in animals (34), it is possible to separately elucidate the microbial metabolome. This determination of interacting host and microbial metabolomes we refer to as dual metabolomics.

2. Materials

1. *Arabidopsis* cell cultures (see Note 1).
2. AT3 medium: Murashige–Skoog salts with vitamins 4.41 g/L, sucrose 30 g/L, NAA 0.5 mg/L, kinetin 0.05 mg/L. Sterilised by autoclaving.

3. Bacterial strains: *Pseudomonas syringae* pathovar *tomato* DC3000 (*Pst*), *Pst avrRpm1*, and *Pst hrpA* (see Note 1).
4. Nutrient Agar (NA) 5 g/L peptone, 3 g/L beef extract, 15 g/L agar; adjusted pH to 7.0. Sterilised by autoclaving.
5. Nutrient Broth (NB 5 g/L peptone, 3 g/L beef extract, adjusted pH to 7.0). Sterilised by autoclaving.
6. Whatman No 1 filter paper (pore size ~11 πm) or equivalent.
7. Vacuum Pump.
8. Laminar Flow Cabinet.
9. Antibiotics. The antibiotics used will reflect the bacterial strains and plasmids under selection. In the experiments described here, the following antibiotics and concentrations were used: rifampicin (100 πg/mL) and kanamycin (10 πg/mL).
10. $MgCl_2$ used at 10 mM concentration in de-ionised (for example, by distillation) water. Sterilised by autoclaving.
11. Evans Blue strain (Sigma Pharmaceuticals Ltd) is used at 0. 25% (w/v) in water.
12. NaCl was used at 0.85% (w/v) concentration in de-ionised (for example, by distillation) water. Sterilised by autoclaving.
13. A mix of chloroform (Fisher Scientific, UK), methanol (Fisher Scientific, UK), and de-ionised (for example, by distillation) and sterile (sterilised by autoclaving) water in a proportion of 1:2.5:1 (v/v).
14. Commercially prepared ultra-pure dH_2O (Fisher Scientific, UK): Sterilised by autoclaving.
15. Acetonitrile (Fisher Scientific, UK) and 0.2% formic acid (Fisher Scientific, UK) in de-ionised (for example, by distillation) and sterile (sterilised by autoclaving) water (v/v) were mixed in a proportion of 1:1 (v/v).
16. Methanol (Fisher Scientific, UK): used at 80% and 60% (v/v) dilution in de-ionised (for example, by distillation) and sterile (sterilised by autoclaving) water.
17. Propan-2-ol (Fisher Scientific, UK): used at 70 dilution (v/v) in de-ionised (for example, by distillation) and sterile (sterilised by autoclaving) water.
18. Acetonitrile (Fisher Scientific, UK): used at 10 dilution (v/v) in de-ionised (for example, by distillation) and sterile (sterilised by autoclaving) water.

3. Methods

3.1. Establishing the Host and Pathogen Metabolomes

3.1.1. Culture of Arabidopsis Suspension Cells

1. The *Arabidopsis* Landsberg *erecta* (L *er*) suspension was first derived from callused stem cells developed by May and Leaver (35).
2. The plant culture regime should be standardised and well established in the investigators group prior to commencing with dual metabolomic studies. Maintain *Arabidopsis* suspension as 200 mL AT3 medium at 24°C on a long day 16 h light cycle at 25 μmol/m^2/s. Cultures should be aerated by shaking on an orbital shaker at 140 rpm. Subculturing should occur after no more than 7 days by transferring ~3 mL of 7 day culture into 200 mL of fresh AT3 in a laminar flow cabinet. The suspension cells should be free of contamination and exposed to minimal stress (see Notes 2 and 3).
3. Maintain large numbers of 200 mL cultures. For each experiment 15 × 200 mL cultures are pooled (see Subheading 3.2); hence, multiple cultures will allow ready inoculation of large numbers of AT3 cultures (see Note 4).

3.1.2. Culture of Bacterial Strains

Whilst it is perfectly valid to examine metabolite changes within a single bacterial strain interacting with a host, the value of the dual metabolomic approach is increased if the responses of different bacterial strains are compared. However, this requires that either the starting metabolomes of each strain be well defined or ideally, be substantially equivalent. The latter can be achieved by growing each strain in chemostats. However, in many laboratories this may not be possible; hence, the following protocol details a semi-batch approach where *Pseudomonas syringae* pv. *tomato* DC3000 (*Pst*—which is virulent on *Arabidopsis*), *Pst avrRpm1* (which is "avirulent"—in that it can elicit a HR from *Arabidopsis*), and *Pst hrpA* (which is non-virulent and is unable to elicit a HR) were grown.

1. Maintain the bacterial strains on solid nutrient agar (NA) plates. Derive single colonies by streaking across the agar surface using a sterile wire loop. Supplement the medium with appropriate antibiotics to maintain any plasmids within the strains.
2. Add a single colony from the bacterial plate to 10 mL of NB and incubate at 28°C in an orbital shaker at ~200 rpm for ~12 h. Use an aliquot of 5 mL of this culture to inoculate 400 mL NB and incubate at 28°C in an orbital shaker at ~200 rpm (~10 × *g*) (see Note 4).
3. The bacteria used for inoculation of *Arabidopsis* cultures should be in a mid-exponential growth phase. Assess samples (1 mL) from the culture for culture turbidity using a spectrophotometer.

An absorbance of 0.01 at 600 nm with a 1 cm path length represents ~2×10^7 cell/mL and indicates that the bacteria are in mid-exponential phase.

3.2. Inoculation Procedure

When inoculating the *Arabidopsis* cultures, consideration must be given to the time required to subsequently collect and process the samples (see Subheading 3.4). In large experiments, the difference in time between processing the first and last samples can be considerable and can complicate interpretation of the results. In our hands, we staggered the bacterial inoculation of individual cultures by 5 min, but this can differ depending on the experience of the researcher(s) and should be practised and assessed in advance. The speed and repeatability of the sample processing steps can have a massive effect on the analytical reproducibility of replicate samples.

When following the subsequent protocol, readers are also referred to the work flow diagram shown in Fig. 3.

1. Once at mid-exponential phase, pellet the bacteria from the 400 mL cultures by a 3 min centrifugation at $17,000 \times g$ at 20°C. Resuspend the pellet in 40 mL sterile 10 mM $MgCl_2$ and transfer to 50-mL sterile tubes. To remove any contaminating media, the suspension should be centrifuged at $6,000 \times g$ for 3 min at 20°C to pellet the bacteria. Discard the supernatant and resuspend the pellet in 40 mL sterile 10 mM $MgCl_2$. Repeat this pelleting and resuspension step twice, with the final resuspension again being in 10 mM $MgCl_2$. At the final resuspension, the bacterial density should be 1×10^{10} cells/mL. It is imperative that at each step where the supernatant is decanted, the bacteria are not contaminated; hence, these steps should be undertaken in a laminar flow cabinet.
2. To reduce experiment-to-experiment variation in the plant metabolome, large numbers of 7-day-old 200 mL *Arabidopsis* suspension cultures should be pooled. Our approach is to pool 15×200 mL cultures into a sterile 5-L conical flask in a laminar flow cabinet to maintain sterility.
3. Compare the responses of the bacterial strains to AT3 medium that had been used to grow plant cells for 7 days (designated as "spent" medium). Thus, at this stage, split the pooled culture into 2×1.5 L cultures in sterile conical flasks. To isolate spent medium from one 1.5 L subculture, filter through Whatman No.1 filter paper using a Buchner funnel linked to an electric vacuum pump in a laminar flow cabinet.
4. Transfer 20 mL aliquots of both the 1.5 L of *Arabidopsis* suspension cell culture and 1.5 L of spent medium to 50-mL sterile tubes in a laminar flow cabinet. Inoculate these with 200 μL of a given *Pst* strain, to yield a final bacterial cell density of 1×10^8 cells/mL.

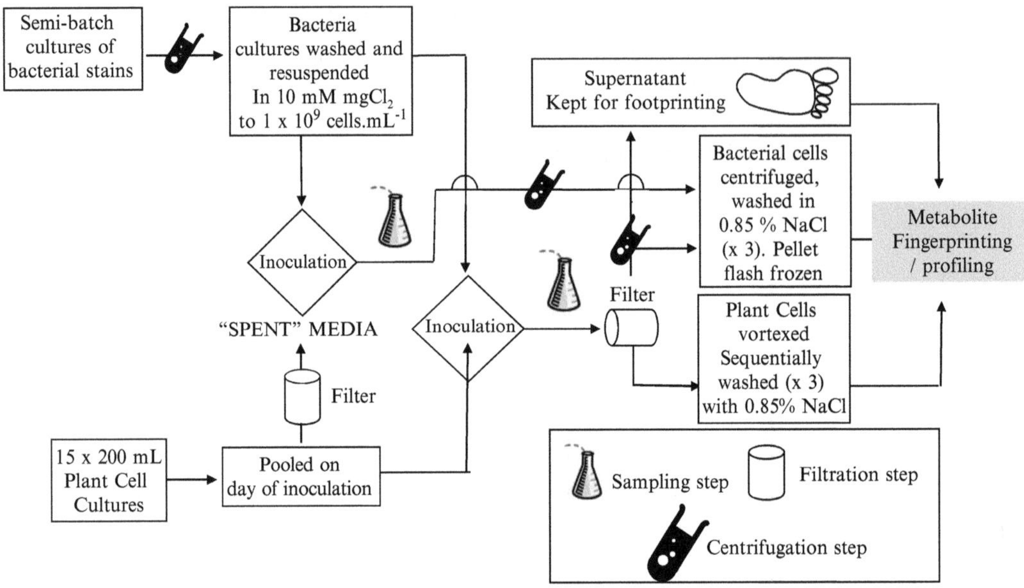

Fig. 3. Work flow for dual metabolomic analyses of the *Arabidopsis thaliana–Pseudomonas syringae* pv. *tomato* interaction. Each of the bacterial strains used in these analyses, the virulent *Pseudomonas syringae* pv. *tomato* (*Pst*), the hypersensitive response (HR) eliciting *Pst avrRpm1* and the non-HR and non-virulent strain *Pst hrpA* were used in an identical manner. The strains were initially grown on nutrient agar plates from which a single colony was used to inoculate 400 mL Nutrient Broth (NB). Once the cell density of the cultures had reached 1×10^9 cells/mL (typically ~ 24 h), 300 μL was used to inoculate 400 mL of fresh NB. This procedure was repeated a further two times once the subcultures had reached the indicated cell density. To prepare the bacteria for inoculation, the cultures were centrifuged, washed in 10 mM $MgCl_2$, re-centrifuged, the supernatant discarded, and finally resuspended in 10 mM $MgCl_2$ to a final concentration of 1×10^{10} cells/mL. *Arabidopsis* cells were continuously maintained as 200 mL cultures of AT3 media and grown at 24°C on a long day 16-h light cycle on an orbital shaker at 140 rpm (~8 × *g*). To prepare *Arabidopsis* cells for bacterial inoculation, ~3 mL of 7 day culture was added to 200 mL. After 7 days, 15 cultures were pooled into one 3 L culture. To provide a source of spent AT3 medium, 1.5 L of the suspension culture was filtered through Whatman No. 1 filter paper using a Buchner funnel and 500 mL side arm flask connected to a vacuum pump. The filtered cells were discarded. Bacterial suspensions were added to 20 mL aliquots of this spent medium in 50-mL centrifuge tubes to give a density of 1×10^8 cell/mL. Sampling of bacteria-spent AT3 cultures or bacteria-*Arabidopsis* cell cultures (sampling stages shown by conical flasks on the Figure) occurred at 12 h post inoculation (hpi). The culture was filtered through Whatman No. 1 paper and the plant cells harvested and sequentially washed with 0.85% (w/v) NaCl. The bacterial pellet was harvested from the filtrate by centrifugation and washed three times in 0.85% NaCl. After the final washing step, plant and bacterial samples were flash-frozen in liquid N_2 and stored at −80°C until metabolomic analysis.

3.3. Validation of the Outcome of Plant Interaction with the Pathogen

A crucial validation step in the dual metabolomic procedure must be establishing that the plant cells are responding in an appropriate manner. For plant–pathogen interactions, we suggest two methods, which in our hands have proven to be robust and easy to perform; the assessment of plant cell death using Evan's Blue Staining and the detection of defence gene expression. Although we highlight these here, the reader may wish to use other suitable indicators. These include the generation of reactive oxygen species (ROS) which may be detected using the indicator stain Amplex Red (36) or NO production detected, for example, using the oxy-haemoglobin method (37), an NO electrode (38), or staining using NO sensitive dyes (39).

3.3.1. Evans Blue Staining

1. Samples of 1 mL of bacterially inoculated plant cell cultures should be taken under sterile conditions.
2. To these samples add 0.5 mL of 0. 25% (w/v) Evans Blue (in water) and leave to absorb for 10 min.
3. Place a drop of each Evans Blue treated sample on a microscope slide with a coverslip. The sample may be viewed under white light using a microscope under 400× magnification. Counts of dead (blue stained cells) should be taken as a proportion of a total number of 100 cells. Cell viability counts should be averaged across three slides.
4. Typically, ~5% of *Arabidopsis* cell clusters should exhibit evidence of Evans Blue retention under unstressed conditions. Plant cultures where >20% of the cell clusters exhibit Evan's Blue staining should be considered to be responding to the bacterial inoculation.

3.3.2. Extracting RNA from Cultured Plant Cells and Assessment of Defence Gene Expression

The selection of suitable marker genes for defence responses should be influenced by the interaction under study. Generally, it can be assumed that responses to biotrophic pathogens can be indicated by increased expression of *pathogenesis related protein 1* (*PR1*, At2g14610), whilst responses to necrotrophic pathogens can be indicated by the defensin gene *PDF1.2* (Ar5g44420). Respectively, these are gene markers for the activation of salicylic acid and jasmonate/ethylene signalling pathways. Increased expression of defence genes will indicate that defence-associated metabolomic reprogramming is occurring. If required, cDNAs for these and other *Arabidopsis* genes may be obtained from http://www.arabidopsis.org.

The techniques of RNA extraction from plant cells are well established and commercial extraction kits are available, so it is not necessary to describe these here. Gene expression can be assessed by either northern blotting and DNA probe hybridisation or quantitative amplification by polymerase chain reaction (qPCR). Suitable protocols for these techniques are described in many places. Our approach follows those described by Sambrook and Russell (40).

In our experiments, *PR1* gene expression is detected 6 h after inoculation with *Pst avrRpm1* and at 12 h with *Pst*. With *PDF1.2* increased expression was detected at 12 h post inoculation with either *Pst avrRpm1* or *Pst*. No significant expression of either gene was detected when inoculated with *Pst hrpA*.

3.4. Sampling Procedure

The time at which the bacterially inoculated *Arabidopsis* cultures may be sampled is very much at the discretion of the investigator. We have concentrated on 12 h post inoculation, as this represents a time when cell death is not prominent in *Pst avrRpm1* challenged samples and yet increased defence gene expression is noted.

1. Filter the 20 mL cultures through Whatman No.1 filter paper to retain the plant cells using a Buchner funnel linked to an electric vacuum pump. Transfer the filtrate containing the bacterial cells to a 50-mL sterile tube and store in ice. This need not be undertaken in a laminar flow cabinet.
2. Resuspend the filtered *Arabidopsis* cells in 20 mL of 0.85% (w/v) NaCl and re-filter as in Subheading 3.4 step 1. Repeat this process twice. The *Arabidopsis* cells (~100 mg fresh weight) should be rapidly transferred to 2-mL microcentrifuge tubes each containing a single 5-mm stainless steel ball bearing (washed in acetone), flash-frozen in liquid N_2, and stored at −80°C.
3. Harvest the bacterial cells in the ice-stored filtrates by centrifugation at 3°C and 6,000 × *g* for 3 min. The supernatant can be transferred to 2-mL microcentrifuge tubes and stored at −80°C, as this represents the co-culture "footprint" (see Note 5). The remaining bacterial pellets should be resuspended in 5 mL 0.85% NaCl, re-centrifuged, and the supernatant discarded. This wash should be repeated twice further prior to flash-freezing the bacterial pellets in liquid N_2 and storage at −80°C. This process should also be followed to assess the effects of spent AT3 medium on bacterial cells.

3.5. Metabolite Profiling by Mass Spectroscopy

The extraction procedure used to extract metabolites from *Arabidopsis* and bacterial samples will depend on the platform used for metabolite profiling and also the chemical diversity across the metabolite groups of interest (41, 42). Our analysis was based on the Direct injection-Electrospray ionisation-Mass Spectroscopy (DI-ESI-MS). The extraction procedure for plant cells should essentially follow that of Fiehn et al. (43). Our approach is to separately extract polar and non-polar metabolite samples.

1. Grind the samples in liquid nitrogen, preferably using a ball mill for high-throughput preparation of large numbers of samples, but a mortar and pestle will suffice.
2. Add 1 mL of chloroform–methanol–sterile dH_2O (1:2.5:1) and vortex thoroughly in a cold room and place onto ice.
3. Add a volume of 0.5 mL of sterile ultra-pure dH_2O to each sample.
4. The polar and non-polar phases should be mixed with a vortex and then centrifuged at 3°C and 17,000 × *g* for 3 min.
5. The polar and non-polar phases are easily separated and the upper aqueous can be removed into a separate microcentrifuge tube using a pipette. Both polar and non-polar phases should be dried down in an environmental speed vacuum concentrator and stored at −80°C. However, it should also be noted that

the non-polar phase can be analysed directly and that lipids are proposed as being more stable over short storage periods when present in solution.

6. Bacterial samples are extracted into acetonitrile: 0.2% formic acid (1:1 (v/v)). The samples are vortexed for 30 s and centrifuged at 17,000 × *g* for 3 min to pellet any debris (44). This represents a rapid bacterial extraction method appropriate for DIMS (see Note 6).
7. Metabolite profiling of both the polar and non-polar plant extracts as well as bacterial extracts is carried out using DI-ESI-MS on a Micromass LCT mass spectrometer. Bacterial extracts may be introduced in their extraction buffer. Non-polar plant extracts should be reconstituted in 100 μL 80% (v/v) methanol and polar extracts in 100 μL 20% (v/v) methanol. Alternatively for non-polar extracts, 100 μL 70% (v/v) propan-2-ol or 10% (v/v) acetonitrile may be used, although in our hands these increased the signal-to-noise ratio. The actual MS conditions to use are described in the legend of Fig. 4.
8. Subsequent data analysis and comparisons can be performed as described in refs. (30, 45).

4. Notes

1. The described protocol allows for the analysis of bacterial pathogens simultaneously with *Arabidopsis* cell cultures; however, this approach can be readily adapted to study other plant–pathogen interactions. Clearly, suspension cultures of any plant species can be utilised. Well-established plant cell cultures which have been employed to examine plant defences include tobacco (*Nicotiana tabacum cv. BY-2*) BY2 cells (46), parsley (*Petroselinum crispum*) cells (47), and soybean (*Glycine* L. Max) cells (48). Considering potential pathogens, we suggest that our system is most appropriate for bacteria. Besides *Pseudomonas syringae*, *Xanthomonas* pathovars and *Ralstonia solanacearum* could be used, as well as necrotrophic pathogens such as *Erwinia carotovora*. Besides plant pathogens, interactions with endophytic bacteria could be assessed (49). However, we suggest that symbiotic interactions with nitrogen-fixing bacteria are not suited to analysis using this system, as this is governed by highly differentiated root tissue. For example, one of the earliest plant responses to *Rhizobium* is the cork-screwing of a root hair to encompass the interacting bacterium. Such complex responses cannot be adequately mimicked in liquid cultures.

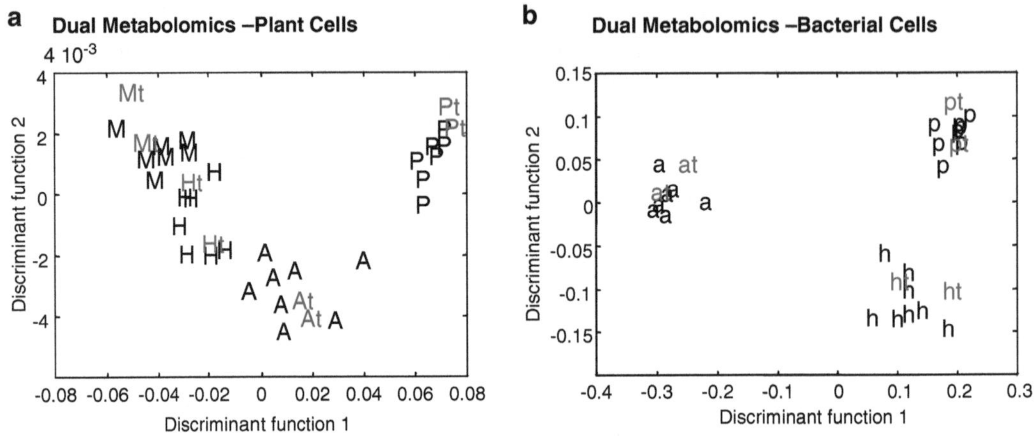

Fig. 4. Metabolomic analyses of *Arabidopsis* and *Pseudomonas syringae* cultures. Principal component-discriminant function analysis (PC-DFA) models of spectra derived from polar extracts of (**a**) *Arabidopsis* and (**b**) *Pseudomonas syringae* pv. *tomato* (*Pst*) strains following Direct injection-Electrospray ionisation-Mass Spectrometry(DI-ESI-MS) in positive ionisation mode. Cultured *Arabidopsis* cells were sampled at 12 h after inoculation (hai) with *Pst* (virulent strain;"**P**"), *Pst avrRpm1* (avirulent strain; "**A**"), *Pst hrpA* (non-avirulent and non-virulent; "**H**"). Control *Arabidopsis* cells were inoculated with 10 mM $MgCl_2$ ("M"). Ten 20 mL plant cultures were sampled per experiment. Sampling involved filtering the cultures through Whatman No.1 filter paper on a Buchner funnel linked to a vacuum pump. The filtered *Arabidopsis* cells were resuspended in 20 mL 0.85% (w/v) NaCl and re-filtered for a further two occasions. The *Arabidopsis* cells were transferred to 2-mL microcentrifuge tubes with stainless steel ball bearings (washed in acetone), flash-frozen, and stored at –80°C. The filtrate gathered, following filtration of the plant cells, contained either *Pst* ("**p**"), *Pst avrRpm1* ("**a**") and *Pst hrpA* ("**h**") These were pelleted by centrifugation at 3°C and 6,000 × *g* for 3 min. The pellets were resuspended in 1 mL 0.85% (w/v) NaCl re-centrifuged as before, the supernatant discarded, and the pellet stored at –80°C. The extraction of *Arabidopsis* and bacteria involved homogenisation in the ball mill in 1 mL of chloroform–methanol–sterile dH_2O (1:2.5:1) was added. The aqueous polar phases were extracted and dried down in a speed vacuum concentrator. Plant extracts were resuspended in 0.5 mL of sterile ultra-pure dH_2O, whilst bacterial extracts were resuspended in 0.5 mL acetonitrile: 0.2% formic acid (1:1 (v/v)). The extracts were introduced by DI at a flow rate of 5 μL/min using a syringe pump in positive ionisation mode ESI-MS. The capillary voltage was always set at +3.0 kV. The desolvation and nebuliser gas flow rate was 400–480 L/h and 50–80 L/h, respectively. The source and desolvation temperatures were 120°C and 250°C, respectively. The cone voltage was 30 V (to minimise in-source fragmentation), the extraction voltage was 5 V, and the radio frequency voltage amplitude was 125 V. Data were acquired over the *m/z* range 65–1,000 Th (Thomson unit; for the physical quantity mass-to-charge ratio) for polar plant extracts and 65–1,500 Th for non-polar plant extracts and bacterial extracts. Data were exported in an ASCII format, binned and each sample aligned to form a data array to employ for PC-DFA and univariate analysis. The derived PC-DFA models were based on 10 PCs and accounted for either (**a**) 99.90% or (**b**) 99.71% of the total variance. Each PC-DFA model was validated by the independent projection of two biological replicates from each experimental class (the test data set (in *grey* and with a "t" suffix)) into the PC-DFA space of their remaining six replicates (the training data set in *black*). Note the discrete metabolomic responses of *Arabidopsis* and bacterial strains during each interaction type.

Conceivably our dual metabolomic approach could be adapted to investigate fungal or oomycete interactions with plants. Necrotrophic fungal pathogens, such as *Botrytis cinerea* (*B. cinerea*), are readily cultured and apparently need no specialised infection structure with which to interact with the host. However, it is more often the case that fungal/oomycete infections involve the formation of specialised infection and/or feeding structures. Hence, any attempt to develop a dual metabolomic model of fungal/oomycete interaction must be carefully and intensively validated.

2. It is essential that the plant suspension cell cultures are not senescent or contaminated in any way, and not stressed by any of the culture conditions, for example by the light levels. It should be noted that unlike many other plant suspension cultures (for example tobacco BY-2 cells), *Arabidopsis* cultures are photosynthetically active and therefore can readily experience light stress.

   Culture contamination will be readily indicated through a change in the colour or consistency of the culture. Cultures may become "milky" as a result of contaminant growth. If there is clumping of plant cells into large balls of tissue, this may also suggest contamination. Alternatively, a loss in the number of viable plant cells could be observed. The latter symptom can be revealed by staining with Evan's Blue prior to inoculation with the pathogen. Evidence of staining in >20% of plant cell clusters should be seen as evidence of either contamination or stress. However, plant stress need not be exhibited by increased cell death, but can still impact on the metabolome. Widely recognised markers for stress are increased generation of ROS and NO. These may be readily assessed using the indicator stain Amplex Red (for ROS) or an NO electrode or staining using NO sensitive dyes.

3. We recommend that, in case of suspension culture contamination, the *Arabidopsis* L *er* cells should be also regularly cultured on solid AT2 agar plates. In our hands, this involves fortnightly culture of *Arabidopsis* cell clusters isolated from liquid culture on Gamborg's B5 basal salts and vitamins 3.06 g/L, 2% (w/v) glucose, 20 g/L, MES 0.5 g/L, 2, 4-D 0.5 mg/L (2,4-Dichlorophenoxyacetic acid), Kinetin 0.05 mg/L (N6-furfuryladenine.), pH 5.5 with 1 M NaOH, 0.8% (w/v) plant tissue culture agar. After preparation of AT2 plates, ~1 cm^2 of a 7-day suspension culture is poured onto the surface under sterile conditions. The plant cells are allowed to settle and the excess liquid decanted away. The cell clusters readily form calli when grown under identical conditions as the liquid cultures. If required, inoculation of the friable calli into liquid culture allows the suspension culture to be established. The calli on solid media should remain green. If they become chlorotic or exhibit signs of microbial growth, they should be discarded.

4. We recommend that the reproducibility of bacterial and plant cultures prior to co-cultivation be exhaustively assessed to avoid misinterpretation of the results of a dual metabolomic experiment. We recommend a rapid metabolite fingerprinting approach, such as Fourier Transform InfraRed spectroscopy (FT-IR) or equivalent, be used. Space precludes a detailed description of the FT-IR approach, but this can be obtained from refs. (45, 50). FT-IR fingerprinting was typically used to

assess variation between cultures of bacterial strains, between plant cultures, and between experiments undertaken at different times.

5. The protocol includes a step where the metabolite footprints of the *Arabidopsis*–bacteria co-cultures are obtained. Metabolite footprinting with FT-IR spectroscopy has been used to monitor the metabolite secretion in bacteria, yeast, and plant cultures (51–53). In the case of this dual metabolomic approach, metabolites secreted from interacting plant and microbe cells will be revealed.
6. For GC- and LC-MS profiling, we recommend the extraction optimised by Winder et al. (54).

Acknowledgements

The authors would like to thank the UK BBSRC for partly funding this work through studentships to JWA and AJL. JWA and RG are also indebted to the EU Framework VI funded project METAPHOR (FOOD-CT-2006-036220).

References

1. M Heil, IT Baldwin: Fitness costs of induced resistance: emerging experimental support for a slippery concept. *Trends in Plant Science* 2002, 7: 61–67.
2. L Salvaudon, T Giraud, JA Shykoff: Genetic diversity in natural populations: a fundamental component of plant-microbe interactions. *Current Opinion in Plant Biology* 2008, **11**:135–143.
3. EC Oerke, H-W Dehne, F Schönbeck, A Weber: Crop production and crop protection—Estimated losses in major food and cash crops. Amsterdam: Elsevier; 1994.
4. M Parniske: Arbuscular mycorrhiza: the mother of plant root endosymbioses. *Nature Reviews Microbiology* 2008, **6**:763–775.
5. KM Jones, H Kobayashi, BW Davies, ME Taga, GC Walker: How rhizobial symbionts invade plants: the *Sinorhizobium-Medicago* model. *Nature Reviews Microbiology* 2007, **5**:619–633.
6. V Bianciotto, P Bonfante: Arbuscular mycorrhizal fungi: a specialised niche for rhizospheric and endocellular bacteria. *Antonie Van Leeuwenhoek International Journal of General and Molecular Microbiology* 2002, **81**:365–371.
7. RP Ryan, K Germaine, A Franks, DJ Ryan, DN Dowling: Bacterial endophytes: recent developments and applications. *Fems Microbiology Letters* 2008, **278**:1–9.
8. LAJ Mur, P Kenton, AJ Lloyd, H Ougham, E Prats: The hypersensitive response; the centenary is upon us but how much do we know? *Journal of Experimental Botany* 2008, **59**:501–520.
9. C Mille-Lindblom, E von Wachenfeldt, LJ Tranvik: Ergosterol as a measure of living fungal biomass: persistence in environmental samples after fungal death. *Journal of Microbiological Methods* 2004, **59**:253–262.
10. D Choi, RM Bostock, S Avdiushko, DF Hildebrand: Lipid-derived signals that discriminate wound-responsive and pathogen-responsive isoprenoid pathways in plants – methyl jasmonate and the fungal elicitor arachidonic-acid induce different 3-hydroxy-3-methylglutaryl-coenzyme-a reductase genes and antimicrobial isoprenoids in *Solanum-tuberosum* L. *Proceedings of the National Academy of Sciences of the United States of America* 1994, **91**:2329–2333.
11. K Shimizu: Metabolic flux analysis based on C-13-labeling experiments and integration of the information with gene and protein expression patterns. In: *Recent Progress of Biochemical and Biomedical Engineering in Japan Ii,*

vol. 91. pp. 1–49. Berlin: SPRINGER-VERLAG BERLIN; 2004: 1–49.

12. TCR Williams, L Miguet, SK Masakapalli, NJ Kruger, LJ Sweetlove, RG Ratcliffe: Metabolic network fluxes in heterotrophic Arabidopsis cells: Stability of the flux distribution under different oxygenation conditions. *Plant Physiology* 2008, **148**:704–718.
13. MK Hellerstein: New stable isotope-mass spectrometric techniques for measuring fluxes through intact metabolic pathways in mammalian systems: introduction of moving pictures into functional genomics and biochemical phenotyping. *Metabolic Engineering* 2004, **6**:85–100.
14. YH Choi, HK Kim, HJM Linthorst, JG Hollander, AWM Lefeber, C Erkelens, JM Nuzillard, R Verpoorte: NMR metabolomics to revisit the tobacco mosaic virus infection in Nicotiana tabacum leaves. *Journal of Natural Products* 2006, **69**:742–748.
15. J Zhao, LC Davis, R Verpoorte: Elicitor signal transduction leading to production of plant secondary metabolites. *Biotechnology Advances* 2005, **23**:283–333.
16. RA Dietrich, MH Richberg, R Schmidt, C Dean, JL Dangl: A novel zinc finger protein is encoded by the arabidopsis *LSD1* gene and functions as a negative regulator of plant cell death. *Cell* 1997, **88**:685–694.
17. DH Aviv, C Rusterucci, BF Holt, RA Dietrich, JE Parker, JL Dangl: Runaway cell death, but not basal disease resistance, in *lsd1* is SA- and NIM1/NPR1-dependent. *Plant Journal* 2002, **29**:381–391.
18. S Lorrain, F Vailleau, C Balaque, D Roby: Lesion mimic mutants: keys for deciphering cell death and defense pathways in plants? *Trends in Plant Science* 2003, **8**:263–271.
19. JD Clarke, SM Volko, H Ledford, FM Ausubel, XN Dong: Roles of salicylic acid, jasmonic acid, and ethylene in cpr-induced resistance in Arabidopsis. *Plant Cell* 2000, **12**:2175–2190.
20. JR Alfano, AO Charkowski, WL Deng, JL Badel, T Petnicki-Ocwieja, K van Dijk, A Collmer: The *Pseudomonas syringae* Hrp pathogenicity island has a tripartite mosaic structure composed of a cluster of type III secretion genes bounded by exchangeable effector and conserved effector loci that contribute to parasitic fitness and pathogenicity in plants. *Proceedings of the National Academy of Sciences of the United States of America* 2000, **97**:4856–4861.
21. T Kondo, S Sawa, A Kinoshita, S Mizuno, T Kakimoto, H Fukuda, Y Sakagami: A plant peptide encoded by CLV3 identified by in situ MALDI-TOF MS analysis. *Science* 2006, **313**:845–848.
22. AK Mullen, MR Clench, S Crosland, KR Sharples: Determination of agrochemical compounds in soya plants by imaging matrix-assisted laser desorption/ionisation mass spectrometry. *Rapid Communications in Mass Spectrometry* 2005, **19**:2507–2516.
23. S Robinson, K Warburton, M Seymour, M Clench, J Thomas-Oates: Localization of water-soluble carbohydrates in wheat stems using imaging matrix-assisted laser desorption ionization mass spectrometry. *New Phytologist* 2007, **173**:438–444.
24. JJ Jones, S Mariccor, AB Batoy, CL Wilkins: A comprehensive and comparative analysis for MALDI FTMS lipid and phospholipid profiles from biological samples. *Computational Biology and Chemistry* 2005, **29**:294–302.
25. P Heraud, S Caine, G Sanson, R Gleadow, BR Wood, D McNaughton: Focal plane array infrared imaging: a new way to analyse leaf tissue. *New Phytologist* 2007, **173**:216–225.
26. F Jamme, P Robert, B Bouchet, L Saulnier, P Dumas, F Guillon: Aleurone cell walls of wheat grain: High spatial resolution investigation using synchrotron infrared microspectroscopy. *Applied Spectroscopy* 2008, **62**:895–900.
27. BO Budevska, ST Sum, TJ Jones: Application of Multivariate curve resolution for analysis of FT-IR microspectroscopic images of in situ plant tissue. *Applied Spectroscopy* 2003, **57**:124–131.
28. Z Movasaghi, S Rehman, IU Rehman: Raman Spectroscopy of Biological Tissues. *Applied Spectroscopy Reviews* 2007, **42**:493–541.
29. N Gierlinger, M Schwanninger: Chemical imaging of poplar wood cell walls by confocal Raman microscopy. *Plant Physiology* 2006, **140**:1246–1254.
30. AJ Lloyd, JW Allwood, CL Winder, WB Dunn, JK Heald, SM Cristescu, A Sivakumaran, FJM Harren, J Mulema, K Denby, R Goodacre, AR Smith, LAJ Mur: Metabolomic approaches reveal that cell wall modifications play a major role in ethylene-mediated resistance against *Botrytis cinerea*. *Plant Journal* 2011, **67**:852–868.
31. PF McCabe, CJ Leaver: Programmed cell death in cell cultures. *Plant Molecular Biology* 2000, **44**:359–368.
32. NM Cecchini, MI Monteoliva, F Blanco, L Holuigue, ME Alvarez: Features of basal and race-specific defences in photosynthetic Arabidopsis thaliana suspension cultured cells. *Molecular Plant Pathology* 2009, **10**:305–310.

33. A Clarke, R Desikan, RD Hurst, JT Hancock, SJ Neill: NO way back: nitric oxide and programmed cell death in Arabidopsis thaliana suspension cultures. *Plant Journal* 2000, **24**:667–677.
34. P Cossart, PJ Sansonetti: Bacterial invasion: The paradigms of enteroinvasive pathogens. *Science* 2004, **304**:242–248.
35. MJ May, CJ Leaver: Oxidative stimulation of glutathione synthesis in *Arabidopsis-thaliana* Suspension-Cultures. *Plant Physiology* 1993, **103**:621–627.
36. MJ Zhou, ZJ Diwu, N PanchukVoloshina, RP Haugland: A stable nonfluorescent derivative of resorufin for the fluorometric determination of trace hydrogen peroxide: Applications in detecting the activity of phagocyte NADPH oxidase and other oxidases. *Analytical Biochemistry* 1997, **253**:162–168.
37. M Delledonne, YJ Xia, RA Dixon, C Lamb: Nitric oxide functions as a signal in plant disease resistance. *Nature* 1998, **394**:585–588.
38. IR Davies, XJ Zhang: Nitric oxide selective electrodes. *Globins and Other Nitric Oxide-Reactive Proteins, Pt A* 2008, **436**:63–95.
39. E Prats, LAJ Mur, R Sanderson, TLW Carver: Nitric oxide contributes both to papilla-based resistance and the hypersensitive response in barley attacked by Blumeria graminis f. sp hordei. *Molecular Plant Pathology* 2005, **6**:65–78.
40. Sambrook J., Russell D.: Molecular cloning : A laboratory manual: Cold Spring Harbor Laboratory Press; 2006.
41. JW Allwood, DI Ellis, R Goodacre: Biomarker metabolites capturing the metabolite variance present in a rice plant developmental period. *Physiologia Plantarum* 2008, **132**:117–135.
42. RD Hall: Plant metabolomics: from holistic hope, to hype, to hot topic. *New Phytologist* 2006, **169**:453–468.
43. O Fiehn, J Kopka, P Dormann, T Altmann, RN Trethewey, L Willmitzer: Metabolite profiling for plant functional genomics. *Nature Biotechnology* 2000, **18**:1157–1161.
44. S Vaidyanathan, DB Kell, R Goodacre: Flow-injection electrospray ionization mass spectrometry of crude cell extracts for high-throughput bacterial identification. *J Am Soc Mass Spectrom* 2002, **13**:118–128.
45. JW Allwood, DI Ellis, JK Heald, R Goodacre, LAJ Mur: Metabolomic approaches reveal that phosphatidic and phosphatidyl glycerol phospholipids are major discriminatory non-polar metabolites in responses by *Brachypodium distachyon* to challenge by *Magnaporthe grisea*. *Plant Journal* 2006, **46**:351–368.
46. V Houot, P Etienne, A-S Petitot, S Barbier, J-P Blein, L Suty: Hydrogen peroxide induces programmed cell death features in cultured tobacco BY-2 cells, in a dose-dependent manner *J. Exp. Bot.* 2001, **52**:1721–1730.
47. E Kombrink, K Hahlbrock: Responses of cultured parsley cells to elicitors from phytopathogenic fungi : Timing and dose dependency of elicitor-induced reactions *Plant Physiol.* 1986, **81**:216–221.
48. A Levine, R Tenhaken, R Dixon, C Lamb: H_2O_2 from the oxidative burst orchestrates the plant hypersensitive disease resistance response. *Cell* 1994, **79**:583–593.
49. C Lodewyckx, J Vangronsveld, F Porteous, ERB Moore, S Taghavi, M Mezgeay, D van der Lelie: Endophytic bacteria and their potential applications. *Critical Reviews in Plant Sciences* 2002, **21**:583–606.
50. R Goodacre, EM Timmins, R Burton, N Kaderbhai, AM Woodward, DB Kell, PJ Rooney: Rapid identification of urinary tract infection bacteria using hyperspectral whole-organism fingerprinting and artificial neural networks. *Microbiology-Sgm* 1998, **144**:1157–1170.
51. NN Kaderbhai, DI Broadhurst, DI Ellis, R Goodacre, DB Kell: Functional genomics via metabolic footprinting: monitoring metabolite secretion by Escherichia coli tryptophan metabolism mutants using FT-IR and direct injection electrospray mass spectrometry. *Comparative and Functional Genomics* 2003, **4**:376–391.
52. FM Carrau, K Medina, L Farina, E Boido, PA Henschke, E Dellacassa: Production of fermentation aroma compounds by *Saccharomyces cerevisiae* wine yeasts: effects of yeast assimilable nitrogen on two model strains. *Fems Yeast Research* 2008, **8**:1196–1207.
53. R Dowlatabadia, AM Weljie, TA Thorpe, EC Yeung, HJ Vogel: Metabolic footprinting study of white spruce somatic embryogenesis using NMR spectroscopy. *Plant Physiology and Biochemistry* 2009, **47**:343–350.
54. CL Winder, WB Dunn, S Schuler, D Broadhurst, R Jarvis, GM Stephens, R Goodacre: Global metabolic profiling of *Escherichia coli* cultures: An evaluation of methods for quenching and extraction of intracellular metabolites. *Analytical Chemistry* 2008, **80**:2939–2948.
55. J Fliegmann, A Mithofer, G Wanner, J Ebel: An ancient enzyme domain hidden in the putative beta-glucan elicitor receptor of soybean may play an active part in the perception of pathogen-associated molecular patterns during broad host resistance. *Journal of Biological Chemistry* 2004, **279**:1132–1140.

56. D Nennstiel, D Scheel, T Nurnberger: Characterization and partial purification of an oligopeptide elicitor receptor from parsley *(Petroselinum crispum)*. *Febs Letters* 1998, **431**:405–410.
57. L Gomez-Gomez, T Boller: Flagellin perception: a paradigm for innate immunity. *Trends in Plant Science* 2002, 7:PII S1360-1385(02)02261-6.
58. A Meyer, A Puhler, K Niehaus: The lipopolysaccharides of the phytopathogen *Xanthomonas campestris* pv. *campestris* induce an oxidative burst reaction in cell cultures of Nicotiana tabacum. *Planta* 2001, **213**:214–222.
59. J Lee, DF Klessig, T Nurnberger: A harpin binding site in tobacco plasma membranes mediates activation of the pathogenesis-related gene HIN1 independent of extracellular calcium but dependent on mitogen-activated protein kinase activity. *Plant Cell* 2001, **13**:1079–1093.
60. S Bourque, MN Binet, M Ponchet, A Pugin, A Lebrun-Garcia: Characterization of the cryptogein binding sites on plant plasma membranes. *Journal of Biological Chemistry* 1999, **274**:34699–34705.
61. DA Navarre, TJ Wolpert: Victorin induction of an apoptotic/senescence-like response in oats. *Plant Cell* 1999, **11**:237–249.
62. M Rohe, A Gierlich, H Hermann, M Hahn, B Schmidt, S Rosahl, W Knogge: The race-specific elicitor, Nip1, from the barley pathogen, *Rhynchosporium-Secalis*, determines avirulence on host plants of the *Rrs1* resistance genotype. *Embo Journal* 1995, **14**:4168–4177.
63. W D'Haeze, M Holsters: Nod factor structures, responses, and perception during initiation of nodule development. *Glycobiology* 2002, **12**:79R–105R.
64. M Parniske, JA Downie: Plant biology – Locks, keys and symbioses. *Nature* 2003, **425**:569–570.
65. M de Torres, JW Mansfield, N Grabov, IR Brown, H Ammouneh, G Tsiamis, A Forsyth, S Robatzek, M Grant, J Boch: *Pseudomonas syringae* effector AvrPtoB suppresses basal defence in Arabidopsis. *Plant Journal* 2006, **47**:368–382.
66. RZ Li, XY Jia, X Mao: Ethanol-inducible gene expression system and its applications in plant functional genomics. *Plant Science* 2005, **169**:463–469.

Chapter 4

Precautions for Harvest, Sampling, Storage, and Transport of Crop Plant Metabolomics Samples

Benoît Biais, Stéphane Bernillon, Catherine Deborde, Cécile Cabasson, Dominique Rolin, Yaakov Tadmor, Joseph Burger, Arthur A. Schaffer, and Annick Moing

Abstract

Plant metabolomics is increasingly a routine option for plant biologists and food scientists. Here, we suggest some precautions for preparation and handling of samples issued from crop plants, in order to ensure sample representativeness and quality before their biochemical analysis. These precautions concern organ harvest either in the greenhouse or in the field, transport to the laboratory, and sampling, as well as sample pooling, storage, and transport to the analytical laboratory. They are in agreement with the recommendations of the "Plant Biology Context" group of the Metabolomics Standards Initiative concerning reporting practices for sample preparation. Some quality checking methods for long-term stability of metabolomics samples are also covered. The corresponding experimental procedures are illustrated using a representative study on melon fruit.

Key words: Metabolite profiling, Melon, Sample harvest, Sample preparation, Sample storage

1. Introduction

Plant metabolomics is increasingly becoming a routine option for plant biologists for use in functional genomics (1, 2) and studies of the biotic or abiotic environment (3–5) as well as for food scientists for the description of organoleptic and nutritional quality changes in food (6), the determination of food authenticity (7), and substantial equivalence studies of GMOs (8, 9). Several examples of potential applications in agriculture were discussed a few years ago (10). Plant metabolomics analyses can be divided into several main steps: (1) choice of experimental design followed by (2) plant growth conditions, (3) harvest of plants or organs,

Nigel W. Hardy and Robert D. Hall (eds.), *Plant Metabolomics: Methods and Protocols*, Methods in Molecular Biology, vol. 860, DOI 10.1007/978-1-61779-594-7_4, © Springer Science+Business Media, LLC 2012

(4) preparation and storage of plant samples, (5) extraction of these samples, (6) biochemical analysis of the extracts, (7) pre-processing of analytical data, (8) statistical analysis and storage of raw or pre-processed data. Several of these steps have been reviewed in detail (11–13). Each step needs to be clearly defined in order to ensure the accuracy of the information generated at the time of the experiment and also for reviewing or reuse of the raw metabolomics data at a later stage. It is therefore necessary to carefully plan the different steps and report the corresponding minimum meta-data as advised by the Metabolomics Standard Initiative (MSI) (http://msi-workgroups.sourceforge.net/bio-metadata/reporting/). The description of meta-data concerning sample growth conditions, harvest and preparation was proposed by the MSI "Plant Biology Context" group (14). The first and second steps concerning the experimental design and the description of culture conditions were reviewed recently (see Chapter 2). Here, we focus on steps 3 and 4 that can be subdivided into organ harvest in the greenhouse or field, transport to the laboratory, and sampling, as well as sample pooling, storage, and transport to the analytical laboratory. Strikingly, such meta-data are rarely reported, especially for crop plants.

The description of crop plants compared to model plants in the context of sample harvest and handling needs particular consideration. Sample preparation for model species such as *Arabidopsis* has been described elsewhere (see Chapter 5). Precise definition of stages of development and organs for model species such as *Arabidopsis* have been provided in the literature (15). Here, we focus on crop plants. For crop plants it may be necessary to rely on description of agricultural practices if precise definitions are not available in scientific publications or ontologies (http://www.plantontology.org/). Here, we detail the experimental procedures for steps 3 and 4 (see Fig. 1) using melon (*Cucumis melo* L.) as a representative species of a fruit crop. These experimental procedures were defined in the context of the META-PHOR European project that aimed to establish a European-based platform for the analysis of plant metabolites based on developing innovative metabolite profiling and identification technologies.

2. Materials

2.1. Harvest and Transport to the Laboratory

1. Air-conditioned car for transport of the harvested organs from the field to the laboratory.
2. Adapted packaging to prevent shocks between the fruits or organs and to allow some air circulation between them.

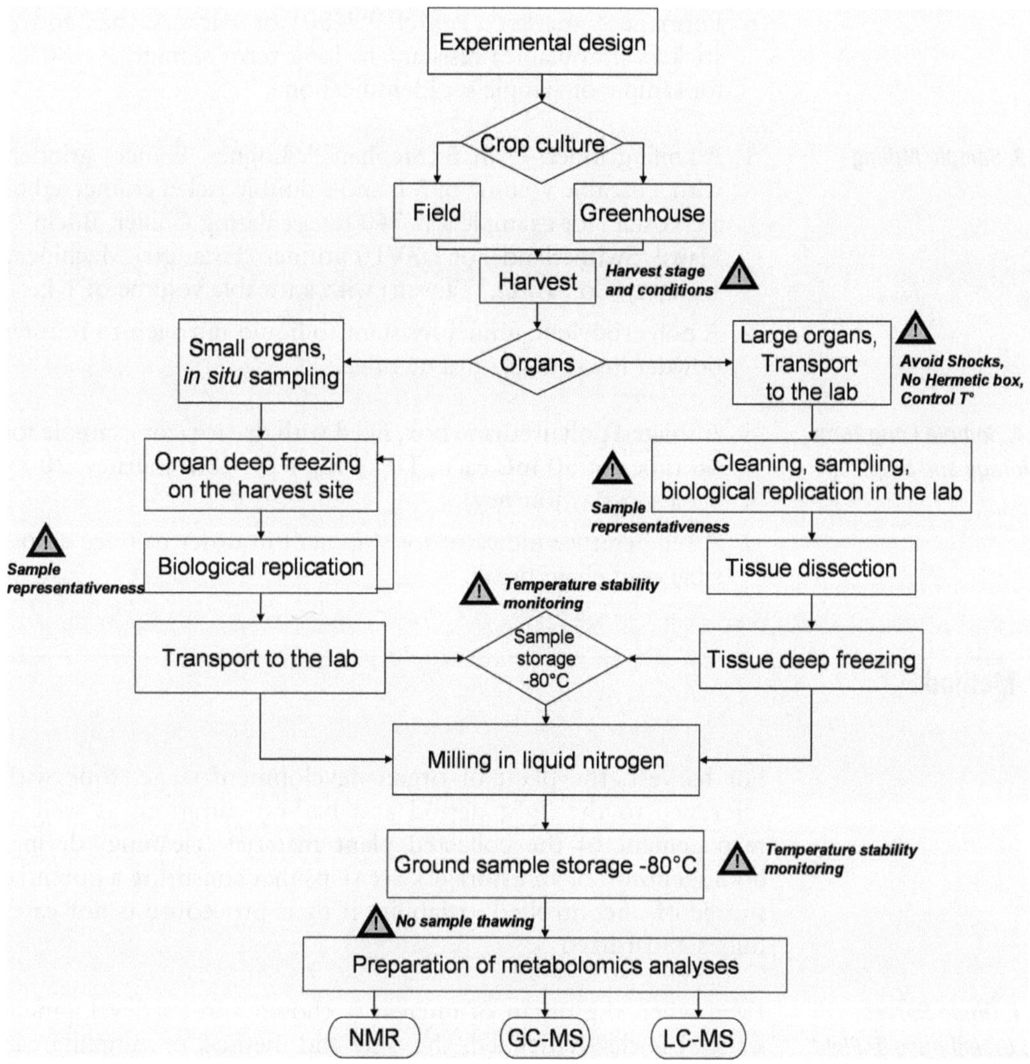

Fig. 1. Schematic representation of the different critical steps for harvest, sampling, storage, and transport of melon samples prepared for analyses with different metabolomics strategies.

2.2. Sampling and Sample Freezing

1. Ultra-pure water (MilliQ™).
2. Ice and liquid nitrogen (handled with protective glasses and protective gloves).
3. Polystyrene box or any container resistant to liquid nitrogen.
4. Stainless steel knife, stainless steel spatula.
5. Storage tubes (50-mL polypropylene tubes with screw cap, Greiner™) or plastic bags resistant to liquid nitrogen and long-term storage at −80°C. Impulse Sealer (TEw™, 300 mm) for rapid sealing of the plastic bags containing the organ pieces or storage tubes.

6. Permanent marker (Pentel™ N50) or stickers (bar coded stickers if available) resistant to long-term storage at −80°C, for sample or sample set identification.

2.3. Sample Milling

1. A knife grinder: UMC5 (Stephan™, Lognes, France) grinder, with a useable volume of 3 L and a double jacket connected to a cryostat (for example a B-740 Recirculating Chiller, Büchi™, Flawil, Switzerland), or D3V10 grinder (Hsiangtai Machinery Industry Co., LTD., Taiwan) with a useable volume of 1 L.
2. A polypropylene funnel resistant to liquid nitrogen to transfer powder into polypropylene tubes.

2.4. Sample Long Term Storage and Shipment

1. Insulated polyurethane box, filled with dry ice (for example for 48 tubes of 50 mL each, 10 kg for a 24–48 h journey, 20 kg for a 4–5 day journey).
2. A temperature indicator for shipment in order to trace a possible cold chain break.

3. Methods

For harvest, the plant or organ development stage, time with reference to the light period and harvest duration, as well as management of the collected plant material (cleaning, drying, homogenization, and storage), are steps that constitute a potential source of uncontrolled variability if their procedure is not carefully standardized.

3.1. Organ Harvest in Greenhouse or Field and Transport to the Laboratory

Even when the organ of interest is chosen and its development stage(s) is clearly defined, the time and method of sampling can still influence the reproducibility of the analysis. Management of the environment during plant cultivation, and/or recording the changes of some of the major environmental variables (e.g. temperature or light) are crucial even in controlled environments such as greenhouses, as small variations (shade and light, diurnal changes, seasonal variations) can cause variations in the biochemical status (16). Thus, special care must be taken to define the relevant harvest time and processes to minimize differences between harvest sessions (14).

1. In our study, for open field melons grown according to usual commercial practice and since pollination was not monitored, visual changes of the skin (beginning of appearance of a network pattern) was used as harvest criteria for the end of the growth stage, and the senescence of the peduncle was used as harvest criteria for commercial maturity (see Note 1 for more precise determination of development stage). Similar precautions

were taken for melons grown in a greenhouse. In both cases the expertise of the harvesting producer was critical.

2. As the composition of the leaf (17) but also fruit (18) metabolome can vary dramatically across a day and night cycle, the time of harvest has to be precisely defined. In the greenhouse, fruits were harvested as early as possible in the morning when the temperature was low and radiation was minimal; of course this does not apply to specific studies of metabolome changes across a day and night cycle. Most of the recommendations indicated below, and concerning harvest in the open field, are also relevant for fruit harvested in the greenhouse, except for the weather issues.
3. For harvest in an open field, weather conditions should be similar for each harvest and without rain or dew (see Note 2). The melons were collected in the morning between 08:30 am and 09:30 am. They were stored in cardboard crates with an individual protective cell for each fruit (see Note 3). This follows commercial practices.
4. After harvest in open field, the melons were transported from the field to the laboratory (180 km), in a special insulated polystyrene box containing ice blocks, in an air-conditioned car within 2 h. From the greenhouse to the laboratory (1 km), the melons were transported immediately after harvest and the samples were prepared in the laboratory within 1 h after harvest (see Note 4).

3.2. Sampling and Sample Pooling

To ensure representativeness of the physiological variations of a given variety of fruit at a certain development stage, several fruits of each variety have to be harvested to constitute a sample set. Several (at least three) sample sets or "supersamples" of each variety have to be independently prepared. The sampling method can greatly influence the reproducibility of the metabolomic analysis (19). Sampling aims to provide samples in a form and amount suitable for metabolomics analysis and which are representative of the total material to be analyzed. Representative means that the sample has to show a high degree of similarity to the total entity to be studied. Extreme care must be taken if the plant material is intrinsically heterogeneous (e.g. an organ made of several tissues). Contamination, loss, metabolism, or any other alteration of the sample properties and composition have to be minimized (20, 21). The main problem for quality assurance in sampling is how to guarantee biological representativeness of the samples (21). Moreover, immediate inhibition of enzymatic activities is required, generally by rapid freezing in liquid nitrogen after harvesting. In raspberry, the levels of vitamin C and phenolics were not affected by freezing in liquid nitrogen (22).

1. The melons harvested in our study were washed and gently brushed for one minute under a tap water flow at room temperature in order to remove every soil particle, then wiped with disposable paper (see Note 5).
2. Forty fruits were collected in each harvest, but only 36 of these were selected for sampling depending on the skin colour and fruit weight, thereby removing less ripe or overripe fruits and melons with extreme weights. Six sets of six fruit each were prepared, each set having the same average and standard deviation of fruit weight (see Note 6).
3. Melons are anisotropic round-shaped fruits, one side lying on the ground. To obtain a representative sampling of such fruits, one approach was to cut each fruit length ways into two halves and to select the two opposite eighths from each half. As flesh was the tissue of interest, the skin, a given width of the external part of the flesh (corresponding to the green part of the flesh for ripening cantaloupe melons) and the seeds were removed with a stainless steel knife. A zirconium oxide knife is recommended if microelement analyses are planned in parallel with the metabolomics ones. Each deseeded and peeled fruit eighth was immediately cut into smaller pieces (e.g. 4.5 ± 0.5 cm^3 cubes) and transferred into a clean polystyrene box filled with liquid nitrogen. Ideally, surfaces that are exposed after cutting should be minimized whilst volume should allow fast freezing of the whole piece (see Note 7).
4. The frozen pieces were transferred into tagged plastic bags, placed on ice and then quickly stored at −80°C within 2 min and for up to several weeks until grinding. It is recommended that all samples for a given experiment follow exactly the same procedure before and after grinding (see Note 8).
5. In our large multi-partner study (META-PHOR project involving more than 20 partners and 8 laboratories), two types of grinders were chosen which were able to process at least 300 g of plant material. The first one was a knife grinder (UMC5™, Stephan, Lognes, France) in which approximately 500 g of material was ground after the bowl was cooled using liquid nitrogen. The UMC5 Stephan grinder was connected to a cryostat set at 4°C. Its bowl was cooled with liquid nitrogen, frozen melon flesh pieces were added, and the volume of liquid nitrogen was increased to cover the highest pieces in the bowl. The samples were ground for 1 min at 3,000 rpm. This performed very good homogenization of the samples. The powder particle diameter, when measured on a representative freeze-dried aliquot, was about 200 μm. With the second grinder (D3V10, Hsiangtai Machinery Industry Co., LTD., Taiwan), each sample corresponding to approximately 500 g FW was ground after the bowl was cooled using liquid nitrogen.

The samples were ground for 30 s. The bulk frozen powder was then divided into small aliquots (about 20 g FW) rapidly transferred to 50-mL tubes with screw caps using a frozen plastic funnel ensuring that all partners obtained representative powdered samples.

6. Labelling and traceability of samples is crucial in quality assurance and quality control throughout the sample preparation process, from the field to the sample storage location and through distribution to chemical analysts. At each level (harvest, constitution of biological replicates, and distribution of samples), the samples were clearly identified (with permanent marker or stickers) and referenced in a suitable database with a unique identifier. Within the META-PHOR project, an online database was developed to gather information about each sample, i.e. from growth conditions to standard operating procedures for sample preparation and analysis (see Chapter 19). In this database, each sample was labelled with a coded name and a unique identifier for each aliquot tube was provided to each partner.

3.3. Sample Storage

The conditions and duration of sample storage need to be controlled and recorded. As reminded by Ryan and Robards (12), studies on the effect of long-term storage of plant samples on their metabolites are needed. No effect of sample storage duration (2, 7, 30 days to 12 months) at −60°C was detected on different sample types including several fruits, although not melon, by measuring the stability of 5-methyltetrahydrofolate (23), as a representative metabolite of the folate family. However, when working on several families of compounds, it is difficult to propose a unique compound as a "marker" of good storage conditions and duration.

1. Depending on the intended analyses, samples can be stored as fresh-frozen (liquid nitrogen ultra-rapid freezing) or lyophilized samples (see Note 9). The melons used in the META-PHOR project were stored as fresh-frozen tissue pieces from the beginning to the end of the harvest period of each year, and as fresh-frozen powders until distribution to the different analytical partners. Nonetheless, if lyophilization of fruit samples is required, the use of a freeze drier with temperature control at sample level is recommended. Start at a temperature of −30°C and progressively increase this temperature. Preliminary tests have to be done to ensure that lyophilization duration is sufficient to obtain constant weight. Care must be taken when transferring the samples out of the freeze drier to avoid water condensing on them.
2. Storage conditions have to be controlled since stability during sample storage is an important factor that is rarely measured. The time in frozen storage was shown to modify some aromatic components in melon (24). Usually samples for metabolomics

are stored at –80°C as are those for transcriptomics analyses (see Note 10 for lyophilized samples). Thawing must be avoided.

3.4. Sample Transport to the Analytical Laboratory

1. In our study, sample storage location and one analytical laboratory were in the same place, but most analyses required sample shipment. When in the same building, rapid transport of samples was done using an ice-chest, a Dewar, and liquid nitrogen. Melon samples were shipped as fresh-frozen tissue pieces or fresh-frozen powder depending on the analyses. The package for sample shipment on dry ice was well insulated, but not hermetically sealed. The quantity of dry ice must be well estimated (see Note 11). If lyophilized samples are shipped, dry conditions need to be ensured using sealed plastic bags containing a desiccant such as silica gel.

3.5. Sample Quality Checking

During storage or after shipment, sample quality has to be verified. A simple visual inspection of colour and powder aspect was used since it can reveal uncontrolled thawing of ground samples. Quality checking can be refined using physico-chemical analyses or biochemical analyses carried out at different times during sample storage (23, 25).

4. Notes

1. The developmental stages of the plants under study and/or of their organs need to be clearly defined relative to standardized growth conditions and/or phenology descriptors, whenever possible by using dedicated ontologies (e.g. Plant Ontology at http://www.plantontology.org/ for phenology). For fruits, and in cases of artificial pollination, harvest might be expressed as Days Post Anthesis (DPA), assuming that maturity should be checked anyway. If fruits are grown under the same temperature conditions, DPA would indicate the developmental stage, otherwise degree-days post-anthesis can be used as a compromise (26). If the exact age of the organ is not known, well-defined criteria (organ aspect, colour, and size) might be used instead, in order to get samples that are as homogeneous as possible.
2. The presence of water on the fruit surface may dilute some metabolites, modify the water–dry matter ratio and generate variability when working on entire organs or on their skin. In addition, free water on the surface greatly increases the risk of disease (e.g. for strawberries, (27)).

3. The harvested organs should be handled with care and stored in packaging which eliminates both shocks between the organs and oxygen limitation. Shocks between the organs can induce stress responses and/or bruising therefore altering the metabolome. Since metabolic processes are fast, enzyme activities need to be quenched to fix the metabolome (28). This is best achieved by freeze clamping or immediately plunging whole organs or tissues into liquid nitrogen at the harvest site. However, this procedure is only suitable for small organs (e.g. leaves or small fruits with a diameter < 5 mm).
4. Transport to the laboratory after harvest has to be done using precautions that depend on the harvested organs and cultivation location. If the plant material is grown close to the laboratory (e.g. greenhouse or experimental field close to a research institute), the transport of organs does not present major issues for the sample quality given that they are processed immediately after harvest. If the plant material is grown far from the laboratory and/or has to be shipped, special care must be taken when packaging the organs to prevent shocks. These organs should be transported to the laboratory as quickly as possible and at a controlled temperature, for instance 20°C for melons harvested in the field in summer. High temperature as well as low temperature stress should be avoided during transport.
5. For sampling of plant organs, it is generally necessary to consider whether the material should be cleaned before chemical analysis. For melon fruit, suitable cleaning procedures have to be carried out since the organ lies on the soil and in this case, the surface contamination of the fruit by earth or fungi may distort analysis results. Cleaning may include purely mechanical steps such as the use of dry or moistened tissues, shaking, blowing, and brushing of the sample material, or recourse to various washing techniques (20). Tap water was chosen because melon peel was removed and not considered for analyses, but deionized or ultra-pure water can be used to avoid mineral contaminations when needed.
6. The constitution of biological replicates has to be defined in the experimental design. As with all biological experiments, replication is essential to incorporate and assess inherent biological or analytical variation and also to provide a sample set representative of the whole population of organs or tissues, thereby allowing the significance and biological relevance of the analytical data to be determined. The number of replicates used should be considered at the biological as well as the extraction and analytical levels (16) and depends on the species, organs and type of experiment and variability. As a compromise, one biological replicate is usually obtained from

several pooled organs or plants. For cherry tomato, 18 fruits from nine plants were pooled to make a representative fruit sample in a study on volatiles (29).

7. If a specific tissue needs to be dissected before freezing, the work surface and every tool (knives, scalpels, spatula, etc.), as well as hands and gloves, should be cleaned and rinsed with distilled water and ethanol according to the usual laboratory practices. However, when studying volatiles including ethanol, care must be taken to avoid contamination with exogenous ethanol. Depending on the studies and if sterile conditions are required, the sampling must be performed using a laminar flow clean bench. The sample must be representative of all of the tissue in the organ, thus material must be taken from several different parts of the organ to avoid local variability. These sampling sites should be defined in the experimental design. The use of knives and scalpels obviously causes wound stress, the effect of which on the metabolism should be kept to a minimum by working as fast as possible. Moreover, it is essential to stop enzyme activities through the use of appropriate treatments such as immediate deep freezing in liquid nitrogen. Other treatments that have to be chosen with respect to the target chemical analyses can be applied at this step. These include acidic treatments with perchloric acid or addition of ethanol or methanol–water mixtures (70% v/v) (19).

8. Sample grinding is the step that provides the increase of surface exchange required to perform an optimal extraction step. It is also used for homogenization of sample material; another important step in the overall analytical procedure. The theoretical and practical approaches to homogenizing samples have been described in detail by Markert (20). Homogenization ensures that the sample that is analyzed has the same mean chemical composition as the original sample. This can be achieved by grinding every organ/tissue of one biological replicate at once. However, possible contamination or volatilization of certain compounds that are to be measured (20) has to be estimated and limited. A large number of grinders exist with different characteristics such as grinding tool (knives, discs, mortar, balls) and construction material (stainless steel, zirconium oxide…). The final degree of powder fineness varies depending on the type of grinder (e.g. from 500 μm for some grinders with disks, to less than 1 μm for planetary ball grinders). The milling technique used will generally depend on the following parameters: overall quantity and number of samples of the material to be homogenized, particle size of the original sample, fineness of the material when ground, physicochemical properties of the sample and of the grinding

equipment, hardness of the material to be homogenized (20). For metabolomics, samples are usually ground whilst still frozen despite the possible modification of the volatile composition upon freezing as shown for strawberry (30). The frequency or speed at which the grinder works must be high enough to achieve good grinding before a significant rise in temperature of the frozen samples can occur. During grinding, loss of frozen fruit material inevitably occurs, e.g. on the wall of the grinder bowl. Therefore, more raw material than needed for analyses must be gathered. If estimation of losses during grinding is not feasible, a compromise is to grind at least twice as much as the quantity needed.

9. Fresh-frozen samples are necessary for subsequent analytical determination of highly volatile compounds as freeze drying will result in a loss of volatiles (31, 32). With fresh-frozen ground samples, precise weighing is tricky and may contribute to increased variability of the absolute quantification data provided by the chemical analysis. However, when relative quantification data are provided by the chemical analysis, the data can be corrected by dividing with the overall signal. Freeze drying provides protection against enzyme activities and microbial decomposition during storage (16) and gives a constant reference value by determining the dry weight, as opposed to the fresh weight which is more difficult to quantify on frozen ground samples. Freeze drying of plant samples is often used when ^{1}H-NMR profiling is required (33). For freeze drying of fruit samples that contain high levels of soluble sugars, especially at maturity, special care must be taken to make sure that the lyophilization duration is sufficient. However, freeze drying can cause a loss of some metabolites through irreversible binding to cell walls or membranes (16).
10. Lyophilized samples can also be stored at 4°C or –20°C, but it has been shown that storage of dried leaves at –20°C is always preferable when studying hydrolysable tannins (34). Whatever the type of the samples stored, recording the full history of storage conditions and duration is compulsory.
11. The quantity of dry ice must ensure a temperature of –60°C for at least 1 week in the package. Indeed, customs issues for plant material (for example if Material Trade Agreement, MTA, or Convention on International Trade in Endangered Species, CITES, is required) can delay the delivery for several days. Therefore devices that enable you to check that the temperature at the sample level did not increase above a given threshold are desirable (for instance RFID temperature sensor from ThermAssureRF™).

Acknowledgements

This work was partially funded by the EU within the plant metabolomics project META-PHOR (FOOD-CT-2006-036220). We gratefully thank Sylvie Bochu, Françoise Leix-Henry from CEFEL (France) for following the cultures and providing the melons, Christel Renaud (France), Uzi Saar, and Fabian Baumkoler (Israel) for technical support, Dr Helen Jenkins for language corrections, and Dr Yves Gibon for critical reading of the manuscript.

References

1. Schauer, N. and Fernie, A.R. (2006) Plant metabolomics: towards biological function and mechanism. *Trends Plant Sci.* **11**, 508–516.
2. Fernie, A.R. (2007) The future of metabolic phytochemistry: Larger numbers of metabolites, higher resolution, greater understanding. *Phytochemistry* **68**, 2861–2880.
3. Pereira, G.E., Gaudillere, J.P., Pieri, P., Hilbert, G., Maucourt, M., Deborde, C., Moing, A. and Rolin, D. (2006) Microclimate influence on mineral and metabolic profiles of grape berries. *J. Agric. Food Chem.* **54**, 6765–6775.
4. Allwood, J.W., Ellis, D.I. and Goodacre, R. (2008) Metabolomic technologies and their application to the study of plants and plant-host interactions. *Physiol. Plant.* **132**, 117–135.
5. Sanchez, D.H., Siahpoosh, M.R., Roessner, U., Udvardi, M. and Kopka, J. (2008) Plant metabolomics reveals conserved and divergent metabolic responses to salinity. *Physiol. Plant.* **132**, 209–219.
6. Hall, R.D., Brouwer, I.D. and Fitzgerald, M.A. (2008) Plant metabolomics and its potential application for human nutrition. *Physiol. Plant.* **132**, 162–175.
7. Cuny, M., Vigneau, E., Le Gall, G., Colquhoun, I., Lees, M. and Rutledge, D.N. (2008) Fruit juice authentication by H-1 NMR spectroscopy in combination with different chemometrics tools. *Anal. Bioanal. Chem.* **390**, 419–427.
8. Catchpole, G.S., Beckmann, M., Enot, D.P., Mondhe, M., Zywicki, B., Taylor, J., Hardy, N., Smith, A., King, R.D., Kell, D.B., Fiehn, O. and Draper, J. (2005) Hierarchical metabolomics demonstrates substantial compositional similarity between genetically modified and conventional potato crops. *Proc. Nat. Acad. Sci. USA* **102**, 14458–14462.
9. Baker, J.M., Hawkins, N.D., Ward, J.L., Lovegrove, A., Napier, J.A., Shewry, P.R. and Beale, M.H. (2006) A metabolomic study of substantial equivalence of field-grown genetically modified wheat. *Plant Biotechnol. J.* **4**, 381–392.
10. Dixon, R.A., Gang, D.R., Charlton, A.J., Fiehn, O., Kuiper, H.A., Reynolds, T.L., Tjeerdema, R.S., Jeffery, E.H., German, J.B., Ridley, W.P. and Seiber, J.N. (2006) Perspective – Applications of metabolomics in agriculture. *J. Agric. Food Chem.* **54**, 8984–8994.
11. Hall, R.D. (2006) Plant metabolomics: from holistic hope, to hype, to hot topic. *New Phytol.* **169**, 453–468.
12. Ryan, D. and Robards, K. (2006) Analytical chemistry considerations in plant metabolomics. *Sep. Purif. Rev.* **35**, 319–356.
13. Saito, K., Dixon, R.A. and Willmitzer, L. (ed.) (2006) *Plant Metabolomics.* Springer, Berlin Heidelberg.
14. Fiehn, O., Sumner, L.W., Rhee, S., Ward, J., Dickerson, J., Lange, B.M., Lane, G., Roessner, U., Last, R. and Nikolau, B. (2007) Minimum reporting standards for plant biology context information in metabolomics studies. *Metabolomics* **3**, 195–201.
15. Boyes, D.C., Zayed, A.M., Ascenzi, R., McCaskill, A.J., Hoffman, N.E., Davis, K.R. and Görlach, J. (2001) Growth stage-based phenotypic analysis of Arabidopsis. A model for high throughput functional genomics in plants. *Plant Cell* **13**, 1499–1510.
16. Dunn, W.B., Bailey, N.J.C. and Johnson, H.E. (2005) Measuring the metabolome: Current analytical technologies. *Analyst* **130**, 606–625.
17. Gibon, Y., Usadel, B., Blaesing, O.E., Kamlage, B., Hoehne, M., Trethewey, R. and Stitt, M. (2006) Integration of metabolite with transcript and enzyme activity profiling during diurnal cycles in *Arabidopsis* rosettes. *Genome Biol.* 7, R76.
18. Ma, F. and Cheng, L. (2003) The sun-exposed peel of apple fruit has higher xanthophyll cycle

dependent thermal dissipation and antioxidants of the ascorbate/glutathione pathway than the shaded peel. *Plant Sci.* **165**, 819–827.

19. Dunn, W.B. and Ellis, D.I. (2005) Metabolomics: Current analytical platforms and methodologies. *Trends Anal. Chem.* **24**, 285–294.
20. Markert, B. (1995) Sample preparation (cleaning, drying, homogenization) for trace element analysis in plant matrices. *Sci. Total Environ.* **176**, 45–61.
21. Wagner, G. (1995) Basic approaches and methods for quality assurance and quality control in sample collection and storage for environmental monitoring. *Sci. Total Environ.* **176**, 63–71.
22. Mullen, W., Stewart, A.J., Lean, M.E.J., Gardner, P., Duthie, G.G. and Crozier, A. (2002) Effect of freezing and storage on the phenolics, ellagitannins, flavonoids, and antioxidant capacity of red raspberries. *J. Agric. Food Chem.* **50**, 5197–5201.
23. Phillips, K.M., Wunderlich, K.M., Holden, J.M., Exler, J., Gebhardt, S.E., Haytowitz, D.B., Beecher, G.R. and Doherty, R.F. (2005) Stability of 5-methyltetrahydrofolate in frozen fresh fruits and vegetables. *Food Chem.* **92**, 587–595.
24. Ma, Y.K., Hu, X.S., Chen, J., Chen, F., Wu, J.H., Zhao, G.H., Liao, X.J. and Wang, Z.F. (2007) The effect of freezing modes and frozen storage on aroma, enzyme and microorganism in Hami melon. *Food Sci.Technol. Internat.* **13**, 259–267.
25. Fish, W. and Davis, A. (2003) The effects of frozen storage conditions on lycopene stability in watermelon tissue. *J. Agric. Food Chem.* **51**, 3582–3585.
26. Bonhomme, R. (2000) Bases and limits to using 'degree.day' units. *Europ. J. Agronomy* **13**, 1–10.
27. Aked, J. (2000) Fruits and vegetables, in *The stability and shelf-life of food* (Kilcast, D, Subramaniam, P, eds), Woodhead Publishing Limited, Cambridge, U.K.
28. AP Rees, T., and Hill, S.A. (1994) Metabolic control analysis of plant metabolism. *Plant Cell Environ.* **17**, 587–599.
29. Tikunov, Y., Lommen, A., de Vos, C.H.R., Verhoeven, H.A., Bino, R.J., Hall, R.D. and Bovy, A.G. (2005) A novel approach for non-targeted data analysis for metabolomics. Large-scale profiling of tomato fruit volatiles. *Plant Physiol.* **139**, 1125–1137.
30. Douillard, C. and Guichard, E. (1990) The aroma of strawberry (*Fragaria ananassa*): Characterisation of some cultivars and influence of freezing. *J. Sci. Food Agric.* **50**, 517–531.
31. Julkunen-Titto, R. and Tahvanaiem, J. (1989) The effect of the sample preparation method of extractable phenolics of *Salicaceae* species. *Planta Med.* **55**, 55–58.
32. Keinänen, K. and Julkunen-Titto, R. (1996) Effect of sample preparation method on birch (*Betula pendula* Roth) leaf phenolics. *J. Agric. Food Chem.* **44**, 2724–2727.
33. Ward, J.L. and Beale, M.H. (2006) NMR spectroscopy in plant metabolomics, in *Plant Metabolomics* (Saito, K, Dixon, RA, Willmitzer, L, eds), Springer, Berlin Heidelberg.
34. Salminem, J.P. (2003) Effects of sample drying and storage, and choice of extraction solvent and analysis method on the yield of birch leaf hydrolyzable tannins. *J. Chem. Ecol.* **29**, 1289–1305.

Chapter 5

Tissue Preparation Using Arabidopsis

Aimee M. Llewellyn, Jennie Lewis, Sonia J. Miller, Delia-Irina Corol, Michael H. Beale, and Jane L. Ward

Abstract

The ability to track changes in the levels of many metabolites in plants has great utility in a number of biological contexts. A metabolomics experiment usually requires the comparison of different varieties in either a functional genomics context or in response to perturbation by an external treatment. Such treatments can result in subtle changes in the final chemical signature of the plant tissue, and therefore, any unwanted variance produced in the generation of that tissue must be minimised. Procedures for plant growth, harvesting, preparation of extracts, and the subsequent collection of data have been optimised to minimise experimental variation within the dataset. This chapter describes in detail how to generate reproducible *Arabidopsis* tissue suitable for a typical plant metabolomics experiment. Issues concerned with tissue sampling, harvesting, and storage are also discussed.

Key words: *Arabidopsis*, Plant growth, Metabolomics, Metabolite profiling

1. Introduction

The science of metabolomics aims to measure simultaneously as many metabolites as possible in the system under study. The technology is rapidly maturing and is now used across multiple areas of plant biology to address a multitude of problems and questions (1). Examples of specific use include the analysis of natural diversity (2), determination of gene function (3), food quality authentication (4), quality trait localisation (5) and in the understanding of plant defence whether under stress (6), pathogen or insect attack (7). Reliable chemical analysis of plants, which are living, multicompartmentalised, developing systems, is not straightforward. A "snapshot" approach can be taken, whereby an attempt is made to freeze metabolism and examine the metabolome at a single time-point. Alternatively, the dynamic nature of plant growth and metabolism can be harnessed and time-course approaches—profiling

Nigel W. Hardy and Robert D. Hall (eds.), *Plant Metabolomics: Methods and Protocols*, Methods in Molecular Biology, vol. 860, DOI 10.1007/978-1-61779-594-7_5, © Springer Science+Business Media, LLC 2012

plant-responses to external factors—can yield metabolite trajectories which not only describe changes in responses to treatments but also set them in context with the general modulation of metabolite levels during plant growth (8, 9).

In order to collect data describing metabolite changes in the plant, a solvent extract is usually made and the solutes can then be introduced into a range of analytical systems (10–13) to produce either quantitative data or in many cases, fingerprint data which can be analysed, using statistics, to discern differences between samples or even highlight similar metabolite behaviour across a dataset. In order to maximise the possibility of generating a high-quality final dataset and thus report on changes in metabolites that are relevant to the study in question, it is necessary to understand and minimise any variation caused by the plant growth and handling regime. Factors such as lighting, temperature, watering schedule and humidity will affect the growth of the plant (14–16) whilst choices made in relation to growth stage, tissue sampling and harvest time will have further impact on the chemical signature recorded (17, 18).

Whilst there are many ways to grow *Arabidopsis*, including tissue generated via hydroponic systems (19, 20), we present here a robust protocol for plant growth on soil that has been utilised routinely for metabolomic tissue generation for a number of years. This chapter deals with the many classic problems encountered in carrying out large-scale growth experiments and the associated collection and storage issues which can impact on post-harvest tissue quality.

2. Materials

2.1. Germination of Arabidopsis on Agar Plates

1. Murashige–Skoog Basal Medium with Gamborg's Vitamins M0404.
2. Sucrose AR grade.
3. Agarose Type PGP.
4. 1 M KOH—(potassium hydroxide).
5. Pure deionised water (polished water, 18 MΩ conductance).
6. Ethanol AR.
7. Sterile Tap Water.
8. Autoclave tape.
9. Thick Bleach (e.g. Parazone Jeyes Active Power).
10. Sterile petri dishes (triple vent 90 mm—Sterilin).
11. Eppendorf polypropylene tubes, 1.5 ml (Eppendorf UK, Cambridge, UK).

12. Parafilm "M", 38 m × 10 cm.
13. pH monitor.
14. Laminar flow hood.
15. p1000 pipette and tips.
16. Tissue culture cabinet.
17. Autoclave.
18. Eppendorf fine tip pen.
19. Scissors, 12.5 cm, dissecting, stainless steel, fine point.
20. Disposable paper towelling.

2.2. Sowing on Soil

1. Levington seed and modular compost: F2 + Sand.
2. Soil trays divided into 12 cells (size P24).
3. Black gravel trays, plastic, 14 × 9 × 3″, no holes (FARGRO).
4. Incubation lids.
5. Plastic plant labels, pointed end (Improved Marking & Label Company Ltd.).
6. Eppendorf fine tip pen (Fisher Scientific UK Ltd.).
7. Tweezers, with a fine, rounded or smooth pincer.
8. Felt growth mats, cut to fit the bottom of the black gravel trays.
9. Disposable balance weighing boat, 100 ml size or smaller (Scientific Laboratory Supplies).
10. Controlled Environment Room/Cabinet.

2.3. Transfer of Arabidopsis Seedlings to Soil

1. Levington seed and modular compost: F2 + Sand.
2. Soil trays divided into 12 cells (size P24).
3. Black gravel trays, plastic, 14 × 9 × 3″, no holes.
4. Incubation lids.
5. Plastic plant labels.
6. Eppendorf fine tip pen.
7. Tweezers.
8. Controlled Environment Room/Cabinet.
9. Felt growth mats.

2.4. Harvesting, Freeze-Drying, and Grinding

1. 50 ml, polypropylene, conical, free-standing (skirted) centrifuge tubes (Sterilin).
2. Cryoware pen, any colour(s), for plastic.
3. Plastic thumb-tacks.
4. Liquid nitrogen, in Dewar.

5. Metal tongs, for use in Liquid Nitrogen, long handled with curved grippers (as for domestic cooking).
6. Forceps, straight, fine, 12.5 cm, stainless steel, smooth internal grip.
7. Scissors, 12.5 cm, dissecting, stainless steel, fine point.
8. Autoclave bags.
9. Heavy-weight clear plastic bags.
10. Freeze-drier.
11. Parafilm "M", 38 m × 10 cm.
12. Mortar and pestle, porcelain.
13. Clean white printer or photocopy paper.
14. 1.7 ml tall glass vials with Bakelite screw caps with tinfoil-faced cork wads, 11.25 × 36 mm (Camlab).
15. Tubees, Hi-Low strips, 32.5 × 13 mm self-adhesive labels (Fisher Scientific).
16. Acetone, AR grade.
17. Deionised water.
18. Disposable paper towelling.

3. Methods

3.1. Experimental Design

Good experimental design ensures that the correct data can be collected over the duration of an experiment to answer a specific hypothesis. Key to this is the design of the experimental approach that not only addresses the hypothesis posed, but that can be undertaken with confidence of quality and reproducibility. Importantly, the experiment should allow for reproducible tissue collection within the limits of the resources available. Inflexible limitations must be carefully considered and include physical space available, time, and the in-built variability of living plant tissue. Additional limitations can include equipment, staff, and consumables (see Note 1). Limitations posed by these items can, in many instances, be addressed through careful planning. Certain activities such as plating of seed and transferring of seedlings to soil should only be done in a single day to limit variability in growth stage and rate (see Note 2). Three biological replicates (Trays) and three technical replicates provides evidence that sample production and analysis is accurate; thus, it must be possible within the design to produce a final sample of sufficient material for analysis (see Note 3). The inclusion of an appropriate number of control or statistical tracking samples should also be included for larger experiments where tissue is to be generated across several different growth experiments and which is later to be compared across experiments. Harvesting generally should not take longer than 2 h per harvesting day (see Note 4). In sample preparation, homogenising material from several plants (1 tray = 18–24 plants) averages out any natural variations between individual plants. This method has been optimised for the sample collection of *Arabidopsis thaliana* sp. ecotypes, but can be applied across many variations of *Arabidopsis* sp. with minimal changes.

3.2. Sterilising Arabidopsis Seeds

1. Stratify the *Arabidopsis* seeds (see Note 5).
2. Transfer the required number of seeds to an eppendorf tube labelled with the seed line number.
3. Add 10% bleach (1 ml, v/v in water) (see Note 6), to sterilise the seeds.
4. Invert each eppendorf tube five times in 3 min, to ensure all the seeds are exposed to the bleach. Do not leave seeds in bleach for longer than 3 min, as this will cause seed coat breakdown.
5. Working in a laminar flow hood, using sterile tips, pipette off the bleach solution and add sterile tap water (1 ml), to remove residual bleach from the seeds.
6. Invert the tube five times to ensure thorough washing of the seeds.
7. Remove the water with a pipette and repeat the sterile water wash.
8. Remove the majority of the water, leaving a small volume in the base of the tube to facilitate plating or sowing of seeds.

3.3. Germination on Agar Plates

Germination on agar plates is routinely used as our Standard Operating Procedure for consistent, good quality plant material production. Germination on plates allows the scientist to judge the viability of seed and observe the general health of the seedlings. It also allows for selection of seedlings for transfer so that the scientist can select good quality seedlings of the same growth stage. The method provides the best control of early-stage seedlings, and more control over experimental parameters such as longer days for quicker germination, or shorter days for more root growth.

3.3.1. Preparation of Growth Medium

1. The growth medium should be freshly prepared when required.
2. Label a 1-L Duran bottle with M + S + 3% Sucrose.
3. Weigh Murashige-Skoog medium (4.4 g), sucrose (30 g, 3% final concentration), Agarose Type PGP, (7 g, 0.7% final concentration) into the Duran bottle.
4. Add polished water (1 L) to the bottle.

5. Mix the growth medium solution by inverting the bottle three times.
6. Adjust the solution to pH 5.6 with 1 M KOH. If pH 5.6 is exceeded in error, do not adjust back to pH 5.6; instead, discard medium and prepare fresh (see Note 7).
7. Slightly loosen the lid of the Duran bottle and place a strip of autoclave tape over the top of the lid.
8. Autoclave the bottle, tap water, and pipette tips.
9. Once the autoclave programme has finished remove the Duran bottle; tighten the lid and place in the Laminar Flow Hood to cool (see Note 8).

3.3.2. Pouring Agarose Plates

1. Using a black marker pen, label the outside of the bottom half of sterile petri dishes, with: M + S + 3% Sucrose, the line number of the seed to be plated and the date and the name of the operator.
2. Once the agarose is cool enough to handle (usually 30–60 min after removal from the autoclave) pour it directly from the bottle into the labelled petri dishes until the agarose is 5 mm thick.
3. Push the petri dishes to the back of the flow hood and leave the lids of the petri dishes slightly open (offset by approximately 5 mm), with the opened lid facing the filter at the back of the flow hood (see Note 9).
4. Allow the plates to stand for 1 h by which time the agar has set and then close the lids (see Note 10).

3.3.3. Plating of Arabidopsis Seeds

Lift the lid of the agarose plate to be plated (see Note 11). Using a p1000 pipette fitted with a sterile pipette tip, aliquot seeds from the appropriate eppendorf tube. Insert the pipette tip with the seeds under the lid and distribute the seeds individually at regular intervals over the plate (see Note 12).

1. Add a maximum of 50 seeds per plate.
2. Seal the plates by wrapping a 1-cm strip of parafilm around the petri dishes.
3. Place the petri dishes flat, in the tissue culture cabinet, ensuring the conditions are set at 24 h light and 22°C.
4. Leave the plates in the tissue culture cabinet for 10 days or until the seedlings have four leaves.

3.3.4. Transfer to Soil

Transfer seedlings to soil 10 days after plating. The plants are at this time usually at the four-leaf stage. The preferred soil mix used is Levington seed and modular compost: F2 + sand. We have found this to be ideal for both transferring from plates and sowing direct to soil. Soil should be pre-treated by deep freezing at −20°C for at least 2 weeks. This substantially reduces or eliminates the occurrence of scarid fly and their larvae. Other methods of pest control, such as chemical root drenches and pesticide application are not appropriate for tissue generation for metabolomic analysis as the chemicals involved can appear in the final metabolite fingerprint. When potting up the plant trays, the soil is removed from the freezer, taken immediately to the controlled environment rooms to thaw, and only opened once inside. Any unused soil is not reused.

1. On the day of transferal make up soil trays according to the number required. Fill with soil and place into gravel trays lined with growth mats.
2. Water the soil from below using a hose. Put sufficient water in each tray to cover the mats on the bottom. Allow the water to soak up through the soil (see Note 13).
3. Ensure that the soil is wet through. Do this by disturbing the soil, with a label, in one cell. If the soil is dry below the surface, add some more water to the trays.
4. Label each tray or a set of 12 cells with the following: the line number, the date of plating, the date of soil transfer, and the tray number.
5. Collect petri dishes containing seedlings from the tissue culture cabinet.
6. Carefully pick the seedlings off of the agarose plates with a pair of tweezers.
7. Lay the seedlings on the soil surface, two per cell.
8. Use the tweezers to gently push all the roots under the surface of the soil, leaving the seedling's leaves resting on top of the soil.
9. Place trays in a randomised array across the growth room (see Note 14).
10. Cover the trays with an incubation lid (Note 15), ensuring all vents are closed and transfer the trays to controlled environment, with conditions specific for *Arabidopsis* growth (see Note 6).
11. Leave the incubation lids on for 2–3 days, until the seedlings have grown and look healthy, then open the vents and leave for a further couple of days (see Note 5). After 5 days, remove the lids and grow the plants until the required developmental stage for harvest.
12. Record the date of transferal of seedlings to soil in the lab journal.

3.4. Germination Directly onto Soil

Germination on soil is an alternative to plating when some varieties of *Arabidopsis* are unable to be germinated on agar plates, or when the plants do not thrive when plated, and either die whilst on plates or when transferred to soil. It is also helpful when very large amounts of *Arabidopsis* tissue need to be produced as a time-saving measure. Care must be taken, however, as it is not possible to judge seed viability when sowing directly onto soil. The following section includes protocols used in our laboratory for both approaches.

All seed should be stratified (see Note 5), clean, dry, and in good condition.

1. On the day of sowing, make up soil trays according to the number required. Fill with soil and place into trays lined with growth mats.
2. Water the soil from below using a hose. Put sufficient water in each tray to cover the mats in the bottom. Allow the water to soak up through the soil (see Note 13).
3. Ensure the soil is wet through. Do this by disturbing the soil, using a plastic label, in one cell. If the soil is dry below the surface, add some more water to the trays.
4. Label each tray, or set of 12 cells with: the line number, the date of sowing, the tray number (if the plants are being grown for analysis), and the name of the operator.
5. Pour a few seeds into the base of a disposable balance weighing boat. Aim for about 100 seeds.
6. Dip the tweezers into some water in the tray to form a water droplet on the end of one pincer.
7. Touch the water droplet to a single seed and then carefully place into 1 cell.
8. Repeat to allow 2 seeds per cell (see Note 17).
9. Place the trays in a randomised array across the growth room (see Note 14).
10. Cover the trays with an incubation lid, ensuring all vents are closed, and transfer the trays to a controlled environment, with conditions specific for *Arabidopsis* growth (see Note 16).
11. Leave the incubation lids on until the first signs of germination and then open the vents slightly to allow air circulation.
12. Record the date of sowing in a Laboratory Notebook.

3.5. Harvesting Arabidopsis Samples

1. Prepare a table (Harvest Sheet) to record data during the harvesting procedure (see Note 18).
2. Label a 50-ml centrifuge tube using the cryopen with the sample name/line, the tray number, the date of harvest, and the operator name. Use separate tubes for different ecotypes and different trays of the same ecotype.
3. Using a thumb tack, or push-pin, make a minimum of three holes in the centrifuge tube caps.

4. At 14 h into the growth daylight, start to harvest plants that have reached the bolt stage, (see Note 19) or the growth stage specified by the experimental design.
5. In the Controlled Environment room, using tongs, dip a labelled 50-ml centrifuge tube into a dewar of liquid nitrogen to fill.
6. Record the harvest time on the Harvest Sheet.
7. Hold the plant sample using the forceps, immediately below the rosette, and cut using scissors as close to the soil surface as possible.
8. Immediately transfer into the tube containing liquid nitrogen to freeze instantly and thereby minimise further metabolic changes.
9. Repeat steps 6 and 7 for up to five plants per tube.
10. Cap the tube with the already perforated cap and immerse in the liquid nitrogen to keep frozen.
11. Record the number of plants harvested per line on the Harvest Sheet.
12. When harvesting is completed, dispose of any remaining plant material and soil into autoclave bags for autoclaving and disposal.
13. In the laboratory, using the tongs remove the centrifuge tubes from the dewar, and drain off any liquid nitrogen back into the dewar. Transfer the tubes into heavy-weight plastic bags labelled with the experiment number, the date, and the operator name.
14. Store immediately at –80°C until further preparation for analysis.
15. Record all details and observations in your laboratory notebook.

3.6. Freeze-Drying Arabidopsis Samples

Depending on the type of metabolomics analysis required and the analytes of interest, harvested tissue can be used "fresh-frozen" or it can be lyophilised. For protocols requiring weighing and solvent extraction for metabolome analysis of for example polar metabolites, we have found that freeze-drying samples gives a more reproducible final dataset with no significant loss of metabolome coverage. The ability to work with a dry powder which can be accurately weighed is in many cases advantageous compared to working with frozen tissue which is hard to weigh accurately and which would contain variable levels of moisture across different tissue types.

1. Turn on the freeze drier and pre-chill the chamber, following the manufacturer's instructions.
2. Remove the centrifuge tubes containing the plant material from the –80°C storage, and transfer directly to the pre-chilled freeze drier.

3. Leave the plant material to freeze-dry completely, usually for 48–72 h.
4. When the plant tissue is dried, remove from the freeze-drier.
5. Carefully replace the perforated caps with caps without holes. Seal the join between cap and tube with a small strip of parafilm.
6. Store immediately at −80°C until required for analysis.

3.7. Sample Preparation: Grinding

Thorough grinding of *Arabidopsis* tissue is essential to ensure that the analysis undertaken is carried out on a homogeneous sample. In our experience, aerial arabidopsis tissue can be milled very finely with the final powder resembling that of icing sugar. Grinding can be performed by hand for smaller quantites of tissue or via utilisation of a mechanical grinding machine (see Note 20).

1. Remove the plant samples to be ground from −80°C storage and allow them to come to room temperature in the laboratory.
2. Group the dry plant samples of the same sample name/line and tray number.
3. Empty all centrifuge tubes containing the same type of material into a mortar or similar grinding receptacle.
4. Keep the material not being ground at −80°C until grinding can be carried out.
5. Use a pestle and mortar (or mechanical grinder) to mill and homogenise the material until it resembles the consistancy of icing sugar. Use scissors if necessary to first chop up large pieces of plant material.
6. Using an appropriate label and cryopen, label a 1.7-ml glass vial with the sample name/line, the tray number, the job number, the date, and the name of the operator.
7. Empty the mortar onto a clean piece of white paper, that has been folded down the middle.
8. Using the paper as a "funnel", transfer the plant material to an appropriate cryogenic storage vial. Cap the vial.
9. Store the vial(s) at −80°C in labelled boxes and record the material location in a lab book.
10. Between grinding different samples, clean all the equipment thoroughly.

3.8. Expected Biomass

Depending upon the design of the experiment and the required final tissue requirements, the amount of biomass will vary from one application to another. In general, when execution of the above protocol is carried out one can expect to generate around 22 g fresh tissue from a tray of 24 arabidopsis plants when harvested

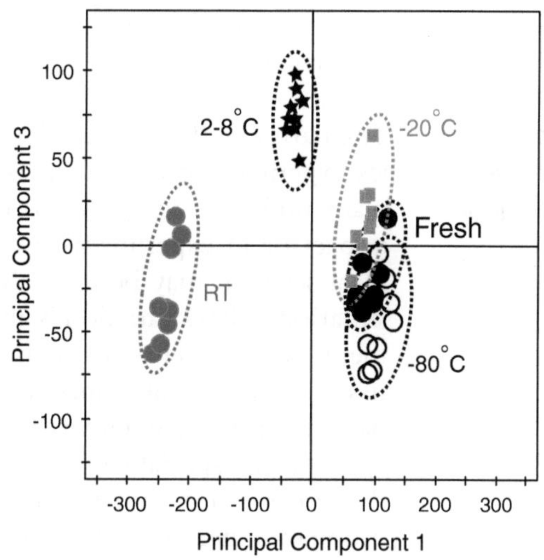

Fig. 1. Scores plot obtained from Principal Components Analysis of NMR data using polar solvent extracts of *Arabidopsis* tissue which had been previously stored under different temperature conditions. The data represented is that analysed 12 months after storage conditions were implemented.

at the large rosette stage, just prior to bolting. Lyophilisation of this material typically yields around 1.9 g of homogeneous freeze-dried powder.

3.9. Tissue Storage

Post harvest processing of tissue and subsequent storage can affect the final metabolome signature. Whilst this is less of a problem for smaller experiments where the tissue can be analysed immediately, it is nevertheless a major problem for larger experiments where tissue needs to be stored prior to sample analysis. Any change in metabolite levels as a consequence of storage could therefore create unnecessary and unwanted trends across a sample dataset, especially in experiments where samples from different harvest times are under study. As an example, data is shown in Fig. 1 from a tissue storage experiment whereby freshly harvested *Arabidopsis* aerial tissue was subdivided at the beginning of the experiment and subjected to different storage conditions. Analysis at regular intervals (9, 12, and 24 months) showed the rapid deterioration of freeze-dried *Arabidopsis* not undergoing refrigeration. Although stable initially, tissue stored in a regular refrigerator (2–8°C) had also deteriorated at 9 months. By 12 months, as demonstrated in the PCA scores plot, tissue maintained at room temperature or at 2–8°C showed distinct clustering away from the data collected from fresh tissue. Whilst good overlap with fresh material was maintained using tissue stored at –80°C, it was also evident that tissue maintained in a chest freezer (–20°C) was after 12 months storage, also beginning to become detectably different. Following this experiment, we recommend that all *Arabidopsis* tissue, whether fresh-frozen or freeze-dried, is stored at −80°C until analysis can be carried out.

3.10. Analysis of Individual Plants Versus Pooled Tissue

The nature of the biological question being asked by a metabolomics experiment can sometimes dictate the sampling regime of the plant under study. To assess natural variation and environmental impact or in experiments involving segregation of seed lots, the ideal scenario may be to analyse tissue from individual plants. In situations where variation is an undesirable influence in the experiment one may choose to smooth out the variability by pooling tissue from several plants to create a more homogeneous tissue base that is representative of all the individual plants. This pooling approach is especially useful if trying to elucidate gene function or the effect of a specific treatment or environmental condition. The problem of heterogeneity across tissue samples manifests itself to a greater extent in the analysis of seed material, e.g. across different ecotypes or mutants where the ratio of seed coat to endosperm can vary greatly. Whilst both sampling approaches are perfectly acceptable, the choice can sometimes come down to feasibility within the overall experimental design. The need for biological replication in any experiment is clear but numbers of required replicates can vary and this impacts not only on the practicalities of collecting high quality, reproducible tissue but also on the cost of the overall experiment and the amount of tissue available for multiple analyses. Figure 2 shows a PCA scores plot of NMR data obtained from polar solvent extracts of *Arabidopsis* wild type tissue (Col-0) and demonstrates the power of pooling tissue to reduce heterogeneity in the final metabolome signature due to plant-to-plant variation.

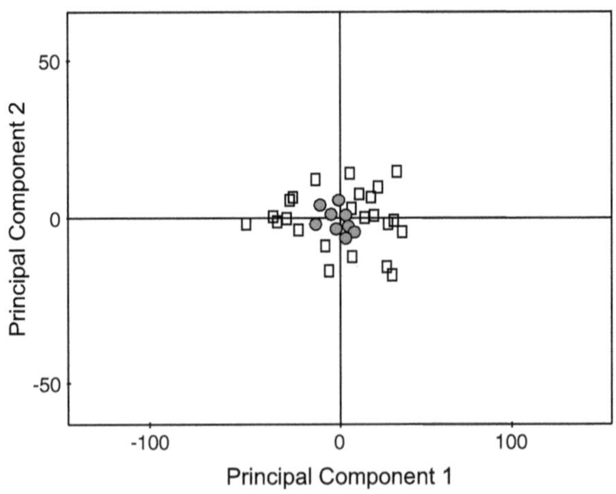

Fig. 2. PCA scores plot demonstrating variability of pooled (*grey circle*) versus single plant (*open box*) samples. Pooled material is made up by combining aerial tissue from 24 single plants which were ground together to give one homogeneous batch of tissue to sample from. Data shown is from NMR spectra of polar solvent extracts using 15 mg freeze-dried *Arabidopsis* tissue.

Data in the centre of the plot (represented by grey circles) is representative of three trays, each containing 24 plants, which have been pooled to give three biological replicates. Each of these have been analysed in triplicate. Surrounding this are triplicate data points generated from tissue of 24 individual plants. It is clearly evident that data from individual plants is more variable than that obtained from pooled tissue. Thus for experiments where individual plant sampling is desired, the need for higher replication to cover this variation is clear but brings with it the practical problems (in a large experiment) which can actually reduce sample quality. As an example, collection of tissue from many individual plants in a large experimental array can be more difficult to achieve within the required time window allowed to minimise diurnal effects. In summary, it is our preferred choice to work on pooled *Arabidopsis* tissue where possible and appropriate.

4. Notes

1. The growth room facilities available to us for metabolomic tissue growth consist of two Sanyo 228 cabinets. These cabinets have a growing area of 1.68 m^2 and growing height of 1.4 m. Temperature control range 5–35°C ± 0.3°C. Humidity control range 65–95% ± 5%. Lighting is provided by T5 49 W colour, 83 fluorescents providing up to 900 μmol/m^2/s at 300 mm from the lights, adjustable from 10 to 100%. A Eurotherm 2704 controller, linked to a SCADA package, provides control. The average maximum plant production under standard operating procedures is 1,440 individual *Arabidopsis* sp. plants per room. For most experiments undertaken, there is therefore, a limitation of 2,880 individual plants.
2. 50 agar plates can be produced reliably each day by a single technician. Best practice number of seeds per plate is 50–60 seeds. 1.5 technician-days per full growth room are suggested for transferring to soil.
3. Sample weights and measures for producing sufficient plant material for analysis: For pre-emergence plant material, assume 60 mg of seed is needed to achieve enough material for each biological replicate, and three technical replicates. For example, "Line A" would require 60 mg × 3 = 180 mg seed. Early stage plantlets (e.g. 2-leaf) have a wet weight of approximately 0.9 mg per single plant. For 60 mg of freeze-dried early stage material, at least 15 agar plates should be produced.
4. Previous studies investigating the effect of diurnal variation of *Arabidopsis* indicated that key metabolic changes occur throughout the day–night cycle. Figure 3 indicates the effect

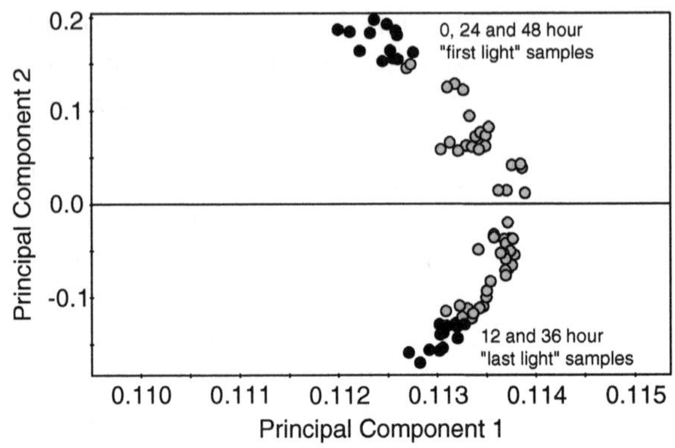

Fig. 3. Effect on diurnal variation. PCA model of NMR data collected from *Arabidopsis* tissue collected at 2 hourly intervals over 48 h. *Plot* shows separation of samples collected at first and last light with data from intermediate time points cycling between these two extremes.

that these changes can have on the final metabolome signature. These effects can be limited by harvesting samples over a shorter period of time—ideally at a point of low diurnal effects. Harvesting at the same time of day in a window no wider than 2 h represents the optimum practical conditions. For other studies, such as looking at the effect of a treatment over time, it may be necessary to sample at regular intervals throughout the diurnal cycle. In these cases, an appreciation of the metabolites varying diurnally throughout the experiment in the wild type or control sample needs to be attained. Subsequent modelling of such an experiment is then best carried out by examining different profiles across the time series rather than by looking at metabolome changes at single time points.

5. Seeds can be stratified by placing in fridge for 4 days (2–4°C) prior to use.
6. Before carrying out sterilisation of seeds make up a 10% bleach solution in a 100-ml Duran bottle using bleach and deionised water (v/v).
7. Make sure the pH meter is calibrated. Submerse the probe in the agar and adjust accordingly making sure the agar solution is mixed to ensure an accurate reading is measured.
8. (Steps in Notes 2–4 to be carried out in a Laminar Flow Cabinet). Before starting the procedure turn on the Laminar Flow Hood and wipe the working surface with Ethanol. Leave the Hood running in LAF purge mode for 10 min and then turn it to the recommended flow for working.
9. To prevent condensation whilst minimising contamination.

10. Transfer all the equipment needed—seeds, eppendorf tubes, bleach solution, sterile water, pipettes (P1000 and P20), pipette tips, parafilm, and scissors—to the Laminar Flow Hood.
11. Gloves should be worn to minimise contamination and the plates should be handled towards the back of the laminar flow hood, closest to the filter.
12. Use water tension to aid pipetting. Ensure that the tip does not dig into the agar.
13. This makes transfer easier as the dry soil does not "stick" to the wet seedling and tweezers, and allows the deposit of the seedling onto the soil surface. It may take up to 30 min before the soil surface is moist.
14. When trays of plants are randomised across the growth room any environmental effects should not cause an unwanted trend in the dataset.
15. This firstly prevents the seedling from scorching and then finally removes excess condensation from inside the lids.
16. Long day conditions: 16 h; Day Hours: 0–16, Night Hours: 16–24; Day Temperature: 23°C, Humidity 75%; Night Temperature: 18°C, Humidity 80%; Light levels: 350–400 μmol/m^2/s.
17. If you are unsure of the quality of the seed, or germination, sow more seeds per cell to ensure that two plants germinate per cell. If you find that more than two seeds germinate well, pick out and remove the weakest additional seedlings using tweezers.
18. Table 1 shows an example of a Basic Harvest Sheet.
19. For ecotypes being grown for NMR metabolomic analysis, we harvest the plants at growth stage 6.0–6.1 (18). This growth stage is very easy to spot as it occurs when the bolt has occurred and the first flower has opened.

Table 1
Typical harvesting log sheet

Date	**3.12.09**		**4.12.09**	
Tray 1 Sample/line	Time	Number plants	Time	Number plants
Tray 2 Sample/line				
Tray 3 Sample/line				

20. A typical mechanical mill suitable for grinding plant tissue is an ultra centrifugal mill such as that produced by Retsch (Type ZM200). Sieves of 0.75 or 1.0 mm are suitable to produce a fine homogeneous powder from freeze-dried material.

Acknowledgements

This work has been funded by the EU Framework VI programme META-PHOR (FOOD-CT-2006-036220) and the UK Biotechnology and Biological Sciences Research Council (BBSRC).

References

1. Sumner, L.W., Mendes, P. and Dixon, R.A. (2003) Plant metabolomics: large-scale phytochemistry in the functional genomics era. *Phytochemistry* **62**, 817–836.
2. Ward, J.L., Harris, C., Lewis, J. and Beale. M. H. (2003) Assessment of ^{1}H NMR spectroscopy and multivariate analysis as a technique for metabolite fingerprinting of *Arabidopsis thaliana*. *Phytochemistry* **62**, 949–957.
3. Schauer, N. and Fernie, A.R. (2006) Plant metabolomics: Towards biological function and mechanism. *Trends Plant Sci.* **11**, 508–516.
4. Cuny, M., Vigneau, E., Le Gall, G., Colquhoun, I. J., Lees, M. and Rutledge, D. N. (2008) Fruit juice authentication by ^{1}H NMR spectroscopy in combination with different chemometrics tools. *Anal Bioanal Chem* **390,** 419–427.
5. Fu, J., Keurentjes, J.J.B., Bouwmeester, H., America, T., Verstappen, F.W.A., Ward, J.L., Beale, M.H., de Vos, R.C.H., Dijkstra, M., Scheltema, R.A., Johannes, F., Koornneef, M., Vreugdenhil, D., Breitling, R. and Jansen, R.C. (2009) System-wide molecular evidence for phenotypic buffering in Arabidopsis. *Nature Genetics* **41,** 166–7.
6. Shulaev, V., Cortes, D., Miller, G. and Mittler, R. (2008) Metabolomics for plant stress response. *Physiol. Plant.* **132**, 199–208.
7. Bezemer, T.M. and van Dam, N.M. (2005) Linking aboveground and belowground interactions via induced plant defences. *Trends Ecol Evol* **20**, 617–624.
8. Mounet, F., Lemaire-Chamley, M., Maucourt, M., Cabasson, C., Giraudel, J.L., Deborde, C., Lessire, R., Gallusci, P., Bertrand, A., Gaudillere, M., Rothan, C., Rolin, D., Moing, A. (2007) Quantitative metabolic profiles of tomato flesh and seeds during fruit development: Complementary analysis with ANN and PCA. *Metabolomics* **3**, 273–288.
9. Bläsing, O.E., Gibon, Y., Günther, M., Höhne, M., Morcuende, R., Osuna, D., Thimm, O., Usadel, B., Scheible, W-R. and Stitt, M. (2005) Sugars and Circadian Regulation Make Major Contributions to the Global Regulation of Diurnal Gene Expression in Arabidopsis. *Plant Cell.* **17**, 3257–3281.
10. Ward, J.L., Baker, J.M. and Beale, M.H. (2007) Recent applications of NMR spectroscopy in plant metabolomics. *FEBS Journal* **274,** 1126–1131.
11. De Vos, R.C.H., Moco, S., Lommen, A., Keurentjes, J.B., Bino, R.B. and Hall, R.D. (2007) Untargeted large-scale plant metabolomics using liquid chromatography coupled to mass spectrometry. *Nature Protocols* **2,** 778–791.
12. Lisec, J., Schauer, N., Kopka, J., Willmitzer, L. and Fernie, A.R. (2006) Gas chromatography mass spectrometry–based metabolite profiling in plants. *Nature Protocols* **1**, 387–396.
13. Sogat, T., Igarashi, K., Ito, C., Mizobuchi, K., Zimmermann, H.P. and Tomita, M. (2009) Metabolomic profiling of anionic metabolites by capillary electrophoresis mass spectrometry. *Anal. Chem.* **81**, 6165–6174.
14. Kaplan, F., Kopka, J., Haskell, D.W., Zhao, W., Schiller, K.C., Gatzke, N., Sung, D.Y. and Guy, C.L. (2004) Exploring the Temperature-Stress Metabolome of Arabidopsis. *Plant Physiol.* **136,** 4159–4168.
15. Lugan, R., Niogret, M.F., Kervazo, L., Larher, F.R., Kopka, J. and Bouchereau, A. (2008) Metabolome and water status phenotyping of Arabidopsis under abiotic stress cues reveals new insight into *ESK1* function. *Plant Cell Environ* **32**, 95–108.

16. Gibon, Y., Usadel, B., Blaesing, O.E., Kamlage, B., Hoehne, M., Trethewey, R. and Stitt, M. (2006) Integration of metabolite with transcript and enzyme activity profiling during diurnal cycles in Arabidopsis rosettes. *Genome Biol* 7, R76.
17. Tarpley, L., Duran, A.L., Kebrom, T.H. and Sumner, L.W. (2005) Biomarker metabolites capturing the metabolite variance present in a rice plant developmental period. *BMC Plant Biol* 5, 8 (31May2005).
18. Boyes, D.C., Zayed, A.M., Ascenzi, R., McCaskill, A.J., Hoffman, N.E. Davis, K.E. and Görlach, J. (2001) Growth Stage–Based Phenotypic Analysis of Arabidopsis. A Model for High Throughput Functional Genomics in Plants. *Plant Cell* **13**, 1499–1510.
19. Norén, H., Svensson, P. and Andersson, B. (2004) A convenient and versatile hydroponic cultivation system for *Arabidopsis thaliana*. *Physiol Plant* **121**, 343–348.
20. Robinson, M.M., Smid, M.P.L and Wolyn, D.J. (2006) High-quality and homogeneous Arabidopsis thaliana plants from a simple and inexpensive method of hydroponic cultivation. *Can J. Bot* **84**, 1009–1012.

Part II

Chemical Analysis Approaches

Chapter 6

Solid Phase Micro-Extraction GC–MS Analysis of Natural Volatile Components in Melon and Rice

Harrie A. Verhoeven, Harry Jonker, Ric C.H. De Vos, and Robert D. Hall

Abstract

The natural fragrance compounds produced by plants play key roles in the long-term fitness and survival of these plants as well as being of direct/indirect benefit to man. Almost all plant fragrances, either pleasant or unpleasant, comprise many different compounds, from different chemical classes and can indeed be highly complex in composition involving several hundred types of volatile molecule. Analyzing these mixtures and identifying their main (bio)active components is of importance in both fundamental and applied science. Gas Chromatography–Mass Spectrometry (GC–MS) plays a central role here. GC–MS has regularly been used for fragrance analysis and different extraction/adsorption and detection protocols have been designed specifically for plant materials. In this chapter, two methods are presented for two highly contrasting plant organs—a melon fruit and rice grains. Metabolomics analyses of these important food crops are already helping us understand better which components are most important in determining the flavour of these important food crops and how we might go about producing new "designer" crops which are even tastier than the existing ones.

Key words: Gas chromatography, Melon, Rice, Natural volatiles

1. Introduction

That which we call a rose; by any other name would smell as sweet

[Shakespeare].

Individual plants are reputed to be nature's most successful organic chemists with the biochemical capacity to produce maybe tens of thousands of primary and secondary metabolites (1). The plant kingdom as a whole may be capable of synthesising a couple of hundred thousand different molecules, from simple, single carbon molecules, to highly complex multi-/heterocyclic ringed structures (2). Such an arsenal of metabolites can be subdivided in many different ways but one key subgroup, separated on the basis of

Nigel W. Hardy and Robert D. Hall (eds.), *Plant Metabolomics: Methods and Protocols*, Methods in Molecular Biology, vol. 860, DOI 10.1007/978-1-61779-594-7_6, © Springer Science+Business Media, LLC 2012

their physico-chemical properties are the volatile components. Such components are generally volatile (present in the gaseous phase) at typical ambient temperatures and are to be detected in the fragrance profiles of plants. The types of volatile components produced by plants are chemically highly diverse comprising major groups such as esters, alcohols, the mono-, di-, and sesquiterpenoids, and the benzoates, as well as many other minor groups (3). Typically, the tastes and fragrances that we associate with plants, the smell of a rose, the intense fragrance/sweet taste of a ripe melon, the stink of fritillary, or the almost sickening odour of the voodoo lily, are not down to just one or two key components but are often highly balanced mixtures of very many different compounds from several biochemical classes (4). This makes their study particularly challenging and the identification of the key components more complex. Furthermore, understanding the importance of individual components—in terms of their contribution to overall fragrance and bioactivity—has an additional level of complexity because for example humans have different odour thresholds for different compounds. While some components might be detectable almost at the level of individual molecules by the human nose, others may require millimolar concentrations or higher before we are able to perceive them (5). This entails that such information must always be borne in mind in the biological interpretation of analytical results, as concentrations alone can be meaningless.

Plants have evolved to produce a wide range of volatile components for a variety of reasons, usually related to the impact they have on the environment around them. Compounds which act as attractants to animals and insects are produced by flowers and fruit to assist in the process of pollination and fruit and seed dispersal. Different fragrance profiles can be used to target specific types of pollinator and diurnal rhythms in volatile production and release are known from nature as a way to target day or night insects (6). Under the influence of man, breeding of food and ornamental crops has often led to significant gains (as in melon) or losses (e.g. in roses) in fragrance composition and intensity. Volatile components, however, often have an even more important role in plant survival. Many such components are known to have a certain bioactivity in that they may act as anti-pest or anti-grazing agents by having an unpleasant (or even lethal) effect on attacking organisms. Other volatiles have been shown to inhibit the germination and growth of spores and hyphae of pathogenic micro-organisms, thus protecting the plant against disease (7). Natural volatiles have also been implicated in signalling between plants and allellopathic effects in warmer countries where plants have been reported to control population density by inhibiting the germination of their own seed or seed from other, potentially competing species through the accumulation of natural volatiles in the soil surrounding growing plants (7).

Metabolomic analyses of natural plant volatiles are based first upon a suitable extraction, entrapment, or adsorption technique to collect and concentrate the components. Separation of these mixtures is then performed using optimized Gas Chromatography (GC) conditions suitable for the specific groups of compounds present. Subsequently, Mass Spectrometry (MS) is used to detect the metabolites and dedicated software and compound databases are then employed to analyze and annotate the results. For naturally released volatiles, so-called headspace trapping followed by GC coupled to electron impact (EI) MS is appropriate (8). Porous polymers specifically designed to adsorb volatile organic compounds are used to collect and concentrate molecules released from the plant material. In Solid Phase Micro-Extraction (SPME), these polymers are used to coat microfibres which are then exposed to the air above the volatile-emitting plant material (headspace) for a specific length of time (9). These fibres can then be directly inserted into the injection port of the GC and the adsorbed compounds are purged at high temperature in order to facilitate their direct release into the machine. Subsequent separation of individual compounds then takes place using an appropriate temperature gradient in the GC. Alternatively, adsorbant polymer powder can be used to fill small columns (e.g. flow-through TENAX cartridges) and clean air drawn over the sample carries any released volatiles across the column during which they are adsorbed onto the polymer filler, thus allowing clean air to pass out the other side (10). Again, such a process must be optimized regarding the amount of material and time in order to collect sufficient amounts of volatiles for accurate GC–MS analysis. Thermal desorption is then used to release the compounds from the column directly into the GC or alternatively, the collection column is washed through using an appropriate organic solvent and the extract obtained is then reconcentrated before analysis. A range of polymers are available which have differing affinities for compounds with contrasting chemical properties and so a guided choice needs to be made for the most suitable one before analysis starts. The plant materials used may be ground or frozen powders or whole living organs or even growing whole plants. Consequently, the whole process can be performed in a fully non-destructive way, thus enabling, for example, time course analyses of organ development, stress response, pathogen or pest attack, etc. to be executed. When using ground samples, additional treatments such as gentle heating or chemical purging can be used to help drive off the volatile compounds for trapping, thus shortening extraction times and enhancing compound concentrations.

Separation in the GC is dependent upon the chemical properties and size of the molecules present and the type and length of the GC capillary column and of course the temperature gradient used during the run. Separated molecules leaving the GC capillary are fragmented on entering the MS and these fragments (radicals and charged molecular ions), which are usually unique for each molecule—"its molecular fingerprint"—are then detected. This molecular fingerprint (spectrum) can then be used to identify the original molecule. A range of molecular databases are available for GC–MS compound libraries such as the on line NIST library (11) or the one commercially available from for example WILEY (Wiley publishers) and comparative analyses between the experimental and reference data present in the databases will yield a table of potential "hits" ordered according to statistical probability. Identifications, however, remain putative until standard reference compounds are run either separately or in a spiked extract in the same GC–MS instrument. Successful sample analysis and compound identification is clearly only possible once all four analytical steps have been optimized: stabilization of the plant material to give biologically relevant profiles, extraction of volatile components in an unbiased way, chromatographic separation gradients for the mixtures under examination, and finally efficient detection and statistically reliable data analysis to reveal the (discriminatory) compounds of interest.

Most of what is listed above is no different to the types of GC–MS analyses which have been performed since the design of the technology. However, the key difference, in terms of *metabolomics*, compared to traditional analytical organic chemistry, lies in the scale at which reliable and uniform analyses can be performed and the data pre-processing and processing software which are now available for complex, statistically reliable multivariate analyses of large sample sets which are required to reveal degrees of similarity and difference (12, 13). In this chapter, we describe reliable approaches to analyze highly contrasting crop plant materials—melon fruit and rice grains. Melon fragrance is a key quality aspect and major selling-point for the development of new fruit varieties (14). It is also an important marketing feature, used by consumers to judge quality and ripeness in the supermarket. Similarly, rice fragrance is not only important because it is used by consumers to distinguish types and varieties of rice associated with superior taste (e.g. Basmati, Jasmine, Pandan, etc.), but also because rice fragrance generally confers a higher market price compared to non-fragrant varieties (15–17).

2. Materials

2.1. Melon: SPME Adsorption of Natural Volatiles

1. Ripe melon fruits.
2. Freezer-proof plastic bags.
3. Screw-top, plastic 50-mL centrifuge tubes (e.g. Corning, NY, USA) for sample storage.

4. Liquid nitrogen for sample quenching and grinding (see Note 1).
5. Protective insulating gloves for handling super-cooled objects.
6. Pestle and mortar or preferably, a metal electric grinder—Basic Analytical mill A11 (IKA, Germany) both pre-cooled with liquid N_2 (see Note 2).
7. Metal spatula or small spoon, pre-cooled with liquid N_2.
8. Freezer between −70°C and −80°C for (long-term) sample storage. Second freezer at −20°C for sample preparation.
9. Milli-Q water or the double-distilled equivalent.
10. An aqueous (Milli-Q) NaOH/EDTA stock solution—200 mM EDTA adjusted to pH 7.5 using 1 M NaOH and a 5 M $CaCl_2$ stock solution in Milli-Q water mixed freshly, on the day of use, in the ratio 1:37.
11. 4-mL glass screw cap vials, 15 × 45 mm (e.g. Bester, Amstelveen, The Netherlands).
12. Polypropylene 10-mL bimetal crimp cap vials (e.g. Bester, Amstelveen, The Netherlands), with silicon/Teflon septa, 20 mm, (Interscience, Breda, The Netherlands).
13. Shaking water bath at 30°C.
14. Ultrasonic bath.
15. Balance for accurate weighing of 100–2,000 mg samples.
16. SPME fibres (65 μm Polydimethylsiloxane-Divinylbenzene (PDMS-DVB)) (Supelco, Code Blue, USA) (see Note 3).
17. Gas chromatograph (e.g. Fisons 8060) coupled to a mass spectrometer (e.g. MD 800, Fisons, Germany) and fitted with a CombiPAL auto-sampler (CTC Analytics, Switzerland) (see Note 4).
18. GC capillary column: HP-5 (50 m × 0.32 mm, film thickness 1.05 μm; Hewlett Packard) (see Note 5).
19. Helium supply linked to the GC–MS for use as carrier gas.

2.2. Rice: SPME Adsorption of Natural Volatiles

1. Uncooked polished rice grains (see Note 6).
2. Screw-top, plastic 50-mL centrifuge tubes (e.g. Corning, NY, USA) for sample storage.
3. Liquid nitrogen for grinding.
4. Protective insulating gloves for handling super-cooled objects.
5. A metal electric grinder—Basic Analytical mill A11 (IKA, Germany) pre-cooled with liquid N_2.
6. Metal spatula or small spoon, pre- cooled with liquid N_2.
7. Freezer at between −70°C and −80°C for (long-term) sample storage.

8. Polypropylene 10-mL bimetal crimp cap vials (e.g. Bester, Amstelveen, The Netherlands), with silicon/Teflon septa, 20 mm, (Interscience, Breda, The Netherlands).
9. Roller bank for sample mixing.
10. Balance for accurate weighing of 100–2,000 mg samples.
11. SPME fibres (65 um Polydimethylsiloxane-Divinylbenzene (PDMS-DVB)) (Supelco, Code Blue, USA) (see Note 3).
12. Gas chromatograph (e.g. Fisons 8060) coupled to a mass spectrometer (e.g. MD 800, Fisons, Germany) and fitted with a CombiPAL auto-sampler (CTC Analytics, Switzerland) (see Note 4).
13. GC capillary column: HP-5 (50 m × 0.32 mm, film thickness 1.05 μm; Hewlett Packard) (see Note 5).
14. Helium supply linked to the GC–MS for use as carrier gas.

3. Methods

The basic procedure for SPME volatile analysis comprises a number of defined steps. First, the fruit material has to be collected and prepared in a uniform and representative way. Second, the samples have to be treated in sealed containers in such a way as to allow the release of the natural volatiles into the container "headspace" after which they are allowed to adsorb onto the SPME fibre, again for a defined "trapping" period. Third, the adsorbed volatiles are measured using a GC–MS after which the data obtained is pre-processed and statistically analyzed. These steps are defined below first for melon samples and second for dry rice grain samples.

3.1. Solid Phase Micro-Extraction/ GC–MS of Melon Volatiles

3.1.1. Melon Fruit Sampling

For any metabolomics analysis, it is critically important to have comparable samples, obtained through identical harvesting techniques. For a full overview of all the "do's and don'ts" specifically for melon, the investigator is referred to Chapter 4 in this volume. Where possible, every aspect of the fruits to be compared should be equivalent in terms of, for example, time of day for harvesting, cultivation history, stage of development or ripeness, position of tissue within the fruit, etc. The challenge here is particularly great because of the considerable size of the melon fruit. It is not feasible to grind whole fruits and because there are gradients of ripening from top to bottom and from inside to outside (14, 18), taking comparable, representative samples is essential. Furthermore, because many melons develop lying on the ground there is also an asymmetry related to upper side/lower side which also needs to be taken into account. These details are all covered in Chapter 4.

1. Preferably, representative tissue sections should be removed from five comparable fruits. These are immediately pooled,

flash-frozen in liquid N_2 and then transferred to a -70°C freezer for storage (see Note 7).

2. Pre-label the required number of 50-mL tubes with a freezer-proof marker pen or freezer-compatible labels.
3. Grind the five frozen pooled fruit segments in liquid N_2 in a metal electric grinder until a uniform powder is obtained (ca. 20 s). Keep the powder deep frozen at all times and transfer immediately to 50-mL storage tubes using a pre-cooled metal spoon.
4. Store the pooled powder samples in the freezer at ca. -70°C until all samples are ready to be analyzed together.

3.1.2. Solid Phase Micro-Extraction: Melon

1. Take the stored sample from the –80°C freezer and transfer to a suitable container containing liquid N_2 and transfer to the laboratory or weighing room. All samples must remain deep frozen (in liquid N_2) at all times during this step.
2. Cool the end of a metal spatula in liquid N_2 and carefully weigh out 200 mg of the frozen melon powder into a pre-frozen 4-mL screw-cap vial.
3. Immediately close the vial with its screw cap and transfer to a -20°C freezer for 24 h.
4. Remove all the samples from the -20°C freezer and incubate with gentle agitation in a pre heated water bath at 30°C for 10 min (see Note 8).
5. Quickly open each vial one by one, and immediately add 3.8 mL of the EDTA–$CaCl_2$ mixture to give a final EDTA concentration of 5 mM and 4.625 M $CaCl_2$ with a sample concentration of 0.05 g/mL. Quickly re-close the vial and shake thoroughly and sonicate for 10 min in an ultrasonic bath (see Note 9).
6. Transfer a 1-mL aliquot of the suspended melon pulp into a 10-mL crimp cap vial and close immediately (see Note 10). Transfer the vials for SPME-GC–MS analysis.

3.1.3. SPME/GC–MS Analysis: Melon

Before beginning with GC–MS analyses, it is important to make sure the instrument is running optimally. Follow the instrument maker's instructions and check for stability by running either a standard sample or the first sample three times at the start of a run series. Run the same sample also in the middle and at the end of a run series. For large sample series, the GC–MS system should preferably be cleaned before starting and if possible, a new SPME fibre and a new conditioned column should be used. All columns should be exclusively used for SPME analysis, and columns (and preferably instruments) used for the analysis of derivatized samples must be avoided at all cost. SPME analyses should be performed with splitless injection and the injector liner should be clean or preferably be replaced between each run series.

1. Set the GC–MS parameters to the following values:
   (a) Helium pressure is maintained at 37 kPa
   (b) The GC interface is set to 260°C
   (c) The MS source temperature is set to 250°C
   (d) The GC temperature gradient programme used starts at 45°C for 2 min, linear gradient raise to 250°C at a rate of 5°C/min and finally maintain for 5 min at 250°C
   (e) Between each sample the column is automatically cooled down from 250°C to the starting temperature of 45°C, ready for the next run
   (f) The total run time is 60 min which includes the cooling step to bring the oven and column back to the starting temperature
2. Prior to volatile adsorption, incubate the vials from step 6 of Subheading 3.1.2 containing the melon suspension at 50°C for 10 min with gentle agitation (see Note 11).
3. Using the CombiPal, insert the SPME fibre (see Fig. 1) through the septum to bring the adsorbant polymer into contact with the headspace of the vial. Expose the coated fibre for 20 min while maintaining the temperature at 50°C and continuing to agitate (see Note 12).
4. Retract the fibre and insert into the injection port of the GC. Drive off the trapped volatiles through temperature desorption at 250°C for 1 min.
5. Run the temperature gradient and record all mass spectra in the 35–400 m/z range with the MS set to scan at 2.8 scans/s and with an ionization energy of 70 eV (see Fig. 2) (see Note 13).
6. Data pre-processing and initial analysis can be performed using the commercial software supplied with the MS instrument used (see Note 14).

3.2. Headspace Trapping of Fragrant Rice Volatiles

3.2.1. Rice Grain Sampling

1. Take a representative sample of rice grains from each source to be analysed. We usually grind 40 g per genotype. If possible, avoid any grains which look different through disease or incomplete polishing.
2. To facilitate the grinding process, pre-cool the metal grinder with liquid N_2, add ca 25 mL liquid N_2, then after evaporation of all liquid N_2, grind the rice grain sample to a fine powder. This usually takes 15 s. From this point onward the sample should not be allowed to thaw out before analysis begins.
3. Using a pre-cooled metal spoon, transfer the rice flour to a pre-cooled 50 mL screw-top plastic tube, seal and transfer immediately to the −70°C freezer for storage.

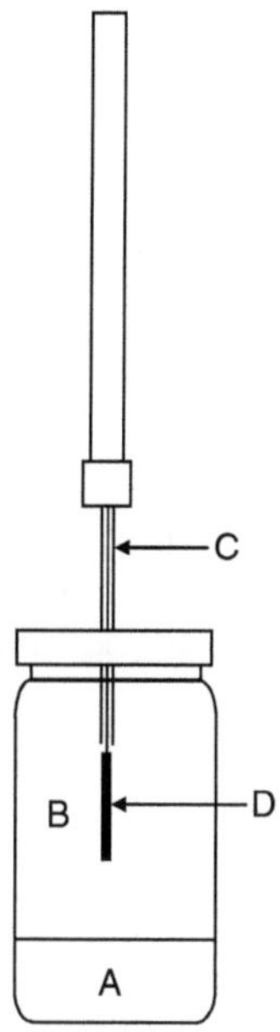

Fig. 1. Schematic representation of an SPME fibre exposed to the headspace in a glass vial containing biological material. The biological material (e.g. frozen melon powder) is transferred to a suitable glass vial to which $CaCl_2$–EDTA solution (**A**) is immediately added and the vial immediately crimp-capped. Care is taken to ensure all liquid + powder is present at the bottom of the vial and that the underside of the crimp cap septum is clean. The vial is warmed to 50°C and agitated for 10 min in order to enhance the release of volatile molecules from the sample. The fibre (**D**), still protected by its metal sheath (**C**), is then injected through the septum into the vial after which the fibre (**B**) is extended to become exposed to the headspace. After a further 20 min, the fibre is retracted into its sheath, removed, and inserted into the injection port of the GC for further analysis. (Reproduced from ref. 10).

3.2.2. Solid Phase Micro-Extraction: Rice

1. Weigh out 1 g of the still-frozen rice flour into the 10-mL vial and immediately seal the vial closed using the crimp cap and septum.
2. Once all samples have been weighed out and sealed into the crimp-cap vials, transfer to a roller bank and incubate at room temperature for 24 h with 50–60 rotations/min (see Note 15).
3. Remove the vials from the roller bank and tap gently on the bench to bring most of the flour to the bottom of the vial.
4. Leave all vials standing for 30 min to allow all the flour to settle at the bottom of the vial to avoid contamination of the SPME fibre.
5. Transfer the vials to the CombiPal for further volatile extraction and analysis.

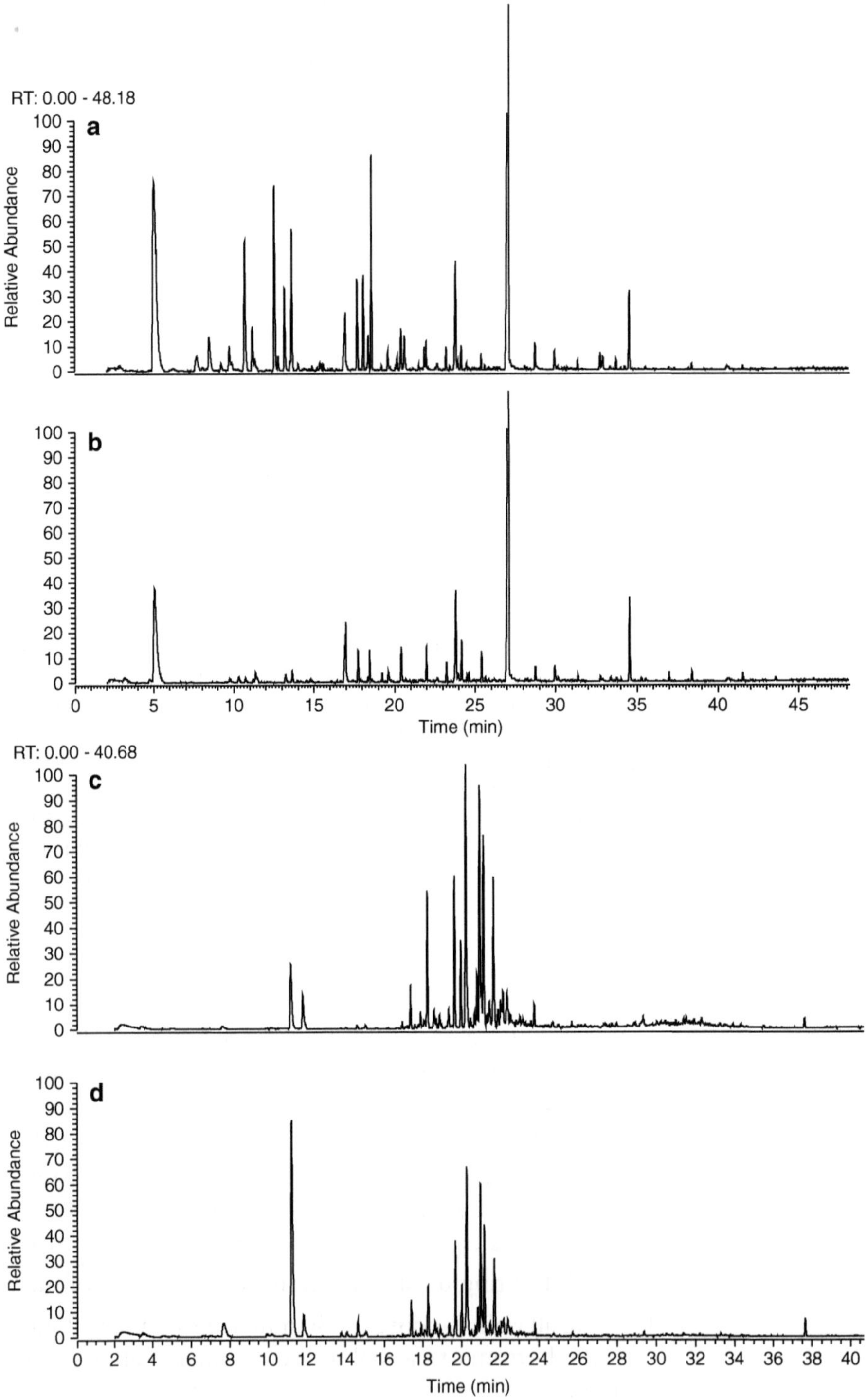

Fig. 2. Typical SPME GC–MS profiles of melon fruit (**a** and **b**) and uncooked rice grain volatiles (**c** and **d**). Ripe melon samples from variety Cezanne (**a**, grown in France) and Nov Yizreél (**b**, grown in Israel); Polished rice grains from variety Taori Basmati (**c**, from India) and PTT1 (**d**, from Thailand).

3.2.3. SPME/GC–MS Analysis: Rice

Please refer above to Subheading 3.1.3 for all precautions required prior to initiating SPME GC–MS analyses.

1. Set the GC–MS parameters to the following values:
   (a) Helium pressure is maintained at 37 kPa
   (b) The GC interface is set to 260°C
   (c) The MS source temperature is set to 250°C
   (d) The GC temperature gradient programme used starts at 45°C for 2 min, linear gradient raise to 250°C at a rate of 4°C/min and finally maintain for 5 min at 250°C
   (e) Between each sample the column is automatically cooled down from 250°C to the starting temperature of 45°C, ready for the next run
   (f) The total run time is 68 min which includes the cooling step to bring the oven and column back to the starting temperature
   (g) The split valve is closed during injection (1 min at 250°C)
2. Using the CombiPal, insert the SPME fibre (see Fig. 1) through the septum to bring the adsorbant polymer into contact with the headspace of the vial. Expose the coated fibre for 20 min while maintaining the temperature at 50°C and continuing to agitate (see Note 12).
3. Retract the fibre and insert into the injection port of the GC. Drive off the trapped volatiles through temperature desorption at 250°C for 1 min with closed split valve.
4. Run the temperature gradient and record all mass spectra in the 35–400 m/z range with the MS set to scan at 2.8 scans/s and with an ionization energy of 70 eV (see Note 13).
5. Data pre-processing and initial analysis can be performed using the commercial software supplied with the MS instrument used (see Note 14).

3.3. Data Analysis

Once the data have been generated from the different GC–MS runs, individual samples can be compared. A good starting point for data analysis is to use the software package(s) supplied with the instrument used. Once differential mass peaks have been identified these are then usually exported into programmes such as AMDIS (19) and databases such as the NIST mass spectral library (11) are then interrogated in order to predict the identity of potentially interesting compounds. The hit lists obtained can then be examined in detail (it is never wise to assume the highest-scoring hit is the most likely identity) and, if possible, commercially available or in-house synthesized standards should be used to confirm the compound identity using the same instrument used for the sample analyses.

For untargeted metabolomic comparisons of large sample numbers, pre-processing of the data is required to enable (semi-) automatic multivariate comparisons. For this, we routinely use the Freeware package metAlign™ (20) for data pre-processing (*see* also Chapter 15 for the details of use). This software performs initial baseline correction and eliminates most noise after which the spectra in different datasets are fully aligned so that every mass peak (m/z) with a defined retention time is lined up across all data sets being analyzed. Once this has been performed, statistical and multivariate software can then be used to search for mass peaks which show differences between different samples. These peaks potentially represent metabolites which have non-uniform distribution within the series of samples and may therefore be linked to phenotypic differences such as fragrance or flavour.

Depending on the tissues used, the ripening stage, etc., the complexity of the metabolomic spectra obtained following GC–MS analysis can vary considerably. In Fig. 2, chromatograms are presented which were obtained for rice grains and melon fruits (ripe stage). Such chromatograms however, may vary considerably between different varieties or even between samples of the same variety but from, for example, different locations/production periods.

4. Notes

1. When liquid nitrogen is being used, standard laboratory practice is required where safety precautions should be taken at all times (eye and hand/skin protection, protective clothing, etc.).
2. For large sample numbers a mill is preferred over a pestle and mortar as this avoids potential risk of developing RSI. The type of analytical mill listed can withstand the extreme temperatures of liquid nitrogen. Other types (such as commercial coffee bean mills) will quickly fail after only a few samples. Whichever is used, both must be thoroughly cooled down with liquid N_2 before the first sample goes in. Liquid N_2 should also be present during grinding to help the milling process and to prevent the sample warming up.
3. A range of SPME fibres are available with different adsorbant properties (see e.g. ref. 21). For metabolomics, fibres with a broad adsorption capacity are desirable. The one listed here has been chosen for its broad affinity to volatiles of different chemical origins and is appropriate for both types of tissue described.
4. GC–MS machines suitable for plant metabolomics are now made by a range of manufacturers (e.g. LECO, Agilent, Waters,

Thermo, Bruker, Perkin Elmer, etc.). Each manufacturer provides a wealth of information to help you chose which machine is most appropriate to meet your needs and budget.

5. Again, several different column types/length, etc. are available from a range of different manufacturers. The HP5 column listed here was found suitable for our needs and its use is frequently reported in the literature. Furthermore, there are much data available on retention index values for this column which will be useful for final compound identification.
6. Polished rice grains are those most frequently eaten. They are generally white in appearance and have lost their outer bran layers and the aleurone and sub-aleurone layers (17). Unpolished or poorly polished rice will still retain (part of) these tissues and will have a distinctly different metabolic profile.
7. Biological variation between different replicate samples, even from the same genotype, must always be taken into account. To reduce the inter-sample variation, we always work with pooled materials taken from several (usually 5–10) plants grown under identical conditions and harvested at the same time or stage. A number of biological replicates of each sample is also recommended in order to get statistically reliable information. A good rule of thumb is to use three to five replicates each comprising three to five pooled tissue samples. Once volatiles related to a specific trait or genotype have been identified, if needed or desired, the variation between plants within a limited number of pools, for instance the most extreme ones, may be determined for these volatiles.
8. All metabolomics analyses are geared to obtain the most "natural" overview of the biochemical composition of plant material. This entails a minimum of sample handling and rapidity in extraction and analysis. However, some volatile compounds are only released after cell disruption through enzymic and non-enzymic processes. As such compounds can be of considerable relevance specifically in flavour and fragrance studies, we usually incorporate a strictly controlled incubation protocol prior to analysis. After this incubation all (enzymic) reactions must be immediately stopped using a saturated $CaCl_2$ solution.
9. A concentrated $CaCl_2$ solution is used here for two main purposes: first, it should stop any further metabolic reactions and thus preserve the biochemical composition; second, $CaCl_2$, is also known to help drive volatile components out of the matrix and into the headspace above (18). EDTA has been included to prevent oxidation of the volatile by its chelating effect while at the same time increasing the pH of the extract to 6.2–6.3. Such a procedure has been found to preserve the volatile composition of for example fruit pulp for at least 12 h.

10. This transfer of a uniform aliquot requires care. We prefer to use an automatic pipette fitted with a 1 mL disposable tip which has had the end cut off to create a wider opening. Also, take care to gently transfer the aliquot fully to the bottom of the vial and not spray it around the edges.
11. Agitation and heating helps to drive off the volatiles from the sample matrix.
12. It is important to check all vials to ensure that there is no liquid or plant tissue deposited on the inside of the septum prior to injecting the SPME fibre. Bringing such material into contact with the fibre polymer will damage it and also give potentially erroneous results.
13. The method described has been shown to give relatively stable profiles for 12 h after sample preparation. Consequently, no more than 12 samples should be prepared at a time. SPME is a semi-quantitative analysis method. It is mostly used for the reliable detection of qualitative differences between samples and for larger quantitative differences. This is because the adsorbent polymer has different affinities for different types of compounds and also because the distribution of volatiles between sample matrix and its headspace is influenced by the specific composition of the mixture being analyzed. Furthermore, the dynamic range for individual components is usually limited.
14. While there are many different approaches for data analysis, all will eventually yield a "putative" compound identification. Definitive confirmation of structural identity is only possible when an authentic standard compound is available that co-elutes under various chromatographic conditions.
15. Using the EDTA–$CaCl_2$ treatment described for melon has proven not to be suitable for rice flour, as the reaction of $CaCl_2$ with the very high starch content of the rice grain results in gel formation which prevents the extra purging of volatiles from the sample.

Acknowledgements

This work was performed within the EU META-PHOR Project (FOOD-CT-2006-036220). RDH and RdV acknowledge additional funding from the Centre for BioSystems Genomics (CBSG) and the Netherlands Metabolomics Centre (NMC), both part of the Netherlands Genomics Initiative (NGI).

References

1. Baxter, I.R. and Borevitz, J.O. (2006) Mapping a plant's chemical vocabulary. *Nat Genet* **38**, 737–738.
2. Oksman-Caldentey, K-M. and Inzé, D. (2004) Plant cell factories in the post-genomic era: new ways to produce designer secondary metabolites. *Trends in Plant Sci* **9**, 433–440.
3. Dudareva, N., Negre, F., Nagegowda, D.A., and Orlova, I. (2006) Plant Volatiles: Recent Advances and Future Perspectives. *Crit. Rev. Plant Sci.* **25**, 417–440.
4. Baldwin, E.A., Scott, W.J., Shewmaker, C.K. and Schuch, W. (2000) Flavor trivia and tomato aroma: biochemistry and possible mechanisms for control of important aroma components. *HortSci.* **35**, 1013–1022.
5. Wilkie, K., Wootton, M. and Paton, J.E. (2004) Sensory testing of Australian fragrant, imported fragrant and non-fragrant rice aroma. *Int. J. Food Prop.* 7, 27–36.
6. Verhoeven, H.A., Beuerle, T. and Schwab, W. (1997) Solid-phase micro extraction: artefact formation and its avoidance. *Chromatographia* **46**, 63–66.
7. Mallik, A.U. (2002) (Ed.). *Chemical ecology of plants: allelopathy in aquatic and terrestrial ecosystems.* Birkhäuser Verlag, Basel.
8. Song, J., Fan, L. and Beaudry, R.M. (1998) Application of solid phase microextraction and gas chromatography/time-of-flight mass spectrometry for rapid analysis of flavor volatiles in tomato and strawberry fruits. *J. Agric. Food Chem.* **46**, 3721–3726.
9. Augusto, F., Valente, A.L.P., Tada, E.S. and Rivellino, S.R. (2000) Screening of Brazilian fruit aromas using solid-phase microextraction–gas chromatography–mass spectrometry. *J. Chromat. A* **873**, 117–127.
10. Tikunov, Y.M., Verstappen, F.W.A. and Hall, R.D. (2007) Metabolomic profiling of natural volatiles: headspace trapping: GC-MS, in *Metabolomics. Methods and Protocols Vol 358* (Weckwerth, W., ed.) Humana Press, Totowa, USA, pp. 39–53.
11. http://www.nist.gov/srd/nist1a.htm.
12. Hall, R.D. (2006) Plant metabolomics: from holistic hope, to hype to hot topic. *New Phytol.* **169**, 453–468.
13. Keurentjes, J.J.B., Fu, Y, De Vos C.H.R., Lommen, A, Hall, R.D., Bino, R.J., Van Der Plas, L.H.W., Jansen R.C., Vreugdenhil, D. and Koornneef, M. (2006) The genetics of plant metabolism. *Nat. Genet.* **38**, 842–849.
14. Biais, B., Allwood, J.W., Deborde, C., Xu, Y., Maurcourt, M., Beauvoit, B., Dunn, W.B., Jacob, D., Goodacre, R., Rolin, D., and Moing, A. (2009) H NMR, GC-EI-TOF MS and dataset correlation for fruit metabolomics: application to spatial metabolite analysis in melon. *Anal. Chem.* **81**, 2884–2894.
15. Fitzgerald, M.A., Sackville-Hamilton, N.R., Calingacion, M.N., Verhoeven, H.A. and Butardo, V.M. (2008) Is there a second fragrance gene in rice? *Plant Biotech. J.* **6**, 416–423.
16. Hall, R.D., Brouwer, I.D. and Fitzgerald, M.A. (2008) Plant metabolomics and its potential application for human nutrition. *Physiol. Plant.* **132**, 162–175.
17. Fitzgerald, M.A., McCouch, S and Hall, R.D. (2009) More than just a grain of rice: the search for quality. *Trends Plant Sci.* **14**, 133–139.
18. Bezman, Y., Mayer, F., Takeoka, G.R., Buttery, R.G., Ben-Oliel, G., Rabinowitch, H.D. and Naim, M. (2003) Differential effects of tomato (*Lycopersicon esculentum* Mill) matrix on the volatility of important aroma compounds. *J. Agric. Food. Chem.* **51**, 722–726.
19. http://www.amdis.net.
20. http://www.metalign.nl/.
21. http://www.sigmaaldrich.com/Brands/Supelco_Home.

Chapter 7

Profiling Primary Metabolites of Tomato Fruit with Gas Chromatography/Mass Spectrometry*

Sonia Osorio, Phuc Thi Do, and Alisdair R. Fernie

Abstract

Metabolite profiling is a rapidly expanding technology which aims to quantify the entire metabolome of biological samples. Gas Chromatography Mass Spectrometry (GC-MS) is one of the most widely used analytical tools for profiling highly complex mixtures of primary metabolites, such as organic and amino acids, sugars, sugars alcohols, phosphorylated intermediates, and lipophilic compounds. This chapter summarizes all of the preparatory steps for metabolite profiling of polar compounds by GC-MS in tomato fruit, from the sampling of plant material to the derivatization procedures required to render the metabolites volatile.

Key words: GC-MS, Metabolite profiling, Primary metabolism, Tomato fruit, Derivatization

1. Introduction

The metabolome is defined as the total small-molecule complement of a cell, and metabolomics is therefore the study of all the low-molecular-weight molecules or metabolites of a cell or organism (1, 2). Technological developments have considerably extended our ability to describe complex biological systems, facilitating the simultaneous detection of individual compound classes with a complex diversity of chemical properties. Gas chromatography hyphenated to mass spectrometry (GC-MS) is one of the most versatile and widely applied technology platforms in modern metabolomic studies (the others being LC-MS and NMR). It combines two strongly complementary technologies: GC which can separate metabolites that have almost identical mass spectra (such as isomers), while MS provides fragmentation patterns that

*Sonia Osorio and Phuc Thi Do contributed equally to this work.

Nigel W. Hardy and Robert D. Hall (eds.), *Plant Metabolomics: Methods and Protocols*, Methods in Molecular Biology, vol. 860, DOI 10.1007/978-1-61779-594-7_7, © Springer Science+Business Media, LLC 2012

differentiate between co-eluting, but chemically diverse, metabolites. Therefore, GC-MS can facilitate the identification and robust quantification of a few hundred metabolites within a single plant extract, resulting in fairly comprehensive coverage of the central pathways of primary metabolism (3–8). Although no single analytical system can cover the whole metabolome, GC-MS has a relatively broad coverage of compound classes, including organic and amino acids, sugars, sugar alcohols, phosphorylated intermediates, and lipophilic compounds (4, 8). The main advantages of this technology are that it has long been used for metabolite profiling and thus there are stable protocols for machine set-up and maintenance, and chromatogram evaluation and interpretation. Furthermore, the advent of faster computers, improved algorithms, statistical software packages, and available databases will likely allow for the exploitation of this method and thus enable the capture of more biologically relevant information (9). Metabolite profiling is a rapidly expanding technology and interest in applying it in plant biology continues to grow with fields of application, including diagnostic and descriptive analysis of metabolic responses to various genetic and/or environmental perturbations such as biotic and abiotic stresses, use in gene function annotation and systems biology (10, 11). Since metabolites present in organisms show a great diversity of chemical properties and wide ranges of concentration, in order to be able to identify and quantify as many different metabolites as possible by the most reliable manner, species specific extraction and chemical modification (derivatization) methods applied for the sample preparation must be considered.

In this chapter, an analytical chemistry approach for the detection of polar compounds of tomato primary metabolism is presented. It is based on the use of relatively inexpensive and robust technology—Gas chromatography time-of-flight mass spectrometry (GC-TOF-MS). GC-TOF-MS is used rather than quadrupole technology (GC-quad-MS), since it provides faster scan times, which give rise to either improved deconvolution and higher mass accuracy or reduced run times for complex mixtures, thus facilitating operation at high throughput (12). The measured metabolite levels should reflect the status in vivo. In fact, metabolite turnover is extremely rapid as compared to mRNA or protein turnover. Therefore, immediate inactivation of metabolism to avoid turnover of metabolites is required. Quenching of metabolism is generally achieved by rapidly freezing samples (at a constant temperature of –60°C or less). The combination of methanol and high temperature (70°C) in the extraction procedure render the majority of enzymes inactive (13, 14). In order to make various classes of compounds volatile and thus accessible for analysis by GC, chemical modification of the polar functional groups using derivatization reagents is necessary. Methoxyamination followed by silylation is the most appropriate derivatization procedure used in GC-MS plant metabolomics studies. This two-step derivatization process involves conversion of aldehyde or keto groups into oximes using methoxylamination reagent, followed by conversion of functional groups such as –OH, –COOH, –NH, –SH into trimethylsilyl (TMS) ethers, TMS-esters, TMS-amines, or TMS-sulphides, respectively. Detailed tests have revealed that derivatization time, temperature, and amount of reagent influence the outcome of the results, indicating that this procedure is a crucial step of the protocol (15, 16). Derivatization of compounds often results in more than one peak for a metabolite of interest, owing to either partial silylation or isomerization in the case of methoxyaminated compounds such as sugars. One of two approaches is commonly take into account for this. Either all peaks are identified but quantification is carried out only on the most reliable peak or summation of all the peaks that represent a given metabolite can be carried out.

The basic steps in the process can be summarized as follows:

1. Sampling the plant tissue, homogenizing, and exact weighing of sample aliquots.
2. Extracting metabolites concomitant with enzyme inactivation and the addition of internal standards and/or authentic chemical standards for peak identification or assessment of extraction efficiency.
3. Drying the polar extract.
4. Derivatizing the polar extract by methoxyamination followed by silylation, and adding retention time index standards.
5. Analyzing the derivatized samples by GC-MS.

2. Materials

2.1. Sampling and Extraction

1. Argon.
2. Centrifuge (capable of 3,700×*g*), (e.g. Allegra® x-15R, Beckman Coulter, Krefeld, Germany).
3. Methanol gradient grade for liquid chromatography (see Note 1).
4. MilliQ water approx 0.055 μS/cm.
5. Oscillating ball mill MM200 (e.g. Retsch GmbH and Co.KG, Haan, Germany) or alternatively a pestle and mortar.
6. Ribitol, purity ≥99.0%; 0.2 mg/mL in dH_2O.
7. Speed vacuum concentrator (e.g. SPD111V-230, Thermo-Electron Corporation, Langenselbold, Germany).

8. Schott glass AR-GLASS® culture tubes (soda-lime) (DURAN GMbH, Mainz, Germany).
9. Thermoblock (capable of heating to up to 70°C).
10. Liquid nitrogen supply.
11. Vortex.
12. Scalpel blades, aluminium foil, 6-well plates, spatula, balance, microfuge tubes (2 mL).

2.2. Derivatization

1. Methoxyamine hydrocloride, purity 98% (e.g. Sigma, St. Louis, USA). Store at room temperature under dry atmosphere.
2. *N*-methyl-*N*-trimethylsilyltrifluor(*o*)acetamide (MSTFA) (Macherey and Nagel, Düren, Germany). MSTFA should be stored in opaque glass bottles under nitrogen. Contact with water generates hydrogen fluoride gas which is highly toxic. Store at 4°C (see Note 1).
3. Pyridine, analytic grade (Merck, Darmstadt, Germany). Store at room temperature (see Note 1).
4. Retention time index standard mixture: fatty acids methyl esters (FAMES). All must be of standard grade for GC: Esters included are methylcaprylate, methyl pelargonate, methylcaprate, methyllaurate, methylmyristate, methylpalmitate, methylstearate, methyleicosanoate, methyldocosanoate, lignoceric acid methylester, methylhexacosanoate, methyloctacosanoate, and triacontanoic acid methylester. (All available via e.g. Sigma). The esters are dissolved in $CHCl_3$ at a final concentration of 0.8 μL/mL for liquid; 0.4 mg/mL for solid standards. Mix all well, aliquot into glass vials, and store at −20°C.
5. 1.1 mL Screw Top Tapered Vial—Clear Gold Grade (CHROMACOL LTD, Thermo Fisher Scientific Inc, Herts, UK).
6. Shaker (950 rpm).

2.3. GC-TOF-MS

1. Autosampler system (PAL Agilent, Santa Clara, USA).
2. Capillary column MDN-35, 30 m × 0.32 mm, 0.25 μm film thickness (SUPELCO, USA) (or equivalent).
3. Conical single taper split/splitless liner (Agilent, Böblingen, Germany).
4. Gas chromatograph (Agilent 6890N), split and splitless injector with electronic pressure control up to 150 psi (Agilent, Böblingen, Germany).
5. Helium 5.0 carrier gas.
6. Pegasus III Tof mass analyzer from LECO and corresponding software (LECO, St Joseph, USA) (or equivalent).

3. Methods

3.1. Sample Preparation (see Note 2)

1. Collect the tomato fruit, cut into two using a scalpel blade. Peel away the cuticle/epidermis layers and remove the placental tissue and chop the pericarp into small pieces. Transfer pericarp to pre-cooled 6-well-plates or wrap samples in aluminium foil. Freeze immediately in liquid nitrogen.
2. Pre-cool two steel cylinders and metal balls in liquid nitrogen.
3. Quickly take out two samples and place them into independent steel cylinders together with a metal ball and cover the cylinders.
4. Fix cylinders in the mixer mill and mill at 25 Hz/s for 2 min.
5. Quickly take out the cylinders and place back into liquid nitrogen.
6. Transfer the fine powder into a pre-cooled tube and keep in liquid nitrogen.
7. Repeat steps 2–6 until all samples have been homogenized.
8. Weigh out ~250 mg fine powder of each sample into a pre-cooled 2-mL microcentrifuge tube and keep in liquid nitrogen or store at –80°C until use (see Note 3).

3.2. Extraction

1. Remove the homogenized samples and add 1.5 mL 100% methanol (pre-cooled to –20°C) to each and vortex for 10 s (see Note 4). Also prepare one tube without sample as a control (see Note 5).
2. Transfer the mixture to a Schott glass vial.
3. Add a further 1.5 mL 100% methanol (pre-cooled to –20°C) into the 2-mL microcentrifuge tube to wash it out. Transfer to the same Schott glass vial as in step 2.
4. Add 120 μL ribitol (0.2 mg/mL in dH_2O) as an internal quantitative standard in Schott glass vial and vortex for 10 s.
5. Incubate for 15 min at 70°C in a thermoblock.
6. Allow the samples to cool down to room temperature. Then, add 1.5 mL of dH_2O and vortex for 10 s.
7. Centrifuge for 15 min at 3,500 ×*g*.
8. Aliquot 50 μL and 5 μL into two new 2-mL microcentrifuge tubes (see Note 6). The pellet can now be used for starch, protein, and/or cell wall determination or be discarded.
9. As a backup (in case you lose a sample), transfer a second aliquot to another new 2-mL tube.
10. Dry absolutely in a speed vacuum concentrator without heating for between 3 and 12 h.
11. For storage, fill the tubes with argon gas before closing. The tubes can then be stored at –80°C for up to 3 months (see Note 7).

3.3. Derivatization

1. Take out the dried extracts from freezer and dry them absolutely in a speed vacuum concentrator for 30 min (see Note 8).
2. Prepare fresh methoxyamine solution by dissolving methoxyamine hydrochloride at 30 mg/mL in pure pyridine. Work in a fume hood (see Note 1).
3. Add 60 μL methoxyamine solution as prepared in step 2 to each sample and quickly close the tube.
4. Shake for 2 h at 37°C at 950 rpm.
5. Spin down shortly to collect all drops on the walls and lids of the microcentrifuge tubes.
6. Prepare MSTFA reagent with FAMES (1 mL of MSTFA with 50 μL of FAMES) (see Note 9).
7. Add 120 μL of MSTFA reagent prepared in step 6 to each sample tube and quickly close the tube.
8. Shake for 30 min at 37°C at 950 rpm.
9. Spin down shortly to collect all drops on the walls and lids of the microcentrifuge tubes.
10. Transfer reaction solutions into glass vials suitable for the GC-MS autosampler and quickly close the vials (see Notes 10 and 11).

3.4. Data Acquisition by GC-MS

1. Inject 1 μL of sample in splitless or split mode, depending on the metabolite concentration (see Note 6), with the helium carrier gas flow rate set to 2 mL/min by using the autosampler. The flow rate is kept constant with electronic pressure control enabled. The injection temperature is set to 230°C. Injection programmes must include syringe washing steps before and after each injection.
2. Perform chromatography using a 30-m MDN-35 capillary column. The temperature programme should be isothermal for 2 min at 80°C, followed by a 15°C per min ramp to 330°C, and holding at this temperature for 6 min. Cooling should be as rapid as the instrument specifications allow. Set the transfer line temperature to 250°C and match ion source conditions.
3. Set the ion source to maximum instrument specifications, 250°C. The recorded mass range should be m/z 70 to m/z 600 at 20 scan/s. Proceed the remaining monitored chromatography time with a 170-s solvent delay with filaments turned off. Manual mass defect should be set to 0, filament bias current should be −70 V, and detector voltage should be ~1,700–1,850 V. Automatically tune the instrument according to the manufacturer's instructions.

4. Transfer raw GC-MS profile chromatograms to a powerful server and regularly back up them.
5. Proceed with data (pre)processing and analysis as indicated (see Chapters 15 and 16 in this volume).

4. Notes

1. Reagents are extremely toxic and should be handled in a fume hood with gloves.
2. Once the sample has been frozen in liquid nitrogen it must not thaw out even slightly before extraction. Make sure therefore to keep samples at constant freezing temperature (in liquid nitrogen) to avoid degradation of metabolites and precool all components to be used (spatulas, vials, etc.) in liquid nitrogen before they come into contact with the sample.
3. If lyophilized material is used, scale down the amount of material based on fresh weight to dry weight ratio. An amount of ~25 mg dried powder is suitable.
4. Enzymatic activity stops on adding methanol.
5. It is necessary to include the control tube without metabolite extract in order to identify any contaminants. It is important that all chemicals and containers need to be of the highest available purity. Please consider that autoclaved material, although sterile, may nevertheless be *chemically* contaminated.
6. Tomato fruit accumulate high amounts of sugars such as fructose, glucose, sucrose, as well as some organic acids such as citric acid and malic acid. 50 μL of extract in a splitless mode GC-MS run gives overloaded peaks for these compounds. Therefore, in order to measure accurately both high and low abundant metabolites, two approaches are required. Either two modes of run, one splitless (for low abundant metabolites) and one split (for high abundant metabolites) using 50 μL of extract are performed, or two different extracts of 50 μL (for low abundant metabolites) and 5 μL (for high abundant metabolites) in splitless mode runs are used (see Fig. 1).
7. Argon-filled sample tubes prevent the extract from oxidization and degradation by reactions through components of atmospheric air. Extracts can be stored at −80°C for up to 3 months. A longer storage time has not yet been investigated.
8. The most critical point is to avoid any water or moisture during derivatization. The silylating step is highly especially sensitive and even minor contamination with water will lead to inconsistent results.

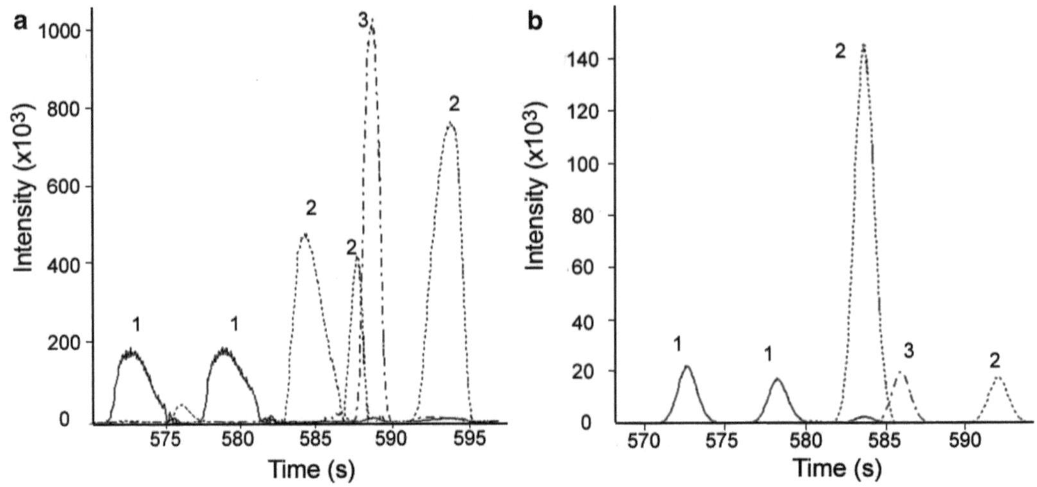

Fig. 1. Example of splitless (**a**) and split (**b**) mode run of tomato fruit of fructose *m/z* 307 (1), glucose *m/z* 160 (2), and citric acid *m/z* 273 (3). Overload peaks were observed in splitless mode (**a**), but not in split mode (**b**).

9. For amount of FAMES, if two modes of run (splitless and split) are applied, 50 μL of FAMES/1 mL MSTFA is required. If only splitless mode run is used, a reduced amount of FAMES to 20–30 μL/mL MSTFA is possible.
10. The rest of derivatized samples can be stored in glass vials (in case something goes wrong with injection or measurement), but always in the dark at room temperature, for up to 2 days. Avoid storage in a cold room.
11. Samples must be injected in statistically valid randomized order to minimize the influence of experiment handling.

Acknowledgements

Recent work in our laboratory on tomato metabolite profiling has been supported by The European Union grant FOOD-CT-2006-036296 "DEVELONUTRI".

References

1. Oliver, S.G., Winson, M.K., Kell, D.B., Baganz, F. (1998) Systematic functional analysis of the yeast genome. *Trends Biotechnol* **16**, 373–378.
2. Tweeddale, H., Notley-McRobb, L., Ferenci, T. (1998) Effect of slow growth on metabolism of *Escherichia coli*, as revealed by global metabolite pool ("metabolome") analysis *J. Bacteriol.* **180**, 5109–5116.
3. Fiehn, O., Kopka, J., Dörmann, P., Altman, T., Trethewey, R.N., Willmitzer, L. (2000) Metabolite profiling for plant functional genomics. *Nature Biotech.* **18**, 1157–1161.
4. Roessner, U., Luedemann, A., Brust, D., Fiehn, O., Linke, T., Willmizer, L., Fernie, A.R. (2001) Metabolic profiling allows comprehensive phenotyping of genetically or

environmentally modified plant systems. *Plant Cell* **13**, 11–29.

5. Roessner-Tunali, U., Hegemann, B., Lytovchenko, A., Carrari, F., Bruedigam, C., Granot, D., Fernie, A.R. (2003) Metabolic profiling of transgenic tomato plants overexpression hexokinase reveals that the influence of hexose phosphorylation diminishes during fruit development. *Plant Physiol.* **133**, 84–99.
6. Weckwerth, W. (2003) Metabolomics in systems biology. *Ann. Rev.Plant Biol.* **54**, 669–689.
7. Kopka, J., Fernie, A., Weckwerth, W., Gibon, Y., Stitt, M. (2004) Metabolite profiling in plant biology: platforms and destinations. *Genome Biol.* **5**, 109.
8. Schauer, N., Semel, Y., Roessner, U., Gur, A., Balbo, I., Carrari, F., Pleban, T., Perez-Melis, A., Bruedigam, C., Kopka, J., Willmitzer, L., Zamir, D., Fernie, A.R. (2006) Comprehensive metabolic profiling and phenotyping of interspecific introgression lines for tomato improvement. *Nat. Biotech.* **24**, 447–454.
9. Fiehn, O. (2008) Extending the breadth of metabolite profiling by gas chromatography coupled to mass spectrometry. *Trends Anal. Chem.* **27**, 261–269.
10. Schauer, N., Fernie, A.R. (2006) Plant metabolomics: towards biological function and mechanism. *Trends Plant Sci.* **11**, 508–516.
11. Guy, C., Kopka, J., Moritz, T. (2008) Plant metabolomics coming of age. *Physiol. Plantarum* **132**, 113–116.
12. Lisec, J., Schauer, N., Kopka, J., Willmitzer, L., Fernie, A.R. (2006) Gas chromatography mass spectrometry-based metabolite profiling in plants. *Nature Protocols* **1**, 387–396.
13. Bligh, E.G., Dyer, W.J. (1959) A rapid method of total lipid extraction and purification. *Can. J. Biochem. Physiol.* **31**, 911–917.
14. Katona, Z.F., Sass, P., Molnar-Perl, I. (1999) Simultaneous determination of sugars, sugar alcohols, acids and amino acids in apricots by gas chromatography-mass spectromety. *J. Chromatogr. A* **847**, 91–102.
15. Gullberg, J., Jonsson, P., Nordstrom, A., Sjostrom, M., Moritz, T. (2004) Design of experiments: an efficient strategy to identify factors influencing extraction and derivatization of *Arabidopsis thaliana* samples in metabolomic studies with gas chromatography/mass spectrometry. *Anal. Biochem.* **331**, 283–295.
16. Erban, A., Schauer, N., Fernie, A.R., Kopka, J. (2007) Nonsupervised construction and application of mass spectral and retention time index libraries from time-of-flight gas chromatography–mass spectrometry metabolite profiles, in *Metabolomics* (Weckwerth, W, ed.), Humana Press, Totowa, NJ, pp. 19–38.

Chapter 8

High-Performance Liquid Chromatography–Mass Spectrometry Analysis of Plant Metabolites in *Brassicaceae*

Ric C.H. De Vos, Bert Schipper, and Robert D. Hall

Abstract

The *Brassicaceae* family comprises a variety of plant species that are of high economic importance as vegetables or industrial crops. This includes crops such as *Brassica rapa* (turnip, Bok Choi), *B. oleracea* (cabbages, broccoli, cauliflower, etc.), and *B. napus* (oil seed rape), and also includes the famous genetic model of plant research, *Arabidopsis thaliana* (thale cress). Brassicaceae plants contain a large variety of interesting secondary metabolites, including glucosinolates, hydroxycinnamic acids, and flavonoids. These metabolites are also of particular importance due to their proposed positive effects on human health. Next to these well-known groups of phytochemicals, many more metabolites are of course also present in crude extracts prepared from *Brassica* and *Arabidopsis* plant material.

High-pressure liquid chromatography coupled to mass spectrometry (HPLC-MS), especially if combined with a high mass resolution instrument such as a QTOF MS, is a powerful approach to separate, detect, and annotate metabolites present in crude aqueous-alcohol plant extracts. Using an essentially unbiased procedure that takes into account all metabolite mass signals from the raw data files, detailed information on the relative abundance of hundreds of both known and, as yet, unknown semipolar metabolites can be obtained. These comprehensive metabolomics data can then be used to, for instance, identify genetic markers regulating metabolic composition, determine effects of (a)biotic stress or specific growth conditions, or establish metabolite changes occurring upon food processing or storage.

This chapter describes in detail a procedure for preparing crude extracts and performing comprehensive HPLC-QTOF MS-based profiling of semi-polar metabolites in Brassicaceae plant material. Compounds present in the extract can be (partially or completely) annotated based on their accurate mass, their MS/MS fragments and on other specific chemical characteristics such as retention time and UV-absorbance spectrum.

Key words: LCMS, Brassica, Arabidopsis, Secondary metabolites, Phenylpropanoids, Flavonoids, Glucosinolates

Nigel W. Hardy and Robert D. Hall (eds.), *Plant Metabolomics: Methods and Protocols*, Methods in Molecular Biology, vol. 860, DOI 10.1007/978-1-61779-594-7_8, © Springer Science+Business Media, LLC 2012

1. Introduction

The *Brassicaceae* family represents a highly interesting and contrasting collection of plants, including a number of major vegetables (e.g., *Brassica oleracea* and *B. rapa*), a source of vegetable proteins and oils (*B. napus*) as well as the model species of plant science, *Arabidopsis thaliana*. Even within *B. oleracea* alone, the diversity of vegetables is remarkable—cabbage, cauliflower, broccoli, Brussels sprouts. This species therefore represents an extremely interesting group of crops to study in the context of their varied biochemical composition and restricted genetic diversity. There is specifically, great interest in a number of secondary metabolite groups and their contrasting profiles within the different sub species of *Brassica*, due to their proposed link with having potentially health-promoting properties (1–3). These secondary metabolite groups include common compounds such as various phenylpropanoids (hydroxycinnamates) and flavonoids, as well as *Brassicaceae*-specific metabolites such as glucosinolates. Within these phytochemicals, there is marked variation between species and cultivars (4–7). However, next to these well-known and frequently analyzed target groups of compounds, *Brassicaceae* plants of course contain many more metabolites. Comprehensive, unbiased metabolomics approaches will reveal much more information on a wider range of metabolites which may also play key roles in other important phenotypic characteristics of *Brassica* plants and their products, such as disease/pest resistance and taste. For example, for this reason, broccoli was chosen as the representative vegetable crop within the EU research project META-PHOR (8) where current technologies and bioinformatics tools are being modified and advanced to increase our ability to generate biologically relevant information relating biochemical profile and phenotype.

The highly rich biochemistry of key secondary metabolites within the *Brassicaceae* covers many compounds that can be efficiently extracted in semi-polar aqueous-alcohol solutions. Of these semi-polar compounds not involved in primary metabolism, quite a number have already been shown to have phenotypic/physiological importance. It is also mainly secondary metabolites through for example their resistance effects, antioxidant properties, and color and flavor characteristics, which are attracting much attention from health, food, and nutrition groups (1–3, 9–11). High Pressure Liquid Chromatography (HPLC), using a C_{18}-reversed phase column, is the most frequently used and favored technique for analyzing semi-polar compounds present in plant extracts (2, 4, 12, 13). Due to the presence of one or more phenolic moieties, (poly)phenolic compounds and their conjugated forms typically exhibit ultraviolet and/or visible (UV/Vis) light absorbance that can be used for their detection. Many phenolic backbone structures

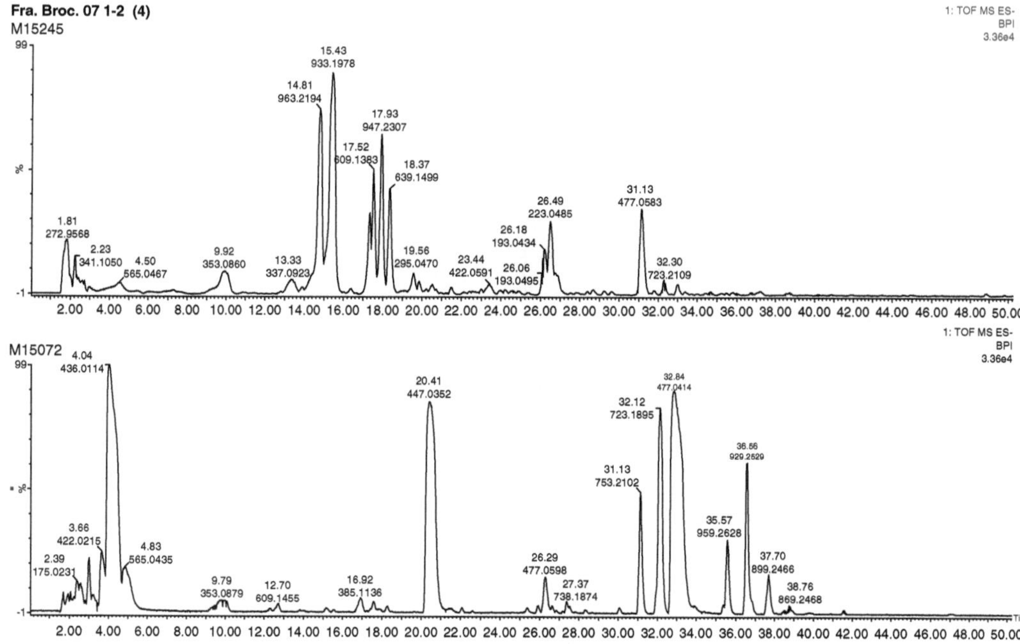

Fig. 1. Representative LC-QTOF MS chromatograms (ESI negative mode) of *B. rapa* Bok Choi leaves (*upper panel*) and *B. oleracea* Broccoli flower head (*lower panel*). The largest peaks represent glucosinolates and flavonoids, which are highly abundant in most *Brassica* species.

show specific absorbance spectra that can be recorded by photodiode array detectors (PDA or DAD) and can be used for (partial) metabolite identification. However, phenolic structures showing similar absorbance spectra, as well as many other secondary metabolites not displaying specific UV/Vis absorbance, need other detection and identification methods. Reversed-phase HPLC, hyphenated with mass spectrometry (HPLC-MS or, in short, LCMS) is nowadays a popular analytical technique for semi-polar compounds and a powerful platform for both targeted and untargeted profiling of plant secondary metabolites (see Fig. 1). High-resolution LCMS with exact mass determination can provide not only comprehensive mass-retention time fingerprints representing hundreds of known and, as yet, unknown metabolites, but in combination with MS-fragmentation abilities (see Fig. 2) and/or UV/Vis detection may also provide valuable structural information of the compounds detected (14–16).

LCMS with electrospray ionization (ESI) can be performed in either negative mode or positive mode, usually depending on the (classes of) compounds of primary interest. Due to their sulfate-glucose core structure, the *Brassicaceae*-specific glucosinolates ionize extremely well in negative electrospray ionization (ESI) mode and, therefore, can be very sensitively detected (17, 18). Other major secondary metabolites usually present in plant materials,

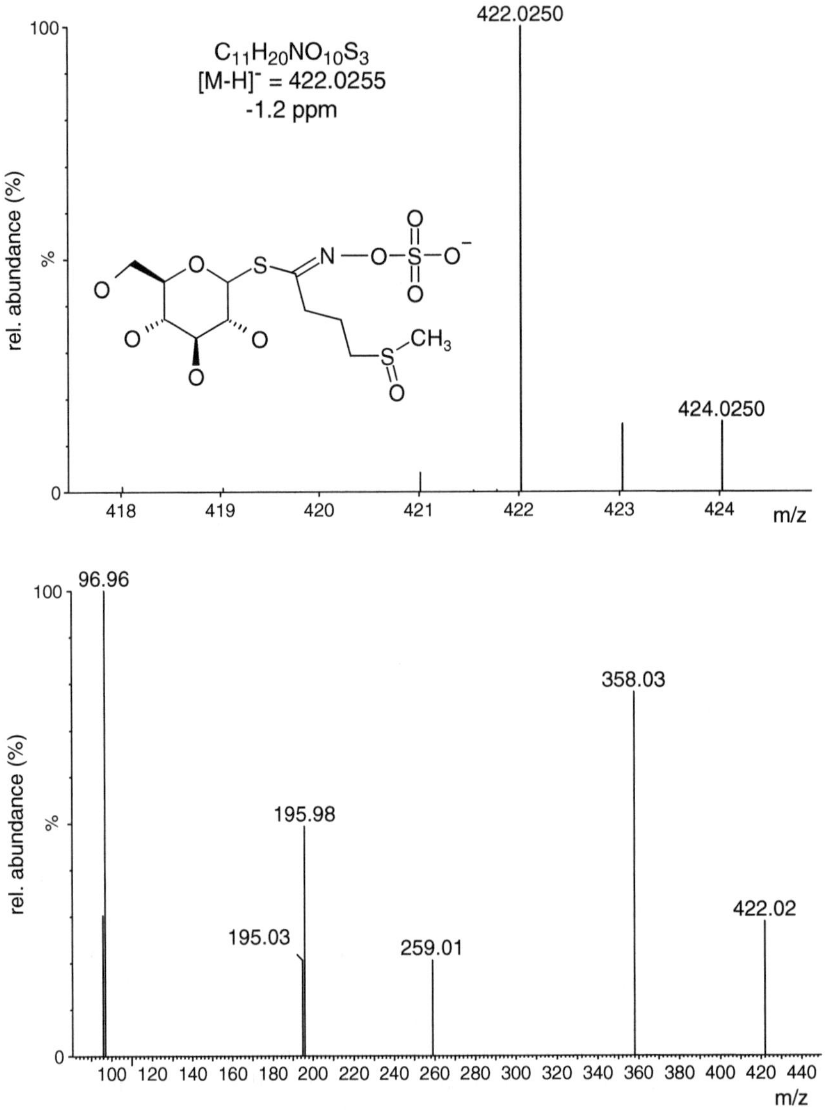

Fig. 2. Accurate mass detection of the molecular ion (*upper panel*) and collision-induced MS/MS fragments (*lower panel*) of glucoiberin (3-methylsulfinylpropylglucosinolate) present in Broccoli florets. Measured masses ([M-H]$^-$) are indicated at the top of the mass peaks. The detected molecular ion deviated –1.2 ppm (0.5 mD) from the calculated mass of the elemental formula of glucoiberin. Glucosinolates show a characteristic HSO_4^- fragment of *m/z* 96.96 upon MS/MS.

such as phenylpropanoids, flavonoids and a range of other glycosylated compounds, can also be readily detected in ESI-negative mode (2, 12, 13). On the other hand, polyamines and other compounds comprising chemical structures that easily form proton adducts, e.g., alkaloids and anthocyanins, can be better detected in ESI-positive mode. Thus, analysis of plant samples in both positive and negative ionization mode will provide the most comprehensive insight into their metabolic composition (16, 19, 20). Nevertheless, profiling in just a single ionization mode may already be sufficient to obtain a global overview of the differences and similarities between samples and to relate metabolic variation to other traits or processes, such as genetic variation (4), gene function analyses (16, 21, 22), plant development (23) and food processing (24). Therefore, this chapter describes a user-friendly protocol to extract and analyze secondary metabolites in *Brassica* and *Arabidopsis* tissues, using accurate mass LCMS in negative mode. This general approach is however, more broadly applicable to most other plant species with the need for only minor modifications.

2. Materials

2.1. Plant Material Sampling

1. Plants, leaves, tissues, etc. of any *Brassicaceae* species.
2. Plastic bags or storage tubes resistant to liquid nitrogen, e.g., polypropylene 50 mL tubes with screw cap (Greiner), Eppendorf micro-test tubes, or 12 mL glass tubes with screw caps (Omnilabo).
3. Liquid nitrogen for sample quenching and grinding (see Note 1).
4. Protective insulating gloves for handling super-cooled objects.
5. Metal spatula or small spoon, precooled with liquid N_2.

2.2. Chemicals (See Note 2)

1. Methanol absolute, HPLC supra-gradient grade.
2. Formic acid for analysis 98–100%.
3. Acetonitrile, HPLC supragradient grade.
4. Ultrapure water (e.g., MilliQ or its double-distilled equivalent).
5. Leucine enkaphaline, ≥95% pure, isolated by HPLC, or equivalent reference compound, to be used for online mass correction (so-called "lock mass").
6. Phosphoric acid p.a. 85% in water solution (w/v) or equivalent compound mixture suitable for mass spectrometer calibration over the mass range of 100–1,500 Da.
7. Liquid nitrogen or nitrogen gas generator for supplying gas to the mass spectrometer ionization source.
8. Argon 5.0, at least 99.999% pure, for supplying gas to the mass spectrometer collision cell.

2.3. Reagents and Solvents

1. *Sample extraction solution*: 0.133% (v/v) formic acid (FA) in pure methanol. Prepare sufficient solution for extraction of the complete series of samples.
2. *HPLC mobile phase*: 0.1% FA (v/v) in ultrapure water (eluent A), and eluent B is 0.1% FA (v/v) in acetonitrile (eluent B). Since

the chromatographic behavior of some *Brassicaceae* metabolites, and especially of intact glucosinolates, is sensitive to even slight variations in the acidity of the mobile phase, prepare the mobile phases freshly by precisely adding 0.1% v/v FA to both water and acetonitrile. Prepare sufficient of both eluents to analyze the entire sample series at one time.

3. *MS calibration solution*: 1 mL of a 0.05% (v/v) phosphoric acid solution in 50% acetonitrile/ultrapure water. Load into the gas-tight glass syringe.
4. *Lock mass solution*: leucine enkaphaline in 50% (v/v) acetonitrile/ultrapure water to obtain a final concentration of 0.1 mg/mL. Prepare sufficient solution for analysis of the complete series of samples.

2.4. Equipment

1. Freezer at –80°C for (long-term) storage of raw and ground plant materials or products.
2. Pipettes and tips suitable for handling organic solvents (Microman, Gilson).
3. Pestle and mortar or, preferably, a ball mill, e.g., Retsch Mixer Mill MM 301 (Retsch, Germany) for small Arabidopsis samples or a metal electric grinder, e.g., IKA A11 Basic Analytical mill (IKA, Germany), for larger samples.
4. Balance for accurate weighing of 100–500 mg frozen sample powder.
5. Ultrasonic bath.
6. Single-use sterile and nonpyrogenic latex-free syringes.
7. Single-use syringe filters free of polymers, such as Anotop 10 (diameter 10 mm, pore size 0.2 μm; Whatman) or Minisart RC4 (diameter 4 mm, pore size 0.45 μm; Sartorius). Filters for MS analyses should be resistant to the extraction solution used (i.e., 75% methanol + 0.1% FA) and free of polyethylene glycol or any other soluble polymer (see Note 3).
8. Crimp cap autosampler vials of 1–2 mL with aluminum crimp caps containing natural rubber/polytetrafluoroethylene septa.
9. Vacuum filtration unit for 96-wells format (see Note 4).
10. Protein filtration plates in 96-well format.
11. 96-well plates (Ritter style) with 700-μL glass inserts (Waters) and a 96-square well PTFE-coated seal (Waters).
12. Analytical column Luna C18(2), 2.0 mm diameter, 150 mm length, 100 Å pore size, spherical particles of 3 μm (Phenomenex).
13. Precolumns: Luna C18(2), 2.0 mm diameter, 4 mm length (Security Guard, Phenomenex).

14. PEEK in-line filter holder with PEEK frit 0.5-μm pore size (UpChurch Scientific).
15. Alliance 2795 HT high performance liquid chromatography system, or comparable system, equipped with an internal degasser, sample cooler, and column heater (Waters).
16. Separate HPLC pump for continuously pumping the lock mass solution at 10 μL/min.
17. Photodiode array detector (PDA) (Waters 2996).
18. High-resolution mass spectrometer: Quadrupole-time-of-flight (QTOF) Ultima V4.00.00 mass spectrometer equipped with an electrospray ionization (ESI) source and separate lock mass spray inlet (Waters) (see Note 5).
19. Syringe pump for injecting calibration solution.
20. Gas-tight glass syringe 0.1–1.0 mL.
21. MS data acquisition software: MassLynx 4.1 (Waters).
22. Mass signal extraction and alignment software such as MetAlign (25).
23. Optional: multivariate analyses software such as GeneMaths (26).

3. Methods

3.1. Plant Growth and Sampling Conditions

Samples to be prepared for metabolomics studies should be as specific and representative as possible for the plant, genotype, tissue, or cell type to be analyzed. For instance, if only specific cell types or tissues are known or suspected to be affected by a certain treatment or mutation, any possible effect of the treatment on the metabolome will be diluted out by other tissues. Thus, if the aim is to detect metabolic changes specifically occurring in root tips, start isolating the root tips from the nonresponding rest of the root system. In studies aiming to link metabolic variation to genetic variation, the epigenetic (biological) variation should be kept as low as possible by means of controlled plant growth and plant pooling. For instance, in the large-scale genetical metabolomics study in Arabidopsis RILs (4), seeds were sown on agar containing a nutrient solution, in Petri dishes with a density of a few hundred seeds per dish. Dishes were temperature-treated to promote uniform germination and were then all randomly placed in a single climate chamber in five blocks where each block contained one replicate dish of each line. After 6 days of controlled growth, the lids of the Petri dishes were removed to ensure that seedlings were free of condensed water on the day of harvest. On day 7, at 7 h into the light period, all seedlings were harvested within 2 h by submerging the complete Petridish briefly in liquid nitrogen and scraping off the seedlings with a razor blade. Finally, per line material from two dishes was pooled to make one of the replicate samples and from the remaining three dishes to make the second. To obtain representative material from larger plants, such as leafy *Brassica* vegetables, a representative number of leaf disks from different leaves or at least three complete leaves should be pooled per plant. In the case of seeds, a large number of seeds (at least 50) should be taken as a representative sample of the genotype, developmental stage, or treatment.

Once harvested, metabolite changes must be kept to a minimum. Therefore, upon harvest, plants or tissues should be snap-frozen in liquid nitrogen, even in the field/greenhouse if at all possible. To obtain homogenous material from the plants, plant parts or products, the frozen material should be ground into a fine powder using liquid nitrogen. Take care that tissues remain fully frozen at all stages from harvest until metabolite extraction; otherwise, throw away the sample. Without knowing the effect of lyophilization on the metabolite profile, lyophilization of tissue is not recommended, unless for specific practical reasons.

3.2. Tissue Sampling

1. Prelabeled bags or tubes with a freezer-proof marker pen or freezer-compatible labels. In the case of seeds or small seedlings (e.g., Arabidopsis) use 1.5- or 2.2-mL Eppendorf tubes; in the case of larger tissues use 50-mL Greiner tubes or plastic bags that are resistant to liquid nitrogen.
2. Harvest a representative amount of tissue (leaf, roots, flower head, etc.) in tubes or bags by rapid freezing in liquid nitrogen (see Note 6).
3. Homogenize the frozen tissue in liquid nitrogen into a fine powder using a pestle and mortar. For large series of samples, preferably use a ball mill for Arabidopsis or an analytical mill for larger tissue amounts. These should be precooled with liquid nitrogen. Homogenize for 20 s. Transfer the homogenized powder into precooled storage containers resistant to liquid nitrogen, using a precooled metal spatula or small spoon.
4. Weigh 100 mg frozen powder of *Arabidopsis* with an accuracy of better than 5% into a precooled Eppendorf tube, or 500 mg in the case of larger amounts of tissue into a 10-mL glass tube with screw cap (see Note 7). Smaller sample amounts can be used as well, but this is not advisable in view of the inherent higher weighing error using frozen material. Also weigh replicate samples of the same plant powder, to be included as quality control samples and technical replicates for extraction and analysis (see Note 8).

3.3. Metabolite Extraction

1. Prepare extracts freshly at the beginning of a series of analyses, after ensuring the LCMS system has been prepared, tested and calibrated properly. Add ice-cold sample extraction solution

(99.867% methanol acidified with 0.133% FA) to the tube containing the weighed frozen powder, in a volume:fresh weight ratio of 3:1 in the case of leaf material. Close the lid and immediately vortex for 10 s. Assuming a tissue-water content of about 95%, this will result in a final concentration of about 75% methanol and 0.1% FA.

2. Extract the metabolites by 15-min sonication at maximum frequency (40 kHz) in a water bath at room temperature.
3. Centrifuge for 10 min at maximum speed (20,000×*g* for Eppendorf tubes; 3,000×*g* for glass tubes) at room temperature.
4. Filter supernatant through a 0.2-μm PTFE filter using a disposable syringe into a 1.8-mL glass vial and close vial with the crimp cap. All filters used should be free of aqueous-methanol soluble polymers, such as polyethylene glycol. In the case of large numbers of samples, if possible use suitable filtration plates in 96-well format and a vacuum filtration unit (see Note 9).
5. If necessary, extracts can be stored at +4 or, preferably at −20°C. After storage, always sonicate and/or filter each sample once more before analysis (see Note 10).

3.4. Conditioning of the HPLC-PDA System

1. Prepare HPLC mobile phase solvents as described in item 2 of Subheading 2.3. Prime HPLC pump and tubing, and degas both solvents for at least for 10 min using the in-line degasser of the Alliance 2795 HT.
2. Install one PEEK in-line solvent filter between the injection system and the precolumn cartridge. Place two precolumns in tandem into the cartridge, place online in front of the analytical column, and place both cartridge and column in the column oven conditioned at 40°C.
3. Place the outlet from the column directly to a waste bottle and precondition the LC system and column system by increasing the percentage of eluent A stepwise (starting at 100% eluent B) until the initial gradient conditions are reached.
4. Program the HPLC method according to the gradient settings given below. In the standard setup, we use relatively long chromatographic runs of 60 min, including a linear gradient from 5 to 35% eluent B for 45 min, column washing at 75% eluent B for 5 min, return to 5% eluent B for 2 min, and reconditioning at 5% eluent B for 5 min, with a mobile phase flow rate of 0.19 mL/min into the analytical column (diameter of 2.0 mm). This flow rate corresponds to 1 mL/min on a 4.6-mm column, which is standard in most HPLC-UV/Vis applications.
5. After preconditioning the LC system, connect the PDA detector and subsequently the QTOF MS with the PEEK tubing. Program the PDA to acquire data every 1 s from 210 nm to 600 nm with a resolution of 4.8 nm. Wavelength range, scan

rate and resolution can be adjusted according to LC run time and the research aims. Always precondition the PDA-lamp, column oven temperature, and analytical column for at least 1 h before starting sample analyses.

6. Check the entire system for air bubbles and all connections for leakage by verifying the LC pressure stability.

3.5. Conditioning of the MS System

Before each series of sample analyses, the mass spectrometer should be well-conditioned and calibrated to obtain good performance in terms of mass accuracy and resolution. In contrast to electron impact ionization, as used in most GC-(TOF) MS applications, detection sensitivity and mass spectra obtained by soft-ionization LCMS are highly dependent on the type of mass spectrometer, ionization source, and chromatographic system used. The procedure and settings described here are for a QTOF Ultima with ESI source and the TOF-tube in V-mode, in combination with the HPLC conditions described above. Depending upon samples and compounds of specific interest, settings and conditions may need specific adaptations.

1. Connect the outlet of the PDA, with an eluent flow rate of 0.19 mL/min, to the inlet of the mass spectrometer and set the capillary voltage to 2.75 kV, cone voltage to 35 V, source temperature to 120°C, and desolvation temperature to 250°C. Use a cone gas flow rate of 50 L/h and desolvation gas flow rate of 600 L/h. Precondition the MS for at least 2 h at these standard settings before sample analysis.
2. Disconnect the LC flow from the MS, and use the syringe pump to inject the MS calibration solution into the ESI source, at an initial flow rate of 10 μL/min.
3. Acquire data from *m/z* 80 to 1,500 at a scan rate of 0.9 s and an interscan delay of 0.1 s. A series of phosphoric acid cluster peaks should appear throughout the entire range of the mass spectrum. To obtain proper calibration and accurate mass calculations, none of the mass calibration peaks should exceed an intensity of 250 counts/s (in continuum mode) and the intensity of the clusters over the mass range should be as uniform as possible. Adjust pump flow, capillary voltage, cone voltage, desolvation gas flow, and/or collision energy until criteria are optimal.
4. Combine the spectra from 50 adjacent scans during acquisition mode at optimal settings in continuum mode, center the mass signals and check mass resolution of the machine at *m/z* 488.8772 (negative ionization mode) or *m/z* 490.8918 (positive ionization mode). Mass resolution is calculated by dividing the *m/z* value of the centered mass signal by the mass difference at half height of the Gaussian-shaped mass peak in

continuum mode, and should be better than 8,500 (with the QTOF Ultima in V-mode). Otherwise, retune the instrument and repeat the procedure.

5. Use the centered mass data for calibration of the instrument using a polynomial-5 fit. Mean residual mass deviation, according to the MassLynx calibration procedure, should be less than 1.0 ppm, otherwise adjust the calibration settings.
6. Reconnect the LC flow to the MS. Check the effluent from the complete LC-PDA system, including mobile phase, tubing, columns and PDA flow cell, by acquiring centroid data from *m/z* 80 to 1,500 under the exact conditions of sample analysis. To prevent excessive ion suppression of sample compounds, individual mass signals at initial gradient conditions should preferably be less than 200 counts per scan (centroid data) in negative mode or less than 500 counts per scan in positive mode.
7. Prepare an MS method file to acquire mass data from *m/z* 80 to 1,500, at a scan rate of 0.9 s and an interscan delay of 0.1 s and in centroid mode. The range of masses to be detected in sample extracts should fall within the range of calibration masses (see Note 11). During sample analyses, set the standard setting of collision energy to10 eV in negative ion mode and to 5 eV in positive ion mode. If needed for optimal ionization of key compounds, the collision energy may be adjusted. The MS is programmed to switch from sample to lock spray every 10 s and to average two scans for lock mass correction (*m/z* 556.2767 in positive mode and 554.2619 in negative mode). The lock mass solution is used for online calibration of the mass accuracy during sample analysis. Adjust the flow rate or concentration of the lock mass solution to obtain a stable intensity of about 600–800 counts per scan (in centroid mode) during LCMS runs (see Note 12).
8. The aqueous-methanol extracts are placed in trays inside the autosampler (20°C) during the analysis series, in a randomized order (see Note 8). Program the injection system to operate in sequential mode and to load the syringe with 5 μL of sample, with 5 μL of air both before and after the sample. The injection needle is washed with 100% methanol between injections.
9. Check for the presence of sufficient eluents, lock mass solution, nitrogen and argon gasses, and computer hard disk space for the entire sample series.
10. Start the sample series with at least four injections of one of the plant extracts, to stabilize the LC and MS systems. Check stability of retention times and mass accuracy of known compounds during these first runs. Deviations of observed known parent masses from their calculated masses should be less than 5 ppm (at signal intensities similar to that of the local lock mass), otherwise stop the series and recalibrate the MS.

11. Once all extracts have been run successfully, transfer data from the LCMS-data acquisition computer to a second computer on which both the acquisition software and data-processing software have been installed.

3.6. Data Processing

Depending on the aim of the research, the raw data may be processed in order to extract metabolite intensity signals is different ways. Relative metabolite intensities may be calculated from their corresponding chromatographic peaks and expressed either as maximal peak height or as area under curve, presuming a more or less Gaussian shape of the chromatographic peak.

In the case of interest in only specific classes of *Brassica* metabolites, e.g., glucosinolates, peak integration tools delivered with the data acquisition software may be used. We use the QuantLynx data processing package delivered with the MassLynx acquisition software (Waters) of the LC-QTOF MS. Since high mass resolution is used, the mass peak integration parameters can be set at a narrow mass window (e.g., 20 ppm) around the exact mass for each compound of interest, enabling specific detection and a high signal to noise ratio. For the untargeted approach, we routinely use the MetAlign software (25). Standard settings for processing of LC-QTOF Ultima MS data from *Brassica* samples, as collected according to the procedure described here, are given in Fig. 3 (see Notes 13 and 14). For further details on the MetAlign software, the reader is referred to the contribution of Dr. Lommen (see Chapter 15).

The MetAlign data output can be cleaned, if needed, for low abundant or misaligned signals and further processed according to the research aim. For instance, metabolite signals significantly differing between samples can be determined, or multivariate analyses techniques such as principal components analyses and hierarchical clustering can be applied to obtain a global view of overall metabolic differences and similarities between samples (4, 12, 22–24).

3.7. LC-MS/MS to Identify Selected Mass Signals

If needed or desired, compounds can be further identified using LC-QTOF MS/MS. For this purpose, masses of interest can be incorporated into a mass inclusion list (data-directed MS/MS) in the MassLynx software. Ten instead of 5 μL of sample containing relatively high amounts of the compounds of interest are now injected, in order to obtain higher intensities of the parent ions and thus also their MS/MS fragments. The collision energy profile is programmed to increase sequentially from 5, 10, 20, to 30 eV (ESI positive mode) or 10, 15, 30, to 50 eV (ESI negative mode). If these settings are insufficient to obtain informative MS/MS information for the masses of interest, the collision energy profile can be adjusted. Also, if the intensities of essential MS/MS fragments are too low for exact mass calculation, the amount of compound

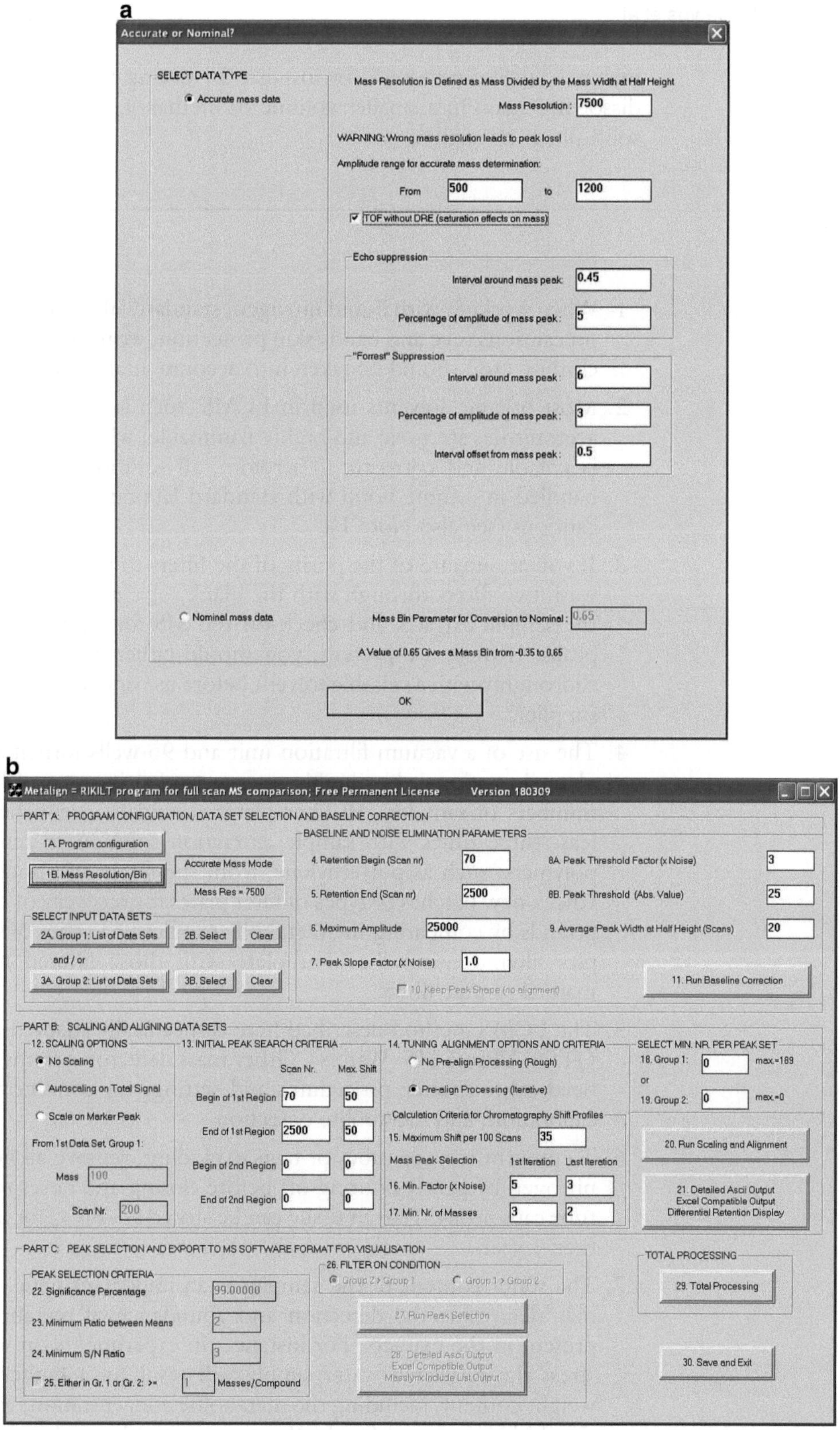

Fig. 3. Interface of MetAlign software with standard settings for peak extraction and alignment of Brassica LC-QTOF Ultima MS data. (**a**) Mass resolution and accurate mass calculation settings (see Note 13). (**b**) Baseline correction and peak alignment settings (see Note 14).

injected can be increased, for instance by drying the extract and dissolving again in a smaller volume of methanol, or by applying solid-phase extraction.

4. Notes

1. When working with liquid nitrogen, standard laboratory safety precautions (eye and hand/skin protection, wearing protective clothes, etc.) should be taken into account at all times.
2. Most organic solvents used in LCMS, such as methanol and acetonitrile, are toxic and highly flammable, while formic acid is volatile and corrosive. Therefore, all solutions should be handled in a fume hood with standard laboratory safety precautions (see also Note 1).
3. If you are unsure of the purity of the filters to be used, always wash two filters through with the blank solvent to be used for the sample extracts and check on the MS for contaminating peaks. If these are present, you should either wash all filters thoroughly with a suitable solvent before use or choose another supplier.
4. The use of a vacuum filtration unit and 96-wells format filter plates is optional, but highly recommended in cases of large numbers of samples. All filter plates should be prewashed at least three times with sample extraction solution to remove polymers, such as polyethylene, from the filters and housing. Check new batches or types of filter plates for recovery of compounds by comparing the metabolic profiles of a series of samples filtered using the filter plate with those filtered using manual syringe filters.
5. The LCMS method described here is specifically adapted for a QTOF Ultima MS (Waters). Other mass detector systems may need other specific procedures and settings for conditioning, calibration, and metabolite detection.
6. To prevent storage tubes or bags exploding, remove all liquid nitrogen by gently pouring off before closing and never screw tube caps firmly! Frozen tissue can be stored at −80°C for more than 1 year.
7. The water content of the samples is an important issue, as it may determine the detection and abundance of metabolites present in the extracts. For instance, in experiments on water stress the restricted water supply will result in a higher dry weight content, including the metabolite concentration, which should be corrected for. In the case of samples such as those with highly variable (predetermined) water contents, freeze-

dried powders or dry seeds, pure water can be added to adjust each sample to always give a final solvent concentration of 75% methanol and 0.1% FA. If the water content is unknown or cannot easily be determined, freeze-dry the samples and prepare the extracts by adding 75% methanol + 0.1% FA. Similarly, if the concentration of metabolites is too high resulting in chromatographic saturation, the extracts can be diluted with 75% methanol + 0.1% FA. Alternatively, plant materials can be extracted with more volume, taking care that the final concentration is 75% methanol + 0.1% FA. For instance, we routinely extract 100 mg Arabidopsis seedlings with 9 volumes of 83.3% methanol + 0.11% FA.

8. It is strongly advised to include a series of extracts from the same plant material as technical replicates, e.g., in order to estimate technical variability, to correct for batch effects, to optimize settings for data processing software, etc. Therefore, prepare a large pooled sample of material from different plants to be analyzed and prepare at least 5 extracts as technical replicates. These technical replicates should be analyzed at least every ten samples, with one replicate at the start and one at the end of the sample entire series. When using multiple 96-well filtration plates, divide five technical replicates over each plate in order to correct for plate variability.

9. We use a TECAN Genesis Workstation 150 equipped with a four-channel pipetting robot and a TeVacS 96-wells filtration unit. Prewash the filtration plates (Captiva 0.45 μm, Ansys Technologies) at least three times with 700 μL 75% methanol containing 0.1% FA. Dry the points of the filter tips by blotting onto filter paper. Place a 96-well plate with 700-μL glass inserts (Waters) in the filtration unit under the prewashed filtration plate. Load each well with 700 μL of extract and vacuum-filtrate until all filters are dry (2 times 20 s). Carefully remove air bubbles trapped at the bottom of the inserts and cover the plate with a 96-square well PTFE-coated seal.

10. The samples should be analyzed as soon as possible after extract preparation, to minimize loss of labile compounds. This, however, is not always practically feasible, for instance due to sudden malfunction of the LCMS system. In these cases, the extracts should be stored at −20°C or +4°C. Before analyzing, sonicate the vials or inserts for 15 min to dissolve possible precipitates before analysis. If needed, filter the samples again.

11. The present procedure describes the mass calibration of compounds within the mass range of *m/z* 80–1,500, using a polynomial function. Extrapolation of this polynomial function towards higher or lower *m/z* values is not valid and results in an incorrect mass detection. So, if compounds outside this

mass range are of specific interest, ensure that the calibration of the machine covers the entire desired mass window.

12. Since the mass detected by the QTOF Ultima is dependent upon the intensity of the signal and most accurate at an intensity corresponding to that of the lock mass (11, 15), the lock mass signal should be as stable as possible during the analysis of the entire sample series. This intensity of the lock mass is used during data MetAlign-processing to calculate accurate masses within a user-defined intensity window (see *also* Note 13).
13. Within the MetAlign software (see Fig. 3b, button 1B), the resolution and amplitude (=intensity) range for accurate mass calculation have to be specified. The optimal settings for the accurate mass calculation are dependent upon the dynamic range of the MS with regard to accurate mass detection. For the QTOF Ultima MS, the most accurate range is between −50% and +50% of the recorded intensity of the lock mass. The more stable the lock mass is during analyses of the sample series, the more sample data points will fall within the selected amplitude range and thus the more reliable the accurate mass output will be. Variation in measured accurate mass across a chromatographic peak may also result in splitting of the metabolite signal into two or more accurate mass peaks. If this occurs, slightly lower the mass resolution value. For example, we routinely analyze Brassica samples by QTOF MS at a resolution of about 8,500, but use 7,500 as the setting for MetAlign.
14. We recommend selecting the sample that has been analyzed just in the middle of the entire LCMS series as the reference file for alignment by MetAlign, i.e., as the first sample in the entire sample list (button 2), to minimize the extent of retention profile correction between first and last samples analyzed. Always perform a test baseline correction and alignment on a few variable samples, to check whether the default settings are at least correct to extract and align mass peaks that are of specific interest (if any). Set parameters for peak extraction and noise (buttons 4–9) and run baseline correction (button 11). Manually check selected mass peaks at the beginning, middle and at the end of the baseline-corrected chromatograms and in the original raw data. If it is obvious that some mass signals from relatively broad chromatographic peaks are missing in the baseline-corrected data, set parameter 9 at a slightly higher value and rerun baseline correction. On the other hand, if closely eluting peaks of compounds with similar accurate mass have been extracted as single peaks, lower the value at button 9. Once peak extraction and baseline correction settings are satisfactory, run baseline correction for all samples. After baseline correction of the entire series, inspect retention shifts in the baseline-corrected data files of the reference sample and of

the first and last sample of the entire data set. Set maximum shift at initial peak searching criteria (button 13) according to default settings, or to a value at least a factor of 2 higher than visually observed retention shifts and higher than that set in parameter 9. After running the alignment (button 20), create the data output file (button 21). Check technical replicates for variation in mass signal intensities and misalignments, e.g., by making scatter plots and frequency distribution tables of signals detected in the replicate extracts. Adapt alignment settings if needed or filter out misaligned or other inappropriate signals from the dataset.

Acknowledgements

This work was financed by the EU Framework VI program project META-PHOR (2006-FOODCT-036220) and additional financing from the Centre for Biosystems Genomics and The Netherlands Metabolomics Centre, both initiatives under the auspices of the Netherlands Genomics Initiative.

References

1. Jahangir, M., Kim, H.K., Choi, Y.H., and Verpoorte, R. (2009) Health-affecting compounds in Brassicaceae. *Comprehensive Reviews in Food Science and Food Safety* **8**, 31–43.
2. Olsen, H., Aaby, K., and Borge, G.I.A. (2009) Characterization and quantification of flavonoids and hydroxycinnamic acids in curly kale (Brassica oleracea L. Convar. acephala Var. sabellica) by HPLC-DAD-ESI-MSn. *J. Agric. Food Chem.* **57**, 2816–2825.
3. Malíková, J., Swaczynová, J., Kolár, Z., and Strnad, M. (2008) Anticancer and antiproliferative activity of natural brassinosteroids. *Phytochemistry* **69**, 418–426.
4. Keurentjes, J.J.B., Fu, J.Y., De Vos, R.C.H., Lommen, A., Hall, R.D., Bino, R.J., Van der Plas, L.H., Jansen, R.C., Vreugdenhil, D., and Koornneef, M. (2006). The genetics of plant metabolism. *Nature Genetics* **38**, 842–849.
5. Bennett, R.N., Rosa, E.A.S., Mellon, F.A., and Kroon, P.A. (2006) Ontogenic profiling of glucosinolates, flavonoids, and other secondary metabolites in Eruca sativa (salad rocket), Diplotaxis erucoides (wall rocket), Diplotaxis tenuifolia (wild rocket), and Bunias orientalis (Turkish rocket). *J. Agric. Food Chem.* **54**, 4005–4015.
6. Jeffery, E.H., Brown, A.F., Kurilich, A.C., Keck, A. S., Matusheski, N., Klein, B.P., and Juvik, J. A. (2003). Variation in content of bioactive components in broccoli. *Journal of Food Composition and Analysis* **16**, 323–330.
7. Kurilich, A.C., Jeffery, E.H., Juvik, J.A., Wallig, M.A., and Klein, B.P. (2002) Antioxidant capacity of different broccoli (*Brassica oleracea*) genotypes using the oxygen radical absorbance capacity (ORAC) assay. *J. Agric. Food Chem.* **50**, 5053–5057.
8. http://www.meta-phor.eu.
9. Ferreres, F., Sousa, C., Pereira, D. M., Valentao, P., Taveira, M., Martins, A., Pereira, J. A., Seabra, R. M., and Andrade, P. B (2009) Screening of antioxidant phenolic compounds produced by in vitro shoots of Brassica oleracea L. var. Costata DC. *Combinatorial Chemistry & High Throughput Screening* **12**, 230–240.
10. Lopez-Berenguer, C., Carvajal, M., Moreno, D.A., and Garcia-Viguera, C. (2007) Effects of microwave cooking conditions on bioactive compounds present in broccoli inflorescences. *J. Agric. Food Chem.* **55**, 10001–10007.
11. Verkerk, R. and Dekker, M. (2004) Glucosinolates and myrosinase activity in red cabbage (*Brassica oleracea* L. var. Capitata f. rubra DC.) after various microwave treatments. *J. Agric. Food Chem.* **52**, 7318–7323.
12. De Vos, R.C.H., Moco, S., Lommen, A., Keurentjes, J.J.B., Bino, R.J. and Hall R.D.

(2007) Untargeted large-scale plant metabolomics using liquid chromatography coupled to mass spectrometry. *Nature Protocols* **2**, 778–791.

13. Bottcher, C., von Roepenack-Lahaye, E., Schmidt, J., Schmotz, C., Neumann, S., Scheel, D. and Clemens, S. (2008) Metabolome analysis of biosynthetic mutants reveals a diversity of metabolic changes and allows identification of a large number of new compounds in Arabidopsis. *Plant Physiol.* **147**, 2107–2120.
14. Matsuda, F., Yonekura-Sakakibara, K., Niida, R., Kuromori, T., Shinozaki, K. and Saito, K. (2009) MS/MS spectral tag-based annotation of non-targeted profile of plant secondary metabolites. *Plant J.* **57**, 555–577.
15. Moco, S., Bino, R. J., Vorst, O., Verhoeven, H. A., De Groot, J., Van Beek, T. A., Vervoort, J. and De Vos, R. C. H. (2006) A liquid chromatography-mass spectrometry-based metabolome database for tomato. *Plant Physiol.* **141,** 1205–1218.
16. Von Roepenack-Lahaye, E., Degenkolb, T., Zerjeski, M., Franz, M., Roth, U., Wessjohann, L., Schmidt, J., Scheel, D. and Clemens, S. (2004) Profiling of Arabidopsis secondary metabolites by capillary liquid chromatography coupled to electrospray ionization quadrupole time-of-flight mass spectrometry. *Plant Physiol.* **134**, 548–559.
17. Rochfort, S.J., Trenerry, V.C., Imsic, M., Panozzo, J. and Jones, R. (2008) Class targeted metabolomics: ESI ion trap screening methods for glucosinolates based on MSn fragmentation. *Phytochemistry* **69**, 1671–1679.
18. Mellon, F.A., Bennett, R.N., Holst, B. and Williamson, G. (2002) Intact glucosinolate analysis in plant extracts by programmed cone voltage electrospray LC/MS: Performance and comparison with LC/MS/MS methods. *Anal. Biochem.* **306**, 83–91.
19. Fait, A., Hanhineva, K., Beleggia, R., Dai, N., Rogachev, I., Nikiforova, V. J., Fernie, A. R. and Aharoni, A. (2008) Reconfiguration of the achene and receptacle metabolic networks during strawberry fruit development. *Plant Physiol.* **148**, 730–750.
20. Hanhineva, K., Rogachev, I., Kokko, H., Mintz-Oron, S., Venger, I., Karenlampi, S., and Aharoni, A. (2008) Non-targeted analysis of spatial metabolite composition in strawberry (*Fragaria x ananassa*) flowers. *Phytochemistry* **69**, 2463–2481.
21. Malitsky, S., Blum, E., Less, H., Venger, I., Elbaz, M., Morin, S., Eshed, Y., and Aharoni, A. (2008) The transcript and metabolite networks affected by the two clades of Arabidopsis glucosinolate biosynthesis regulators. *Plant Physiol.* **148**, 2021–2049.
22. Bino R.J., De Vos, R.C.H., Lieberman, M., Hall, R.D., Bovy, A., Jonker, H. H., Tikunov, Y., Lommen, A., Moco, S. and Levin, I. (2005) The light-hyperresponsive *high pigment-2^{dg}* mutation of tomato: alterations in the fruit metabolome. *New Phytol.* **166**, 427–438.
23. Moco, S., Capanoglu, E., Tikunov, Y., Bino, R. J., Boyacioglu, D., Hall, R. D., Vervoort, J. and De Vos, R. C. H. (2007) Tissue specialization at the metabolite level is perceived during the development of tomato fruit. *J. Exp. Bot.* **58**, 4131–4146.
24. Capanoglu, E., Beekwilder, J., Boyacioglu, D., Hall R.D. and De Vos R. C. H. (2008) Changes in antioxidant and metabolite profiles during production of tomato paste. *J. Agric. Food Chem.* **56**, 964–973.
25. http://www.metalign.wur.nl.
26. http://www.applied-maths.com/genemaths/genemaths.htm.

Chapter 9

UPLC-MS-Based Metabolite Analysis in Tomato

Ilana Rogachev and Asaph Aharoni

Abstract

Recent advances in the performance of hyphenated technologies based on ultrapressure chromatography and high-sensitivity mass spectrometry have set the stage for a myriad of metabolomics studies in plants and other organisms. In this chapter, we describe the use of a UPLC (Ultraperformance Liquid Chromatography)-qTOF (quadrupole time-of-flight) system for profiling semipolar metabolites in the model fruit plant tomato. An optimized extraction method, instrument parameters and data treatment procedures are provided. The value of UPLC instruments, which use small particle size chromatographic columns, in terms of resolution, separation, and short injection times are presented. When coupled to a TOF mass spectrometer with high resolution and mass accuracy, good dynamic range, and a fast spectral acquisition capacity, this system is most suitable for the extensive profiling of hundreds of plant metabolites.

Key words: Tomato, Fruit, UPLC, Mass spectrometry, qTOF, Metabolomics

1. Introduction

Tomatoes and tomato-based products are eaten throughout the world; their consumption is believed to benefit human health (1). Tomato metabolites, both primary and secondary, are responsible for variations in fruit nutritional quality; therefore, the analysis of tomato fruit constituents is highly important. Another benefit from the analysis of the tomato metabolome is that the metabolite data obtained can be interpreted in combination with new data arising from the on-going tomato genome project (International Tomato Sequencing Project (2)), which will lead to better understanding of gene functions.

Tomato fruit extracts contain carotenoids such as lycopene, β-carotene, and vitamin E, which are known as effective antioxidants (3). Beside these lipophilic compounds, tomato tissues comprise numerous semipolar compounds: organic acids (mostly cinnamic acids), flavonoids (mostly naringenin chalcone and

Nigel W. Hardy and Robert D. Hall (eds.), *Plant Metabolomics: Methods and Protocols*, Methods in Molecular Biology, vol. 860, DOI 10.1007/978-1-61779-594-7_9, © Springer Science+Business Media, LLC 2012

glycosilated and acylated derivatives of quercetin and kaempferol) and glycoalcaloids (tomatine, esculeosides) (4–6). HPLC and capillary electrophoresis are the most widely used techniques for the separation of these classes of compound (7, 8). UPLC (Ultra Performance Liquid Chromatography) instruments are based on the use of small particle size chromatographic columns (less than 2 µm), offer substantial resolution enhancement (9), and, hence, more effective separation of the compounds, reduction of both injection time and matrix effects. MS-based techniques, particularly in combination with chromatographic technologies are most popular as these combine very high analytical precision with equally high detection sensitivity (10). A TOF (time-of-flight) mass spectrometer, with high resolution and mass accuracy, good dynamic range, and a fast spectral acquisition capacity (11), is most suitable in combination with UPLC for the extensive profiling of hundreds of plant metabolites typically present, e.g., in tissue extracts (12, 13). In this chapter, we present the use of UPLC-qTOF for profiling semipolar metabolites in tomato tissues.

2. Materials

2.1. Reagents and Equipment

1. Water, double deionized, from the Milli-Q purification system (Millipore, Bedford, MA), resistivity 18.2 MΩ-cm, filtered through a 0.22-µm membrane filter (see Note 1).
2. Acetonitrile, ultra gradient HPLC grade or LC-MS grade (e.g., JT Baker).
3. Liquid nitrogen for grinding and freezing tomato samples.
4. Standards for QC (quality control) samples: L-Tryptophan (Sigma), L-Phenylalanine (Sigma), Chlorogenic acid (Fluka), Caffeic acid (Sigma), *p*-Coumaric acid (Sigma), Ferulic acid (Aldrich), Sinapic acid (Sigma), Rutin hydrate (Sigma), Quercetin dihydrate (Sigma), Tomatine (Apin), Naringenin (Fluka), Kaempferol (Fluka).
5. IKA A11 basic grinder or a mortar and a pestle.
6. Screw-cap polypropylene (PP) tubes (50 ml) for storage of frozen samples (e.g., Greiner, Greiner bio-one Inc.).
7. Screw-cap PP tubes (15 ml; e.g., Greiner) or 2-ml PP safe-lock eppendorf tubes for sample extraction.
8. Ultrasonic bath.
9. Vortex.
10. Centrifuge suitable for 15 ml tubes (3,000 × *g*) and/or Centrifuge suitable for 2 ml Eppendorf tubes (15,000 × *g*).
11. Single-use sterile latex-free syringes, 1 or 3-ml volume.

12. Single-use, 0.22 μm membrane syringe filters (e.g., diameter 4 mm PVDF (Millex-GV)) or diameter 12 mm PTFE (PALL) (see Note 2).
13. Amber-glass 2-ml auto-sampler vials and caps with a PTFE/Silicone septum. Use suitable 250-μl glass inserts when you have a small volume of solution for injection (less than 1 ml).

2.2. Instrumentation and Software

1. UPLC-PDA-qTOF system: e.g., a UPLC Waters Acquity instrument connected in-line to an Acquity PDA (photodiode array) detector and a Synapt HDMS detector (tandem quadrupole/time-of-flight mass spectrometer). The MS detector is equipped with an electrospray ion source (ESI). The Synapt HDMS system is operated in the standard qTOF mode, without using the ion mobility capabilities (see Note 3).
2. UPLC BEH C18 column (Waters Acquity), 100×2.1 mm i.d., 1.7 μm, with a column pre-filter.
3. MassLynx 4.1 instrument software (Waters).
4. XCMS program (14) for mass peak extraction and alignment (see Note 4).

2.3. Solutions

2.3.1. For UPLC

1. Mobile phase A: 5% acetonitrile/water (v/v), containing 0.1% formic acid (v/v).
2. Mobile phase B: 100% acetonitrile, containing 0.1% formic acid (v/v).
3. Strong needle wash solution: 80% methanol (a strong organic solution that dissolves most components of the sample matrix).
4. Weak needle wash solution: 5% acetonitrile/water (v/v) (composition similar to the initial conditions of the gradient).
5. Seal wash solution: 10% methanol/water (v/v).

2.3.2. Standard Mixture for a Quality Control Sample (QC-Mix-12)

1. Prepare individual stock solutions of standard compounds (phenylalanine, chlorogenic acid, caffeic acid, coumaric acid, ferulic acid, sinapic acid, rutin, quercetin, naringenin and kaempferol) at a concentration of 1 mg/ml in methanol. Sonicate all stock solutions for several minutes for better solubility of the compounds.
2. Prepare a tomatine stock solution at a concentration of 0.5 mg/ml in methanol. Sonicate for several minutes.
3. Prepare tryptophan stock solution at a concentration of 1 mg/ml in 80% methanol (v/v) containing 2% formic acid (v/v) (see Note 5). Sonicate for several minutes for better solubility of the compounds.
4. Prepare a stock mixture of standards by combining equal amounts of the individual stock solutions. Each compound will

then be at the concentration of 83 μg/ml (tomatine—42 μg/ml). Aliquots of this solution can be stored at -20°C for 3–4 months without significant changes in the concentration of the compounds.

5. Prepare the working standard mixture (QC-Mix-12) by diluting the stock mixture of standards tenfold with methanol. Final concentration of compounds is 8 μg/ml (tomatine—4 μg/ml). Use this solution as QC (quality control) and SST (system suitability test) samples.

3. Methods

3.1. Sample Preparation

The extraction of biological material with aqueous methanol has so far been the most widely used option for LC-MS metabolite profiling schemes (15). Acidified aqueous methanol at a final concentration of 75% methanol (v/v) and 0.1% formic acid (v/v) was considered to be the most suitable solvent for efficient extraction of a wide range of secondary metabolites from different plant species and tissues (16). A detailed description of the sample preparation procedure can be found in (16). Tomato fruits contain relatively high concentrations of organic acids (6), and, hence, relatively low pH of the obtained extract. The water content in the tomato frozen tissues is approximately 85–95%. Therefore, 100% methanol, added to the frozen tomato samples at a 1:3 (w/v) ratio, is the optimal solvent for tomato fruit extraction.

Perform the sample preparation as follows:

1. Grind representative amounts of frozen tomato tissue (see Note 6) using a cooled grinder, or when the amount of frozen tissue is insufficient to be ground in a grinder, use a mortar and a pestle, precooled in liquid nitrogen. Keep the samples frozen at all times during the grinding procedure.
2. Weigh about 0.5 g of the frozen ground sample into the 15-ml PP screw-cap tube. If the amount of sample is limited, weigh the frozen sample (30–350 mg) into the 2-ml PP eppendorf tube using a spatula pre-cooled in Liquid N_2. Keep the sample frozen during the weighing procedure.
3. Add the required amount of 100% MeOH to the frozen samples while keeping the frozen sample–MeOH at a 1:3 (w/v) ratio. Vortex for several seconds until all the powder has been fully resuspended.
4. Sonicate for 20 min.
5. Vortex for several seconds.
6. Centrifuge for 10 min at about 3,000 × *g* for the15-ml tubes or at 15,000 × *g* for the eppendorf tubes.

7. Filter the supernatant through a 0.22-μm PTFE (or PVDF) filter into the 2-ml eppendorf tube (see Note 7) or directly into the autosampler vial and close tightly with a stopper. If sample volume is less than 1 ml, place a 150 μl aliquot (or less, but not less than 50 μl) into the insert of the autosampler vial, close tightly with a stopper and shake briefly/tap the tube to remove air bubbles from the bottom of the insert.
8. Place vials into the UPLC autosampler kept at 12°C.
9. Blank sample preparation: Add the same amount of 100% MeOH as used for samples extraction to the empty tube (the same type as was used for the preparation of samples). Add water while keeping the water:MeOH at a 1:3 (v/v) ratio, vortex for several seconds and continue sample preparation together with the biological samples.

3.2. UPLC-PDA-qTOF MS Analysis

3.2.1. UPLC Parameters

We routinely use a 26-min gradient method for the analysis of tomato samples on the Acquity BEH C18 column. The parameters for the instrument method are:

1. Mobile phase A is 5% acetonitrile containing 0.1% formic acid, mobile phase B is 100% acetonitrile, containing 0.1%formic acid. Linear gradient (see Table 1).
2. Flow rate is 0.3 ml/min.
3. Autosampler temperature is set to 12°C.
4. Column temperature is 35°C.
5. The injection volume is 4 μl.

Table 1
Parameters of the 26-min gradient run used for the analyses of tomato samples by the UPLC-qTOF instrument

Time, min	A, %	B, %	Gradient stage
0.00	100	0	Initial conditions
22.0	72	28	Elution
22.5	60	40	Elution
23.0	0	100	Washing
24.5	0	100	Washing
25.0	100	0	Equilibration
26.0	100	0	Equilibration

Mobile phase A: 5% acetonitrile–water (v/v), containing 0.1% formic acid (v/v); Mobile phase B: 100% acetonitrile, containing 0.1% formic acid (v/v). Gradient curves are linear

6. Injection needle is washed with 200 μl of the strong needle wash solution and then 600 μl of the weak needle wash solution.
7. Prior to use, the new chromatographic column should be equilibrated with the mobile phases (50%A, 50%B) for at least 60 min and subsequently conditioned by several gradient injections of the standard QC-Mix-12 solution and a sample containing the biological matrix in order to stabilize the retention time (RT) and signal intensities. The same combination of the mobile phases A and B is good for the short-term storage of the column (see Note 8).

Typical UPLC-MS chromatograms of samples obtained by the extraction of different tomato plant tissues (old and young leaves, peel of mature green fruit, peel and flesh of red fruit, seeds, flowers and roots) are presented in Fig. 1.

3.2.2. Injection Time Considerations

We apply a gradient shorter than 26-min when the reduction of the run time is desired for screening experiments with large numbers of samples, or for targeted analysis of a specific compound. When shortening the injection time (e.g., less than 5 min) it should be taken into consideration that co-elution of peaks can lead to an increase in the matrix affect (inhibition or enhancement of ionization of specific molecules, non-linear response, etc.). Examination of a longer gradient (extension of the elution part to 60% of acetonitrile) shows that a 23–44-min window of the tomato fruit peel extract chromatogram is less populated by eluting compounds than the one observed in a 2–23-min window. Therefore, a 26-min gradient presents a sort of compromise between the two options (see Fig. 2a).

3.2.3. Injection Volume Considerations

An injection volume of 4 μl was chosen as optimal in the analysis of tomato fruit extracts because it permits the detection of less abundant compounds in the tomato sample while the elution of most of the abundant ones still does not overload the column or the detector. If the compound of key interest is present at low concentrations in the biological sample, it is advisable to concentrate the sample before injection rather than inject larger volumes. Injection of more than 5 μl volume (e.g., 8–10 μl) leads to broadening of the chromatographic peaks eluting at the beginning of chromatogram, adjustment of their shape and loss of resolution. This is of particular importance for tomato fruit tissue extracts due to the large content of polar cinnamic acid derivatives, eluting at the first several minutes of the chromatogram acquired during a 26-min run.

3.2.4. Analysis of Abundant Compounds (e.g., Tomatine in Green Tomato Fruits, Leaves, and Flowers)

Tomatine is one of the most abundant compounds in the green stage tomato fruits, tomato leaves and flowers. Injection of 4 μl tissue extract (obtained with the extraction method described above) leads to overloading of tomatine. To quantify tomatine

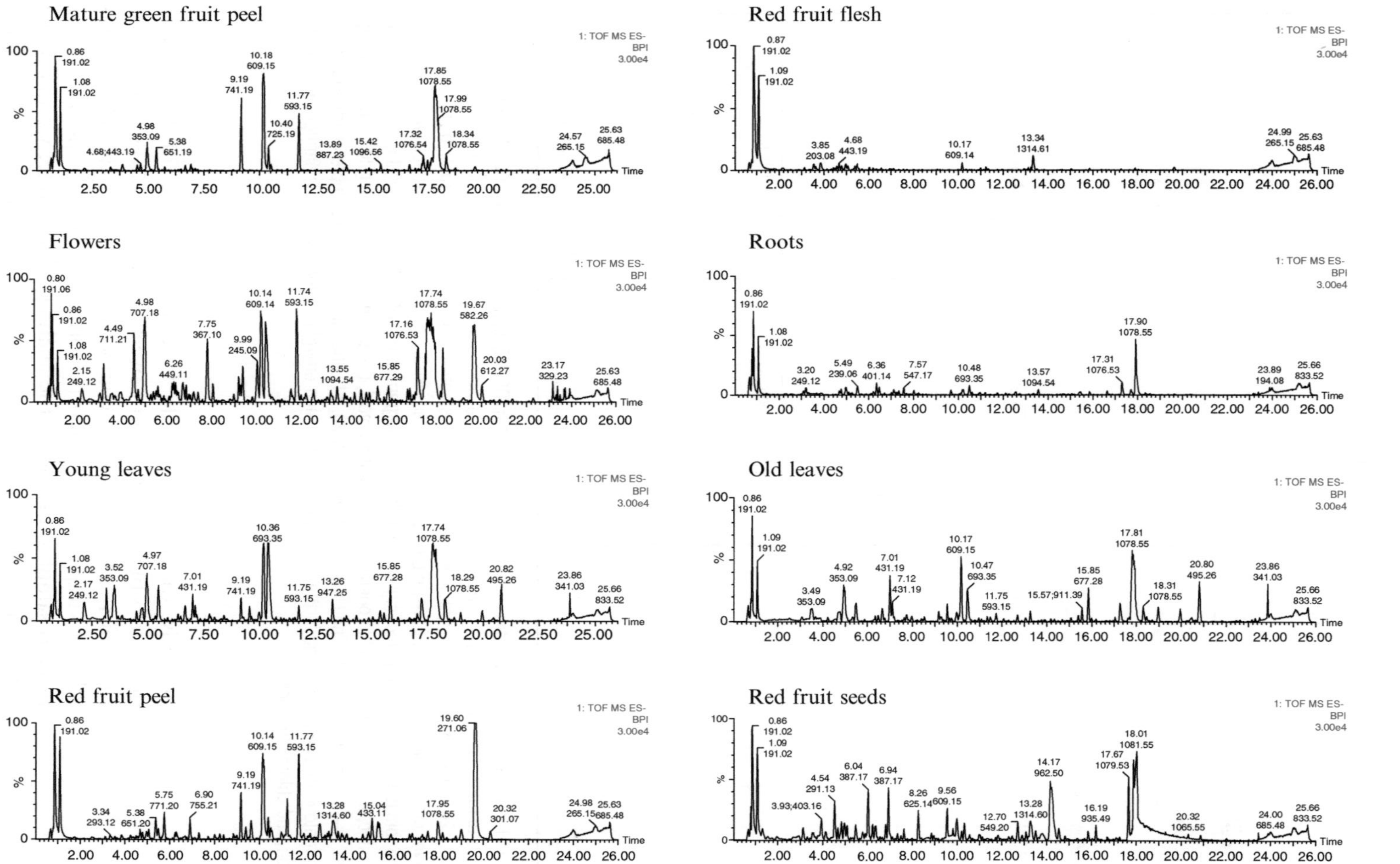

Fig. 1. BPI (Base Peak Intensity) chromatograms of different tomato tissues extracts, injected in the ESI(−) mode. Tomato tissues were extracted using the same procedure and injected at the same LC-MS conditions. Injection volume was 3 μl. The *Y* axes of the chromatograms are linked.

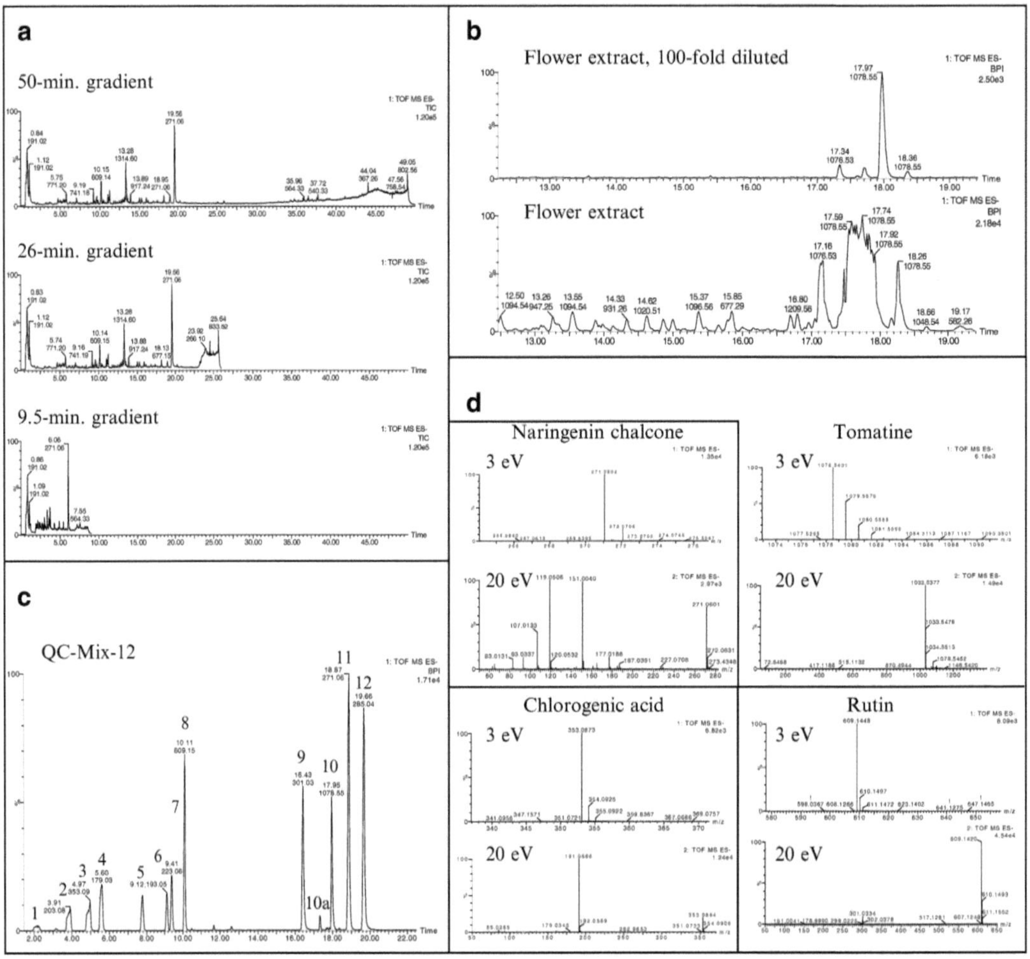

Fig. 2. Typical chromatograms and spectra acquired by the UPLC-qTOF-MS instrument. In (**a**), TIC (Total Ion Current) chromatograms of tomato red peel sample, acquired in the ES(−) mode: 50-min gradient (gradient slope is the same as for the 26-min run), 26-min gradient, 9.5-min gradient (gradient slope is steeper than for the 26-min. run). (**b**), BPI chromatograms of flower extracts, injected in the ES(−) mode: 100-fold diluted extract and non-diluted extract. *m/z* 1078.55 Da represents the tomatine-formic acid adduct signal. (**c**), BPI chromatogram of the QC-Mix-12 acquired in the ES(−) mode. 1—phenylalanine, 2—tryptophan, 3—chlorogenic acid, 4—caffeic acid, 5—coumaric acid, 6—ferulic acid, 7—synapic acid, 8—rutin, 9—quercetin, 10—tomatine (10a—dehydrotomatine), 11—naringenin, 12—kaempferol. (**d**), MS spectra of selected tomato compounds, acquired in the ES(−) mode at collision energies 4 eV and 20 eV: naringenin chalcone, tomatine, rutin, and chlorogenic acid.

properly, the samples should be diluted at least 50–100-fold in order to obtain a sharp, not-overloaded chromatographic peak. Most of the low-abundant compounds are not detectable in this diluted sample. Thus, when performing metabolomic analysis of tomato samples, including the analysis of tomatine, we inject the extract and run with a 26-min gradient before subsequently diluting the same samples 50–100-fold and running them again with a short gradient for the targeted analysis of tomatine (see Fig. 2b).

3.2.5. Photo Diode Array Detector Parameters

The photodiode array (PDA) detector is located between the chromatographic column and the MS detector. UV-Visible absorbance spectra provide valuable complementary information to the MS data which is often extremely helpful for compound identification. We set the Acquity UPLC PDA detector to acquire spectra in the range of 210–500 nm.

3.2.6. qTOF Parameters

We use the Synapt HDMS detector with an ESI source for the analysis of semipolar compounds. The TOF part is operating in the V-mode with mass resolution of 9,000. MS spectra are acquired from 50 until 1,500 Da with scan duration of 0.4 s. and an interscan delay of 0.02 s, in the centroid mode. Acquisition in the centroid mode is essential for further data treatment with peak picking programs. Argon is used as a collision gas and leucine enkaphalin for lock mass calibration.

The following MS parameters have proven suitable for the analysis of tomato tissue samples: capillary voltage—2.4 kV, cone voltage—28 eV, source temperature—125°C, desolvation temperature—275°C, desolvation gas flow rate—650 L/h, collision energy—4 eV. For the acquisition of MS/MS spectra, collision energies are set from 10 till 50 eV, depending on the nature of the compound. The approximate values for the collision energy settings during the MS/MS analysis for tomato metabolites can be found in the Supplementary Table S6 in ref. (6).

3.2.7. System Suitability Test, Quality Controls and Order of Injections

A System Suitability Test (SST) solution is used for checking the performance of the chromatographic column before starting the injections of samples. Quality Control (QC) samples are used to check the chromatographic performance during the injection of samples sequence. The QC sample should be injected at the beginning of each sequence of samples, at the end of sequence and one or several times during the sequence (each 3–5 h).

Two main approaches for the QC samples can be followed. The first, using the biological sample (or pool of biological samples) as the QC sample (17). A second option is to use a standard solution or a mixture of standards. The first approach has several advantages including the possibility to follow the behavior of all compounds during the injection of the samples and avoid possible de-stabilization of the column after injection of the sample that does not contain the biological matrix. The main disadvantage of utilizing a biological sample as QC is that in some cases it is difficult to detect the known compounds in the sample matrix and calculate their mass accuracies and retention time (RT) (the "unknown" matrix should be spiked with the "known" standards in this case). Our laboratory uses the second option: a mixture of standards as QC samples. The QC-Mix-12 consists of a mixture of 12 standard compounds belonging to different chemical classes (see above). These represent various classes of metabolites present in tomato fruit including: amino acids (phenylalanine, tryptophan); organic acids (chlorogenic acid, caffeic acid, coumaric acid, ferulic acid, synapic acid); flavonols (quercetin, rutin, kaempferol); flavanones (naringenin), and glycoalkaloids (tomatine), (see Note 9). The standard compounds mentioned above cover a wide lipophilicity range (peaks are located throughout the chromatogram, see Fig. 2c), are detected in both ionization modes and their mass signals cover a range of the m/z ratios, from 120.08 Da (the immonium ion of phenylalanine, detected in the positive ionization mode) to 1,034.5 Da (the pseudomolecular ion of tomatine, detected in the positive ionization mode) or 1,078.5 Da (the tomatine-formic acid adduct, detected in the negative ionization mode).

We use the same QC-Mix-12 as the SST solution for performing a system suitability test prior to beginning the analysis of the batch of samples. The chromatographic peak shape of caffeic acid provides information about the performance of the Acquity BEH C18 column. A caffeic acid-methanol solution, injected in the 26-min gradient, produces a sharp and symmetric peak. When its peak becomes wide or split, the column needs to be replaced (tryptophan, phenylalanine and chlorogenic acid elute earlier than caffeic acid and produce non-symmetric peaks even on a new and equilibrated column (see Fig. 2c)). The closely-eluting ferulic and sinapic acids also provide some information about the column performance: they have a baseline resolution in the 26-min run. When this disappears and the peaks begin to merge, the column should be replaced.

Errors in mass accuracy measurements (in ppm, parts per million) that are easily calculated for the known QC-Mix-12 compounds allow the determination of the mass accuracy parameter (mass tolerance) that is used for the calculation of metabolite elemental compositions (see Note 10).

Perform injections according to the following sequence:

1. Inject SST (system suitability test) solution at the beginning of sequence using the QC-Mix-12 solution. Check that the UPLC column is properly conditioned: retention times of compounds correspond to the values obtained previously on a fully-functioning (or new) column, and the peaks of ferulic and sinapic acids are resolved. Check that the MS system is properly calibrated: the leucine enkaphalin signal is stable and mass error is not more than 2–3 ppm (see Note 11). Check that the accuracy of mass determination for the mixture of standards is within the 2–3 ppm error.
2. Inject the quality control samples at the beginning and at the end of sequence and also after each tenth sample injection (each 5 h for the 26 min gradient).
3. Inject a blank sample once.

4. Inject the first biological sample three times to check the reproducibility of the injections.
5. Inject the remaining biological samples once, in a randomized order or in appropriate, statistically-compatible block designs.

3.3. Data Analysis

While targeted analysis is focused on one or several metabolites, non-targeted metabolite analysis provides information about all possible metabolites present in the analyzed sample (18). Here we will discuss data processing obtained by non-targeted metabolomics experiments of tomato samples. Data analysis (from LC-MS) can be roughly divided into three main steps: peak picking and peak alignment, statistical treatment of the data and peak assignment.

3.3.1. Peak Picking and Peak Alignment

A number of peak picking and peak alignment programs can be used for the non-targeted processing of UPLC-qTOF MS data such as MarkerLynx (Waters), MZ-mine (19), MetAlign (20) or XCMS (14). The main goal of these programs is to construct the mass-RT and mass-intensity matrix, aligned across all the samples. Our laboratory uses XCMS for peak picking and alignment; thus, the points below refer to this program.

A few steps should be performed after receiving the UPLC-MS raw data:

1. Check the validity of the chromatographic data obtained using the QC samples by examining the reproducibility of peak intensities and mass accuracy (see Note 12).
2. Convert the MassLynx raw data files to NetCDF format using the DataBridge toolbox of the MassLynx program (see Note 13).
3. Group NetCDF files per folders and subfolders according to the experimental relevance. For example, if two plant genotypes (G1 and G2) were analyzed in positive and negative modes in five replicates, prepare two main folders: "Pos" and "Neg," and two subfolders, containing the biological replicates, "G1" and "G2."
4. Prepare XCMS parameter files (treat the data acquired in positive and negative modes separately). The following main XCMS parameters are suitable for the analysis of tomato fruit peel samples injected in a 26 min run (negative mode): fwhm = 10.8, step = 0.05, steps = 4, mzdiff = 0.07, snthresh = 8, max = 1,000.
5. Run the XCMS program. The program produces several sorts of files as an output including a table containing mass—RT—mass intensity values, and images of the aligned mass peaks.
6. Check the quality of the XCMS results: look through the aligned mass signals and check profiles of several masses (see Note 14).

7. It is recommended to remove from further analysis the beginning of the chromatogram representing the column void volume and the last minutes corresponding to the column washing and equilibration. There is co-elution of many compounds at these chromatographic windows, which can lead to matrix effects and a non-linear intensity-concentration response.

3.3.2. Statistical Treatment of the Data

Differential mass peaks can be sorted by applying the appropriate statistical and multivariate analysis. PCA (Principal Component Analysis) is a convenient tool for the visualization of the results. We use PCA for primary filtering of possible outlier samples. Masses belonging to the same metabolite (i.e., fragments, adducts, isotopes, pseudomolecular ions) may be clustered together at this stage of data analysis. The most common strategy for clustering is to cluster according to the similarity in the abundance profiles of masses across different samples (6, 21).

3.3.3. Peaks Assignment

The final step in the non-targeted metabolite analysis is the putative identification of compounds. This procedure is rather complex since only relatively few standards for secondary metabolites are available. The following workflow is recommended for peak assignment:

1. Combine information that can be obtained from the UPLC-MS runs:
   (a) Predict elemental composition of the mass peak of interest using accurate mass and isotopic pattern with the Elemental Composition toolbox of the Masslynx software.
   (b) Retrieve the UV-Visible absorbance spectrum.
   (c) Predict lipophilicity of the compound by its position in the chromatographic column.
2. Compare the obtained data with the information of the standard compound injected under the same conditions for unambiguous identification of the metabolite.
3. For putative identification of the metabolites, search predicted elemental composition in the available databases (e.g., MOTO database (http://appliedbioinformatics.wur.nl/moto; (5)); KNApSAcK metabolite database (http://prime.psc.riken.jp/KNApSAcK, (22, 23)); KOMOCS (Kazusa OMICS, http://webs2.kazusa.or.jp/komics/); MassBank (http://www.massbank.jp/); Madison Metabolomics Consortium Database (http://mmcd.nmrfam.wisc.edu/); ARMeC: High Mass resolution annotation database (http://www.armec.org/MetaboliteLibrary/index.html)).
4. When no suitable candidate is found, search in more comprehensive chemical databases such as the Dictionary of Natural Products (Chapman & Hall/CRC) and SciFinder tool (SciFinder Scholar).

5. Perform an additional UPLC-MS/MS analysis in order to obtain the fragmentation pattern of the metabolite (see also Chapter 10).
6. Combine all the above information (i.e., UV-Visible and MS/MS spectra, predicted elemental composition and lipophilicity) and compare to those found in the literature (4–6, 24, 25).

Typical MS spectra of several metabolites, present in tomato fruit tissues are depicted in Fig. 2d (see Note 15).

4. Notes

1. It is very important that water, used for the preparation of the UPLC solutions does not contain sodium or potassium ions, otherwise a large amount of sodium or potassium adducts will be obtained during the analysis of metabolites in the positive ionization mode.
2. Membrane filters must be resistant to 75% methanol/0.1% formic acid and free of polyethylene glycol and soluble plasticizers.
3. The Waters Synapt HDMS system uses a traveling wave to accomplish ion mobility separation prior to time-of-flight (TOF) *m/z* analysis, thus providing an additional, orthogonal to the UPLC dimension of separation. When running in the ion mobility mode, the Synapt HDMS allows the deconvolution of overlapping isotopic patterns for co-eluting compounds, measures comprehensive MS/MS/MS spectra, and provides additional information which can be utilized for structure elucidation of unknown compounds. When running in the standard mode, this instrument is very good for routine qualitative and quantitative exact mass UPLC-MS and MS/MS measurements.
4. The XCMS program (14) is software from the Bioconductor package in the R statistical language. It can be freely downloaded from the Internet site http://metlin.scripps.edu/download/.
5. Tryptophan is sparingly soluble in methanol. To prepare a solution in 80% methanol:2% formic acid, start by dissolving in water and formic acid, vortex vigorously and then add the corresponding amount of methanol.
6. Harvest several tomato plant materials (i.e., fruit, leaves, flowers, etc.) preferably as a pool from a minimum of three tomato plants to obtain a representative biological sample. Wherever possible, samples should be as uniform as possible. They should be taken from plants grown together under the same conditions, at the same time of day, the same position on the plant etc. Separate the desired tissues (e.g., peel from the flesh and gel) and immediately freeze in liquid nitrogen (in the PP tubes

of the corresponding size). Cut large pieces into smaller ones before freezing.

7. If you consider further use of the samples extracts (e.g., for MS/MS analysis), place the eppendorf tubes with the remaining amount of extract in the freezer as soon as possible (−20°C). The samples can be stored in this way for several weeks without major changes in their composition. However, minor changes to specific compounds might occur (e.g., partial isomerization of naringenin chalcone to naringenin). Vortex vigorously or sonicate the stored tubes after thawing and filter the extract again before injection into the UPLC-MS.

8. Using these chromatographic conditions, we injected more that 3,000 plant samples (extracted by the method described above) without notable reduction of the column performance.

9. A commercial tomatine standard usually contains some amount of dehydrotomatine as impurity, which elutes about 1 min earlier than tomatine in the 26-min run and therefore can be detected in the chromatogram of the tomatine standard. Dehydrotomatine is also a native constituent of tomato fruits and leaves.

10. Elemental composition of the compounds can be predicted with the Elemental Composition tool (MassLynx 4.1) using accurate mass and isotope pattern. The mass tolerance parameter (mass error) should be introduced as the elemental composition calculation parameter.

11. The Waters TOF mass spectrometer instrument is equipped with a lockspray ionization source utilizing an online lock mass correction. Use a leucine enkaphalin solution as a lock-spray reference for on-line calibration of the TOF instrument ($(M+H)^+ = 556.2771$ Da, $(M-H)^- = 554.2620$ Da).

12. It is advisable not to use the data if intensity variations in QC samples are more than 20–30%, and mass accuracy of known compounds is above 5 ppm.

13. The DataBridge program produces separate NetCDF file for each function (channel) of the MassLynx file. For example, a standard MS injection on a UPLC-PDA-qTOF instrument produces a MassLynx raw data file with three functions: (a) MS data, (b) MS data for the Lock Mass injection, and (c) PDA data. NetCDF files obtained from the functions 2 and 3 are not needed for XCMS analysis and should be removed.

14. Quality control (QC) of the XCMS output can be performed by the MetaboQC program (26), or manually, by evaluating 3 main features: (a) the similarity of replicates. If the XCMS profiles of two replicates are dissimilar (i.e., a scatter plot of the values from the two replicates is not linear) but the corresponding chromatograms look similar, it is likely that the

integration/alignment was not correct; (b) the superposition of extracted ion chromatograms, generated by the XCMS visualization tool. The XCMS visualization tool generates extracted ion chromatograms for a chosen number of mass signals and shows the RT regions used for integration. These figures provide information about the correctness of peaks integration and alignment; (c) the total amount of mass signals in the XCMS output. Too little or many mass signals in the XCMS output indicate the inappropriate choice of threshold values in the XCMS parameter file.

15. Naringenin and naringenin chalcone have the same MS spectra, they differ only in their UV-Visible spectra and retention times on the column. Naringenin chalcone elutes half a minute later than naringenin in the 26-min gradient, but these compounds reverse their elution order in the 9.5-min gradient run.

Acknowledgements

We are grateful to Arye Tishbee for operating the UPLC-qTOF instrument, Max Itkin and Sagit Meir for the assistance in the preparation of tomato extracts and Ilya Venger with XCMS analysis. A.A. is an incumbent of the Adolfo and Evelyn Blum Career Development Chair. The research in the A.A. laboratory was supported by the William Z. and Eda Bess Novick Young Scientist Fund, the Y. Leon Benoziyo Institute for Molecular Medicine and the EU project "META-PHOR" contract number FOODCT-2006-036220.

References

1. Dorais, M., Ehret, D. L., and Papadopoulos, A. P. (2008) Tomato (*Solanum lycopersicum*) health components: from the seed to the consumer. *Phytochem Rev* 7, 231–250.
2. Mueller, L. A., Solow, T. H., Taylor, N., Skwarecki, B., Buels, R., Binns, J., Lin, C., Wright, M. H., Ahrens, R., Wang, Y., Herbst, E. V., Keyder, E. R., Menda, N., Zamir, D., and Tanksley, S. D. (2005) The SOL Genomics Network: a comparative resource for Solanaceae biology and beyond. *Plant Physiol* **138**, 1310–1317. http://www.sgn.cornell.edu/about/tomato_sequencing.pl.
3. Engelhard, Y. N., Gazer, B., and Paran, E. (2006) Natural antioxidants from tomato extract reduce blood pressure in patients with grade-1 hypertension: a double-blind, placebo-controlled pilot study. *Am Heart J* **151**, 100.e1–100.e6.
4. Iijima, Y., Nakamura, Y., Ogata, Y., Tanaka, K., Sakurai, N., Suda, K., Suzuki, T., Suzuki, H., Okazaki, K., Kitayama, M., Kanaya, S., Aoki, K., and Shibata, D. (2008) Metabolite annotations based on the integration of mass spectral information. *Plant J* **54**, 949–962.
5. Moco, S., Bino, R. J., Vorst, O., Verhoeven, H. A., de Groot, J., van Beek, T. A., Vervoort, J., and de Vos, C. H. (2006) A liquid chromatography mass spectrometry based Metabolome database for tomato. *Plant Physiol* **141**, 1205–1218.
6. Mintz-Oron, S., Mandel, T., Rogachev, I., Feldberg, L., Lotan, O., Yativ, M., Wang, Z., Jetter, R., Venger, I., Adato, A., and Aharoni, A. (2008). Gene expression and metabolism in tomato fruit surface tissues. *Plant Physiol* **147**, 823–851.
7. von Roepenack-Lahaye, E., Degenkolb, T., Zerjeski, M., Franz, M., Roth, U., Wessjohann,

L., Schmidt, J., Schee, D., and Clemens, S. (2004) Profiling of Arabidopsis secondary metabolites by capillary liquid chromatography coupled to electrospray ionization quadrupole time-of-flight mass spectrometry. *Plant Physiol* **134**, 548–559.

8. Moco, S., Bino, R., De Vos, R. C. H., and Vervoort, J. (2007) Metabolomics technologies and metabolite identification. *Trends in Analytical Chemistry* **26**, 855–866.
9. Wilson, I., Nicholson, J., Castro-Perez, J., Granger, J., Johnson, K., Smith, B., and Plumb, R. (2005) High resolution "ultra performance" liquid chromatography coupled to oa-TOF mass spectrometry as a tool for differential metabolic pathway profiling in functional genomic studies. *J Proteome Res* **4**, 591–598.
10. Verhoeven, H. A., de Vos, C. H., Bino, R. J., and Hall, R. D. (2006). Plant metabolomics strategies based upon quadrupole time of flight mass spectrometry (QTOF-MS), in *Plant Metabolomics – Biotechnology in Agriculture and Forestry* (Saito, K., Dixon, R. A. and Willmitzer, L., eds.) Springer-Verlag, Berlin, Heidelberg Vol. **57** pp. 33–48.
11. Niessen, W. M. (2006) *Liquid chromatography-mass spectrometry*, 3rd edition. Taylor and Francis Group, LLC, CRC Press.
12. Fait, A., Hanhineva, K., Belleggia, R., Dai, N., Rogachev, I., Fernie, A. R., and Aharoni, A. (2008) Reconfiguration of the achene and receptacle metabolic networks during strawberry fruit development. *Plant Physiol* **148**, 730–750.
13. Hanhineva, K., Rogachev, I., Kokko, H., Mintz-Oron S., Venger, I., Kärenlampi, S., and Aharoni, A. (2008) Non-targeted analysis of spatial metabolite composition in strawberry (Fragaria × ananassa) flowers. *Phytochemistry*, **69**, 2463–2481.
14. Smith, C.A., Want, E.J., O'Maille, G., Abagyan, R. and Siuzdak, G. (2006) XCMS: processing mass spectrometry data for metabolite profiling using nonlinear peak alignment, matching, and identification. *Anal Chem* **78**, 779–787.
15. Clemens, S., Böttcher, C., Franz, M., Willscher, E., Roeoenack-Lahaye, E. V., and Scheel, D. (2006) Capillary HPLC coupled to electrospray ionization quadrupole time-of-flight mass spectrometry. In *Plant Metabolomics – Biotechnology in Agriculture and Forestry* (Saito, K., Dixon, R. A. and Willmitzer, L., eds.) Springer-Verlag, Berlin, Heidelberg Vol. **57**, pp. 65–79.
16. De Vos, R. C. H., Moco, S., Lommen, A., Keurentjes, J. J. B., Bino, R. J., and Hall, R. D. (2007) Untargeted large-scale plant metabolomics using liquid chromatography coupled to mass spectrometry. *Nature protocols* **2**, 778–791.
17. Sangster, T., Major, H., Plumb, R., Wilson, A. J., and Wilson, I. D. (2006) A pragmatic and readily implemented quality control strategy for HPLC-MS and GC-MS-based metabonomic analysis. *Analyst* **131(10)**, 1075–1078.
18. Aharoni, A., Keizer, L. C. P., Bouwmeester, H. J., Sun, Z., Huerta, M. A., Verhoeven, H. A., Blaas, J., van Houwelingen, A. M. M. L., De Vos, R. C. H., van der Voet, H., Jansen, R. C., Guis, M., Mol, J., Davis, R. W., Schena, M., van Tunen, A. J., and O'Connell, A. P. (2000) Identification of the SAAT Gene Involved in Strawberry Flavor Biogenesis by Use of DNA Microarrays. *The Plant Cell* **12**, 647–662.
19. Katajamaa, M., and Oresic, M. (2005) Processing methods for differential analysis of LC/MS profile data. *BMC Bioinformatics* **6**, 179.1–179.12.
20. Lommen, A. (2009) MetAlign: an interface-driven, versatile metabolomics tool for hyphenated full-scan MS data pre-processing. *Anal Chem* **81**, 3079–3086. http://www.metalign.nl.
21. Malitsky, S., Blum, E., Less, H., Venger, I., Elbaz, M., Morin, S., Eshed, Y., and Aharoni, A. (2008) The "inner" and "outer" circles of the transcriptome and metabolome effected by the two clades of Arabidopsis glucosinolate biosynthesis regulators. *Plant Physiol* **148**, 2021–2049.
22. Shinbo, Y., Nakamura, Y., Altaf-Ul-Amin, M., Asahi, H., Kurokawa, K., Arita, M., Saito, K., Ohta, D., Shibata, D., and Kanaya, S. (2006) KNApSAcK: A comprehensive species-metabolite relationship database, in: *Plant Metabolomics – Biotechnology in Agriculture and Forestry* (Saito, K., Dixon, R. A. and Willmitzer, L., eds.) Springer-Verlag, Berlin, Heidelberg Vol. **57**, pp. 165–181.
23. Akiyama, K., Chikayama, E., Yuasa, H., Shimada, Y., Tohge, T., Shinozaki, K., Hira, M. Y., Sakurai, T., Kikuchi, J., and Saito K. (2008) PRIMe: a Web site that assembles tools for metabolomics and transcriptomics. *In Silico Biol* **8**(3–4), 339–345.
24. Slimestad, R., Fossen, T., and Verheul, M. J. (2008) The flavonoids of tomatoes. *J Agric Food Chem* **56**, 2436–2441.
25. Yamanaka, T., Vincken, J. P., de Waard, P., Sanders, M., Takada, N., and Gruppen, H. (2008) Isolation, characterization, and surfactant properties of the major triterpenoid glycosides from unripe tomato fruits. *J Agric Food Chem* **56**, 11432–11440.
26. Brodsky, L., Moussaieff, A., Shahaf, N., Aharoni, A., and Rogachev, I. (2010) Evaluation of peak picking quality in LC-MS metabolomics data. *Anal Chem* **82**, 9177–9187.

Chapter 10

High Precision Measurement and Fragmentation Analysis for Metabolite Identification

Madalina Oppermann, Nicolaie Eugen Damoc, Catharina Crone, Thomas Moehring, Helmut Muenster, and Martin Hornshaw

Abstract

The degree of precision in measuring accurate masses in LC MS/MS-based metabolomics experiments is a determinant in the successful identification of the metabolites present in the original extract. Using the methods described here, complex broccoli extracts containing hundreds of small-molecule compounds (mass range 100–1,400 Da) can be profiled at resolutions up to 100,000 (full width half maximum, FWHM), useful for accurate and sensitive relative quantification experiments. Using external instrument calibration, analyte masses can be measured with high (sub-ppm to a maximum of 2 ppm) accuracy, leading to compound identifications based on elemental composition analysis. Unambiguous identification of four analytes (citric acid, chlorogenic acid, phenylalanine, and UDP-D-glucose) is used to validate the performance of the different MS/MS fragmentation regimes. Identifications are carried out either via resonance excitation collision induced dissociation (CID) or via higher energy collision dissociation (HCD) experiments, and validated by infrared multiphoton dissociation (IRMPD) fragmentation of standards. Such results, obtained on both hybrid and non-hybrid systems from metabolite profiling and identification experiments, provide evidence that the strategies selected can be successfully applied to other LC-MS based projects for plant metabolomic studies.

Key words: Metabolomics, Mass spectrometry, CID, HCD, Orbitrap, Exactive, LTQ, LTQ FT

1. Introduction

Metabolomics as an application area has already gained an established place among the 'omics' technologies. While genomics and proteomics have undergone a renaissance with the development of high-throughput technologies and automated database searching, metabolomics has matured at a more even pace, learning from and integrating with these adjacent fields, in approaches that aim to deepen the understanding of how complex biological systems work (1, 2). With analytes of interest in plant extracts typically

Nigel W. Hardy and Robert D. Hall (eds.), *Plant Metabolomics: Methods and Protocols*, Methods in Molecular Biology, vol. 860, DOI 10.1007/978-1-61779-594-7_10,

© Springer Science+Business Media, LLC 2012

belonging to diverse classes of compounds, and their detection sometimes being favoured by different analytical techniques (3), the quantitative analysis of metabolites is a complex process that must handle difficulties associated with both biochemical and biological variability (4). Therefore, stringent criteria are imposed on the design of the experiment (see Note 1) which must include blanks, quality control samples, biological and technical replicates. Demands are set high with regard to both experimental and computational outcomes, while innovative application and technological approaches augment experimental performance (5–9).

"The ultimate goal for any metabolomics experiment is to find patterns in (the fingerprint) data that can describe the biological outcome" (10). The biological outcome under study could be as diverse as metabolic changes induced by a disease, responses to drug therapy, mode of action of a drug, assessment of pharmacotoxicity, or selectivity of desirable traits in cultivars (11–14). The methodologies employed fall into several groups which could be crudely summarized as *metabolic fingerprinting*, a global screening approach which aims to discriminate features that change in response to the condition under study (a proteomic equivalent of this approach is biomarker discovery), and *metabolic profiling*, a targeted approach whereby quantitative analyses are carried out along a metabolic pathway or for a defined subclass of compounds (a proteomic rough equivalent of this approach is biomarker verification).

Food quality, nutritional value, flavour, and resistance to pathogens are among the traits monitored by governments and the food industry alike, in an attempt to promote the creation of robust, healthy, nutrition-rich cultivars that contribute to sustained agronomic development. Metabolomics has been identified as a key mass spectrometry-based approach in the analysis of such traits. Within our organization and as collaborators in research projects (e.g. (15)) we aim to understand and develop the use of highly resolving, highly mass accurate measurement systems and advance the benefits they provide in the identification of small molecules. The general aim within this field is to develop technologies and procedures which are widely applicable so that methodologies developed using one type of plant material can find subsequent application in a wide variety of projects to the broader benefit of crop science and world nutrition programmes.

In this chapter, metabolite profiling and identification procedures using broccoli samples on both hybrid and non-hybrid instruments are presented, along with a general strategy for metabolome analysis and tools supporting these applications. The work presented here describes a general workflow based on broccoli, but it can be used as a template for most other plant metabolomics projects involving other crops or plant materials.

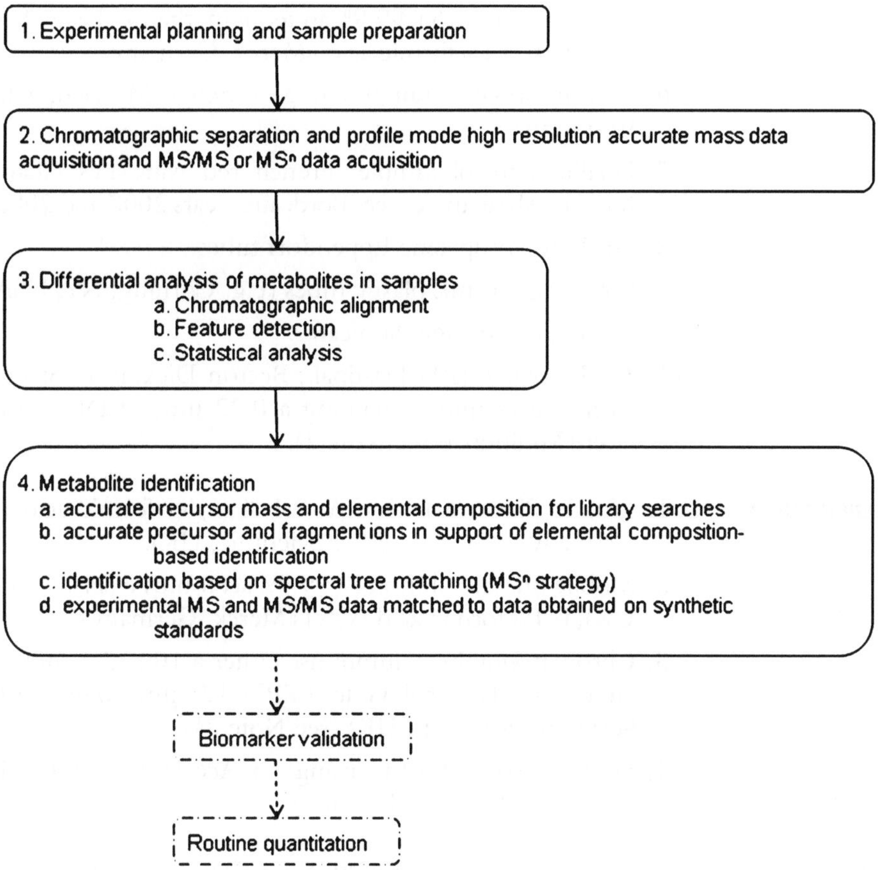

Fig. 1. Metabolomics workflow representing the sequence of events involved in metabolite biomarker discovery (1–4) and progressing towards development of clinical applications.

Figure 1 depicts a typical metabolomics workflow for metabolic fingerprinting, from study design and sample preparation, through chromatographic separation and mass spectrometric data acquisition, to data analysis and validation of results. As part of the experimental design, blanks, quality controls, and randomized injection order are important parameters that must be incorporated into the set-up to be able to assess and standardize chromatographic and analytical robustness.

2. Materials

2.1. Sample Preparation

1. Water (LC-MS Grade e.g. Fisher Scientific, UK).
2. Methanol (LC-MS Grade).
3. Acetonitrile (LC-MS Grade).
4. Formic acid (Suprapur quality).

5. Extraction solution: 99.9% aqueous methanol (Fisher Scientific, UK) with 0.1% formic acid (Merck, Germany).
6. Frozen broccoli samples (e.g. varieties Monaco, Chevalier, Iron Man).
7. Quality control sample: French red wine Les Charmes de Kirwan, Margaux (cuvee, Bordeaux, years 2003 and 2005) (16).
8. 2-mL polypropylene Eppendorf tubes.
9. Screw cap 10-mL plastic tubes (e.g. Corning, NY, USA).
10. Vortex, Centrifuge, Sonicator.
11. 1-mL syringe (BD Plastipak, Becton Dickinson, Spain) fitted with a filter unit containing a 0.22 μm, PVDF Hydrophilic filter (Millipore) (see Note 2).

2.2. Chromatography

1. Mobile Phase: Solution A: 0.1 % (v/v) formic acid (Merck, Germany) in water (Fisher Scientific, UK).
2. Mobile phase: Solution B: 99.9% acetonitrile (Fisher Scientific, UK), 0.1% formic acid (v/v) (Merck, Germany).
3. Chromatography column: use either a 100 × 2.1 mm Hypersil Gold™ or Hypersil Gold PFP™ 1.9 μm column (Thermo Scientific, Runcorn, UK) (see Note 3).
4. High pressure HPLC using an Accela U-HPLC (Thermo Scientific, Bremen, Germany).

2.3. Mass Spectrometry

1. Thermo Scientific Exactive™ (Thermo Scientific, Bremen, Germany).
2. LTQ Orbitrap XL™ hybrid mass spectrometer (Thermo Scientific, Bremen, Germany).
3. LTQ FT Ultra™ hybrid mass spectrometer (Thermo Scientific, Bremen, Germany).

2.4. Software for Data Acquisition and Analysis

1. Data generation: Xcalibur™ software version 2.0.7 (for LTQ Orbitrap XL™ and LTQ FT Ultra™ hybrid mass spectrometers) and version 2.1.0 (for Thermo Scientific Exactive™ mass spectrometer) (Thermo Fisher Scientific).
2. Analysis software: SIEVE™ software (version 1.2.0.477, Thermo Fisher Scientific).
3. Metabolite preliminary identification: ChemSpider (17).
4. Spectra interpretation and metabolite confident identification: Mass Frontier™ software (version 6.0, release candidate 3) (HighChem).

3. Methods

3.1. Sample Preparation

1. Extract 1,500 mg frozen broccoli powder (genotypes Monaco, Chevalier and Iron Man) in 4.5 mL of extraction solution (0.1% formic acid in methanol).
2. Vortex for 15 s for thorough mixing and then sonicate for 15 min.
3. Filter through a PTFE filter and collect the filtrate into a clean 2-mL Eppendorf polypropylene tube.
4. Transfer 500 μL aliquots to fresh Eppendorf tubes and freeze at –80°C until required for analysis.
5. Before injection into the mass spectrometer, stored sample extracts are first quickly thawed at room temperature, vortexed, centrifuged, and then re-filtered (0.22 μm, PVDF Hydrophylic, Milipore; 1-mL syringe from BD Plastipak).

3.2. Blank/Control Injections

1. The blank injection is done using 0.1% formic acid in methanol (see Note 4).
2. The quality control sample used is red wine.

3.3. Chromatography

1. Chromatographic separations can be performed on either a 100×2.1 mm Hypersil Gold™ or Hypersil Gold PFP™ 1.9 μm column (Thermo Scientific, Runcorn, UK) using an Accela U-HPLC (Thermo Scientific, Bremen, Germany) operated in high pressure mode together with mass spectrometric analyses performed on the Exactive™ and the LTQ FT Ultra™ hybrid mass spectrometer. For LTQ Orbitrap XL™ mass analysis a Surveyor MS Pump Plus™ (Thermo Scientific, Bremen, Germany) can be used for liquid chromatographic separations.
2. The column is maintained at 30°C.
3. Either a long or a short gradient can be employed. For the long, 30 minute gradient (flow rate of 300 μL/min for the LTQ FT Ultra™ runs and 150 μl/min for the LTQ Orbitrap XL™ runs) the mobile phase separation starts at 95% A and decreases to 60% over a period of 25 min; then reduces from 60% to 5% A in 1 min and is then maintained at isocratic flow for another 2 min; the gradient then increases from 5% to 95% A in 1 min and is kept for another 1 min prior to the next separation (see Note 5).
4. A short, 15 minute gradient can also be used (flow rate of 500 μL/min) together with mass spectrometric analysis on the Exactive™. Here, the mobile phase separation runs from 95% to 67% A in a period of 11 min, 67–55% A in 0.2 min, 55–5%

A in 0.3 min; maintain at 5% A for 0.9 min; return to 95% A in 0.6 min and maintain for a further 2 min prior to the next separation.

5. 5 μL of sample is injected onto the column and the column eluent is then directed to the mass spectrometer.

3.4. Mass Spectrometry

Mass spectrometry of broccoli samples can be performed on three types of high-resolution, high mass accuracy platforms. The mass spectrometers should be mass calibrated prior to starting any sequence of injections. All data are acquired using external calibration.

1. The Thermo Scientific Exactive™ (Thermo Scientific, Bremen, Germany) mass spectrometer uses a heated electrospray ionization (HESI) source and is operated in negative ion mode. The sheath gas is set to 30 (arbitrary units) at a temperature of 300°C; the auxiliary gas is set to 20 (arbitrary units) and the capillary temperature is set to 250°C. The capillary voltage and spray voltage are set to 50 V and 3.5 kV respectively. The instrument is operated in full scan negative mode, from m/z 120–1,000 at 100,000 resolving power. The data acquisition rate is set to 1 Hz, with 250 ms maximal injection time and the AGC target is set at one million charges. Each full scan is followed by a same-polarity, "all-ion-fragmentation" HCD scan with data acquisition rate of 4 Hz and resolution set at 25,000, m/z 60–1,000, 250 ms maximal injection times, and the AGC target value of three million. The HCD is set at 35 eV.
2. Operate the LTQ Orbitrap XL™ hybrid mass spectrometer (18) in full scan negative mode, from m/z 100–1,400 at 100,000 resolving power. Collision induced dissociation (CID) using a normalized collision energy of 70%, is performed in the linear ion trap in data-dependent fashion, whereby the three most intense analytes detected in the full scan are selected for fragmentation, and fragment ions are detected with the orbitrap analyzer at a resolution of 7,500.
3. Operate the LTQ FT Ultra™ hybrid mass spectrometer in negative ion mode, from m/z 100–1,400 at 100,000 resolving power, without data dependent subsequent fragmentation events. Standards provided are analyzed by direct infusion with either CID or IRMPD as dissociation techniques.

3.5. Data Analysis

1. Process the raw data files generated by Xcalibur™ software using SIEVE™ software for differential analysis based on chromatographic alignment and recursive base peak framing. This enables the distinction of differences that are statistically meaningful. Metabolite identification is performed based on accurate mass elemental composition predictions, via links embedded in SIEVE™ software to ChemSpider (17) and spectral interpretation employing Mass Frontier™ software.

3.6. Validation of Method

1. In the method described, the LTQ FT™ Ultra hybrid mass spectrometer is employed primarily for metabolic fingerprinting. Although the instrument is capable of resolution in excess of 1,000,000 FWHM, in this case the resolution for data acquired in profile mode is limited to 100,000 to assist in an indirect comparison to data obtained on the LTQ Orbitrap XL™ hybrid system and the Exactive™ instrument, despite differences in experimental set-up. We acquire data on the LTQ FT Ultra™ instrument using five biological replicates of three broccoli genotypes and five technical replicates of pooled broccoli samples. On both other systems biological triplicates of two cultivars are usually analyzed in full scan mode followed by HCD fragmentation on the Exactive™ system or trap-based CID fragmentation on the LTQ Orbitrap XL™. Across all samples and instruments, the four standards provided (citric acid, chlorogenic acid, phenylalanine, and UDP-D-glucose) are measured with high mass accuracy employing external calibration to give a set of reference compounds (see Note 6).
2. Implementation of Automatic Gain Control™ (AGC) (19) ensures that the number of ions trapped does not compromise instrument performance by the induction of space charging effects. Fast pre-scans are used to measure the total ion current generated, employed to calculate the optimal injection time. Thus, the number of ions that reach the mass analyser is kept constant, leading to measurements based on reproducible, optimal size ion populations.
3. Data acquisition is done in profile mode at high resolution (20) on all systems and is compatible with HPLC and fast UPLC chromatographic separations. Peak shapes should be well defined by a minimum of ten points and chromatographic widths varying from 4 s (U-HPLC) to under 30 s (HPLC). As seen from Table 1, RMS values for all three systems tested by

Table 1
List of mass measurements, elemental compositions and RMS mass error values obtained for UDP-D-Glucose on the LTQ FT Ultra™, LTQ Orbitrap XL™, and the Exactive™ mass spectrometers

Instrument	Elemental composition (Neg. ion)	Theoretical mass	Number of data points[a]	RMS error (ppm)
LTQ FT ULTRA	C15H23O17N2P2	565.04774	30	0.741
LTQ ORBITRAP XL	C15H23O17N2P2	565.04774	60	0.548
EXACTIVE	C15H23O17N2P2	565.04774	42	1.212

[a]Total number of data points from 3 LCMS runs

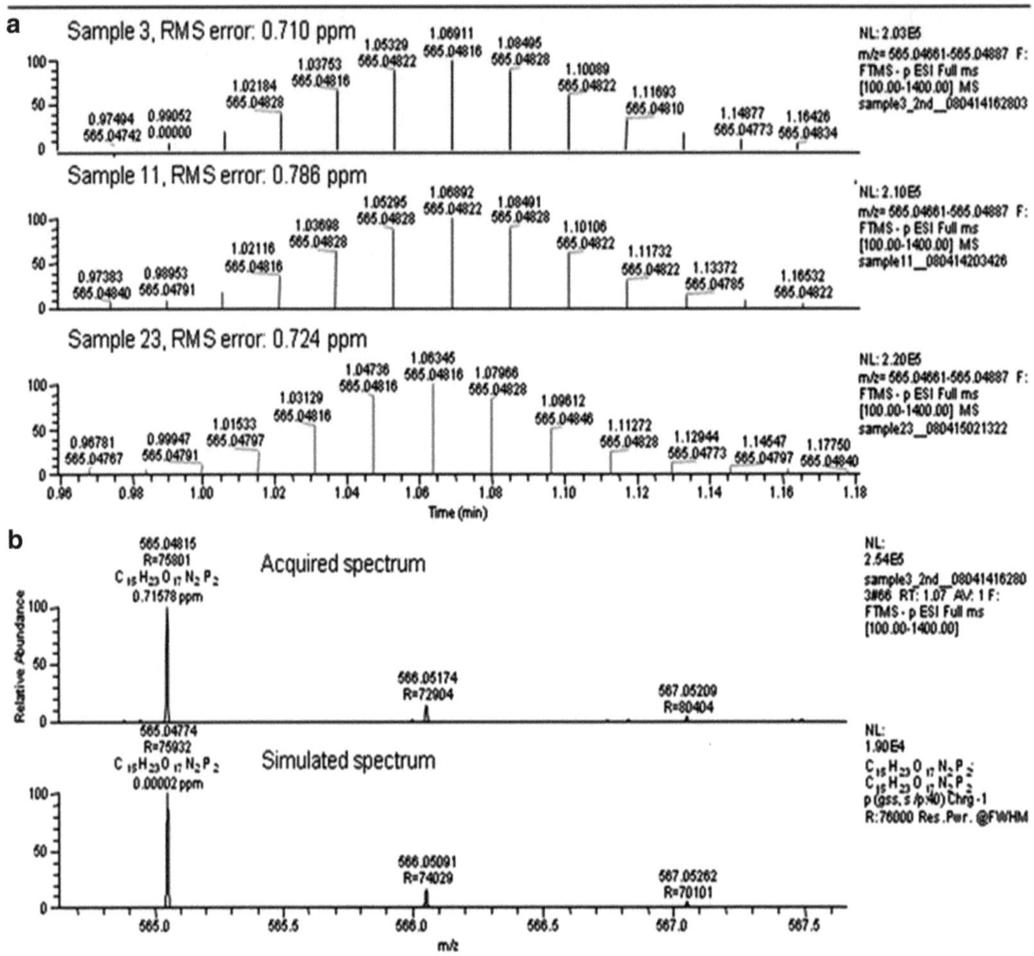

Fig. 2. (**a**) Calculation of RMS error for UDP-D-Glucose measured in three non-consecutive LCMS analyses (analysis 3, 11, 23). (**b**) Theoretical isotopic distribution is close to a perfect match to the observed distribution.

us displayed mass errors well within instrument specifications, with data acquired on the LTQ Orbitra XL™ and the LTQ FT Ultra™ instruments showing results across multiple acquisitions with mass accuracy better than 1 ppm. Figure 2 shows the RMS error value as measured for UDP-D-Glucose in three non-consecutive broccoli analyses. In addition, isotopic distribution of the analyte and intensities should be extremely reproducible from analysis to analysis. High mass accuracy combined with high mass resolution provides powerful data for compound identification: correct isotopic distribution corroborates the evidence provided by elemental composition calculation.

4. Differential analysis is performed with the SIEVE™ software. During alignment, the chromatographic space is divided into frames (m/z versus retention time) where scans from multiple runs are matched. Following feature detection, which sets

cut-off values either as detection thresholds or as a limit on the number of peaks to be detected, peak intensity is extracted for differential expression analysis. For ease of use, features can be tabulated by m/z value and p-value and data linked with correlated MS^n files, supporting the process of metabolite identification. While SIEVE™ provides support for principal components analysis and K-means clustering, software packages such as SIMCA-P™ can be used to process data employing more sophisticated statistical analyses (21). Alternative options which integrate metabolite differential analysis with statistical analysis and which may also provide tools for biological understanding and pathway mapping are provided elsewhere (22).

5. A crucial step and bottleneck in modern metabolite analysis is the identification of metabolites of interest. Here, we have used a two-pronged approach based on accurate mass determination for confirmatory elemental composition matching with a secondary strategy of matching MS/MS product ion data against theoretical fragmentation patterns derived with Mass Frontier™ software (see Note 7). Figure 3 shows a typical fragmentation spectrum from for one of the standards provided, chlorogenic acid. While all three types of fragmentation employed here provided adequate means to identify the compound with good confidence, conclusive evidence is provided by a comparison of accurate precursor mass and isotopic pattern and high quality MS/MS data of a standard to the actual mass spectral data acquired for the sample under analysis.
6. The broad-based orientation of the fingerprinting approach suits experiments where little is known in advance about the effect of the perturbation under study. However, often there are hypotheses for mechanisms and pathways affected, for example, relative to the design of genetically modified organisms. These can be tested more directly by employing class-targeted and/or pathway-targeted approaches (see Note 8).

4. Notes

1. Independent of the approach selected, a number of points which mark the importance of analytical performance in terms of sample and sample preparation, hardware and software processes, merit further discussion: experimental design, from sampling and sample preparation through mass spectrometry and data analysis, must substantiate the validity of the findings; therefore, quality control samples, blanks, and randomized injection order must be planned in the analysis; highly pure, mass spectrometry compatible chemicals must be used throughout

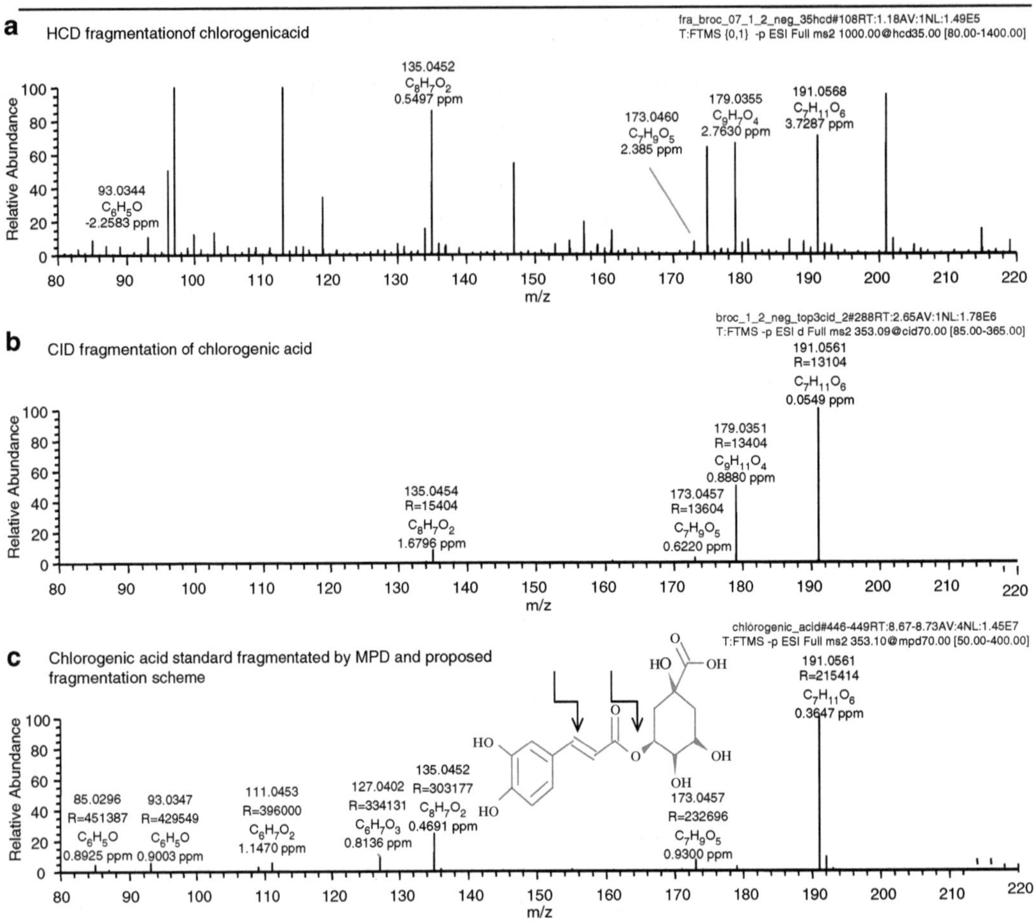

Fig. 3. (**a**) Fragmentation of chlorogenic acid using the high energy collision dissociation, HCD "all ion fragmentation" approach on the Exactive™ instrument; unlabeled fragments were generated by co-fragmentation of other precursor ions. (**b**) CID-like ion trap resonance fragmentation on the LTQ Orbitrap™ XL instrument (**c**). IRMPD fragmentation of standard directly infused in the LTQ FT Ultra™ mass spectrometer; *arrows* indicate potential fragmentation sites.

the experiment; prior to sample analysis, chromatography and mass spectrometry must be tested according to the vendor's procedure to insure that optimal performance is achieved.

2. All filter types used should be checked for contamination before being used, by passing through a blank sample and running it on the LCMS system chosen. Should contaminant peaks be found, the chosen filter type should either be replaced or each one should be thoroughly washed through before use.

3. Compounds with a planar structure (such as steroids) benefit from the retentive properties of Hypersil Gold PFP columns; the two columns tested provided equally good results.

4. A suitable frequency of injecting blank and control samples to monitor carry-over, chromatographic drift and check system performance is every five to ten samples.
5. Profiling feature-rich samples is successfully achieved on longer gradients, where chromatographic conditions have been optimized with respect to sample complexity as well as chemical diversity. Shorter gradients are best employed with samples of limited complexity, also together with chromatography operating in high pressure mode.
6. For example, in our hands, during a set of LC-MS analyses, performed on the Exactive™ mass spectrometer covering approximately 12 h of acquisition time, the mass measurement error for UDP-D-Glucose ranged from 0.30 to 2.03 ppm (data not shown).
7. Metabolite identification carried out either via searches in mass spectral libraries or via alternative strategies, must produce confident results. To achieve this, multiple strategies should be employed, such as combinations of accurate precursor mass/ elemental composition, accurate fragment ion measurement, fragment ion search(es), similarity- or homology-based searches, and MS/MS or MS^n-based identification in spectral libraries.
8. Metabolite quantification must be robust and the results supportive of a pathway and/or mechanistic understanding of the process under study.

Acknowledgements

The authors would like to thank Michael Athanas from VAST Scientific, Robert Mistrik from HighChem, Helen Jenkins and Robert Hall as partners in the META-PHOR project, for advice and useful discussions. This work was in part supported by the EU FP VI project META-PHOR (FOOD-CT-2006-036220).

References

1. Kitano, H. (2002) Systems biology: a brief overview *Science* 295, 1662–1664.
2. van der Greef, J., Stroobant, P., van der Heijden (2004) The role of analytical sciences in medical systems biology. *Curr Opin Chem Biol* **8**, 559–565.
3. Dettmer, K., Aronov, P. A., and Hammock, B. D. (2007) Mass spectrometry-based metabolomics. *Mass Spectrom Rev.* 26, 51–78.
4. Novotny, M. V., Soini, H. A., and Mechref, Y. (2008) Biochemical individuality reflected in chromatographic, electrophoretic and mass-spectrometric profiles. *J Chromatogr B Analyt Technol Biomed Life Sci.* 866, 26–47.
5. Kell, D. B., Brown, M., Davey, H. M., Dunn, W. B., Spasic, I., and Oliver, S. G. (2005) Metabolic footprinting and systems biology: the medium is the message. *Nat Rev Microbiol.* 3, 557–565.
6. Beckmann, M., Parker, D., Enot, D. P., Duval, E., and Draper, J. (2008) High-throughput, nontargeted metabolite fingerprinting using

nominal mass flow injection electrospray mass spectrometry. *Nat Protoc.* 3, 486–504.

7. van der Werf, M. J., Overkamp, K. M., Muilwijk, B., Coulier, L., and Hankemeier, T. (2007) Microbial metabolomics: toward a platform with full metabolome coverage. *Anal Biochem.* 370, 17–25.
8. Scheltema, R. A., Kamleh, A., Wildridge, D., Ebikeme, C., Watson, D. G., Barrett, M. P., Jansen, R. C., and Breitling, R. (2008) Increasing the mass accuracy of high-resolution LC-MS data using background ions: a case study on the LTQ-Orbitrap. *Proteomics* 8, 4647–4656.
9. Kothari, S., Song, Q., Xia, Y., Fico, M., Taylor, D., Amy, J. W., Stafford, G., and Cooks, R. G. (2009) Multiplexed four-channel rectilinear ion trap mass spectrometer. *Anal Chem.* 81, 1570–1579.
10. Enot, D. P., Lin, W., Beckmann, M., Parker, D., Overy, D. P., and Draper, J. (2008) Preprocessing, classification modeling and feature selection using flow injection electrospray mass spectrometry metabolite fingerprint data. *Nat Protoc.* 3, 446–470.
11. Xu, E. Y., Schaefer, W. H., and Xu, Q. (2009) Metabolomics in pharmaceutical research and development: metabolites, mechanisms and pathways. *Curr Opin Drug Discov. Devel.* 12, 40–52.
12. Spratlin, J. L., Serkova, N. J., and Eckhardt, S. G. (2009) Clinical applications of metabolomics in oncology: a review. *Clin Cancer Res.* 15, 431–440.
13. Jacobs, A. (2009) An FDA perspective on the nonclinical use of the X-Omics technologies and the safety of new drugs. *Toxicol Lett.* 186, 32–35.
14. Hall, R. D., Brouwer, I. D., and Fitzgerald, M. A. (2008) Plant metabolomics and its potential application for human nutrition. *Physiol Plant.* 132, 162–175.
15. http://www.meta-phor.eu/
16. Damoc, E., Scigelova, M., Giannakopulos, A. E., Moehring, T., Pehal, F., and Hornshaw, M. (2008) Direct analysis of red wine using ultrafast chromatography and high resolution mass spectrometry. *Thermo Scientific Application Note 30173.*
17. http://www.chemspider.com/
18. Makarov, A., Denisov, E., Lange, O., and Horning, S. (2006) Dynamic range of mass accuracy in LTQ Orbitrap hybrid mass spectrometer. *J Am Soc Mass Spectrom.* 17, 977–982.
19. Stafford, G. C., Taylor, D. M., Bradshaw, S. C., and Syka, J. E. P. (1987) Enhanced sensitivity and dynamic range on an ion trap mass spectrometer with automatic gain control. *Proc. 35th Annual Conference of the American Society for Mass Spectrometry, Denver, CO*, 775–776.
20. http://planetorbitrap.com
21. http://www.umetrics.com
22. http://www.biocyc.org

Chapter 11

Fourier Transform Ion Cyclotron Resonance Mass Spectrometry for Plant Metabolite Profiling and Metabolite Identification

J. William Allwood, David Parker, Manfred Beckmann, John Draper, and Royston Goodacre

Abstract

Mass spectrometry (MS) is usually the technique of choice for metabolomic studies where the volume of sample material is too limited for applications employing nuclear magnetic resonance (NMR) spectroscopy. With the advent of ultra-high accuracy mass spectrometers such as the Orbitrap (resolution ~ 10^5) and the Fourier Transform Ion Cyclotron Resonance (FT-ICR) analysers (resolution potentially in excess of 10^6) there is the opportunity to generate an accurate mass fingerprint (often referred to as a profile since the variables are considered as effectively discrete) of an infused sample extract. In such data representations mass "peaks" are detected in the raw data and the centroid mass intensity calculated. The resolving power and sensitivity of these ultra-high accuracy mass analysers is such that metabolite signals from molecules containing naturally abundant elemental isotopes (e.g. ^{13}C, ^{41}K, ^{15}N, ^{17}O, ^{34}S, and ^{37}Cl) are visible in the data. Such is the instruments precision that it allows for the calculation of highly accurate elemental compositions for the unknown signals, thus aiding greatly in the selection of potential metabolite candidates for the annotation of unknowns prior to their confirmation by comparisons to analytical standards. The application of FT-ICR-MS to plant metabolomics has thus far been limited to a few studies and clear step-by-step methodologies are as yet unavailable. This chapter presents a rigorous method for the extraction and FT-ICR-MS analysis of plant leaf tissues as well as downstream data processing.

Key words: FT-ICR-MS, DI, FI, ESI, CID, Plant metabolomics

Abbreviations

DI	Direct infusion
FI	Flow infusion
FT	Fourier transform
ICR	Ion cyclotron resonance
MS	Mass spectrometry

Nigel W. Hardy and Robert D. Hall (eds.), *Plant Metabolomics: Methods and Protocols*, Methods in Molecular Biology, vol. 860, DOI 10.1007/978-1-61779-594-7_11, © Springer Science+Business Media, LLC 2012

ESI	Electrospray ionisation
LTQ	Linear trap quadrupole
CID	Collision-induced dissociation
QC	Quality control
PCA	Principal components analysis
LDA	Linear discriminant analysis
RF	Random forest

1. Introduction

Fourier transform ion cyclotron resonance mass spectrometry (FT-ICR-MS) in the metabolomics field is currently regarded as an instrument of great potential due to its ultra-high mass accuracy and resolution which permits unequivocal mass assignment and the resolution of ion species currently not possible with alternative mass spectrometry (MS) analysers. The fact that a range of ionisation chemistries, including electrospray ionisation (ESI), atmospheric pressure chemical ionisation (APCI), atmospheric pressure photoionisation (APPI), matrix assisted laser desorption ionisation (MALDI), electron impact and chemical ionisation (EI and CI respectively), can be applied to FT-ICR-MS adds to its value further. In FT-ICR-MS metabolomics, mass resolutions up to several hundred thousand have routinely been achieved, although for analysis over narrow mass ranges, values of greater than one million have been reported (1). High mass accuracy is also a key FT-ICR-MS feature with average errors being less than one part per million (ppm), the limits of detection in MS mode are also comparable to those of alternative MS instrumentation. Studies of complex samples such as crude oils (2) report the resolution of more than 10,000 distinct chemical species in a single FT-ICR mass spectrum without prior chromatographic separation. To discuss all of the theory that is relevant to FT-ICR-MS and how it obtains such high performance is beyond the scope of this chapter. However, at this point we refer the reader to Barrow et al. (3) for an excellent review of the background of the technique.

In one of the first studies where FT-ICR-MS profiling was applied to plant metabolomics, Aharoni et al. (4) traced over 1,000 metabolites by direct infusion (DI) of aqueous methanol and acetone extracts of strawberry fruit collected at four different stages of fruit ripening. In comparison to DIMS analysis of complex plant extracts with conventional electrospray ionisation (ESI) and time-of-flight (ToF) instrumentation (5, 6), it is clear that DI-ESI-FT-ICR-MS is capable of resolving a much greater number and range of metabolites, although with the application of high performance or ultra-high performance liquid chromatography (HPLC and UHPLC respectively) this could potentially be enhanced further since isobaric metabolites would also be resolved. However, with the low scan speeds which are commonly employed by FT-ICR-MS (typically 1 scan/s) it must be considered that the Thermo Hybrid LTQ Orbitrap instrument is more appropriate for the prior application of UHPLC (7).

Several notable FT-ICR-MS based plant studies in recent years have emerged from Japan. Hirai et al. (8) elegantly combined transcriptomics data obtained from Affymetrix *Arabidopsis thaliana* microarrays with FT-ICR-MS metabolomics to investigate the gene-metabolite networks controlling nitrogen and sulphur metabolism. FT-ICR-MS was also applied to investigate the metabolomics behind light/dark regulation in *A. thaliana* cell cultures (9). Here, the authors developed automated data processing software (Dr. DMASS: http://kanaya.naist.jp/DrDMASS/) and an accurate mass metabolite identification system (KNApSAcK: http://Kanaya.aist-nara.ac.jp/KNApSAcK/) (9, 10).

Metabolite putative identification can be performed against KNApSAcK, PubChem (http://pubchem.ncbi.nlm.nih.gov/) or Chemspider (http://www.chemspider.com/) databases as well as many others. However, whilst some of these databases contain information on large numbers of metabolites many of these are not of natural origin. Searching for likely annotation candidates based on accurate mass information in publicly accessible databases is in itself time-consuming as individual database coverage of natural chemistry varies and so a comprehensive search requires query of several if not all relevant databases. Unfortunately, with few exceptions, databases with appropriate metabolite mass information can contain much redundancy, resources for curation are often limited (consequently it is not uncommon to find mistakes relating to mass values, molecular formulae and structure), furthermore some of the entries relate to ionic states, often from interactions with salts, which whilst useful for understanding the metabolites role in metabolism is not so relevant for MS. A further consideration is that publically accessible databases for the most part only contain information regarding the uncharged metabolite, thus to search masses of interest all possible ionisation states must be calculated by the user (taking into account the mass of electrons) and ensuring that the elemental masses used are accurate.

Work carried out at Aberystwyth University has lead to the development of FT-ICR-MS data processing methods as well as two further new resources for the analysis of ESI-MS based data. First, a data analysis package FIEmspro (http://users.aber.ac.uk/jhd) written in the R environment and requiring a moderate knowledge of R command-line usage (11) can be used for all data analysis (i.e. visualisation, normalisation, feature ranking) and thus can be used to narrow down the number of signals of interest for further interpretation. Second, a database constructed using metabolite "structures" harvested from publicly accessible databases and converted into a common format to generate a comprehensive archive known as MZedDB (http://maltese.dbs.aber.ac.uk:8888/hrmet/search/addsearch0.php) (12). This database is based on an archive in a common format of all metabolite "structures" derived from several widely used and publicly accessible databases. Using "rules" derived from structural information and physical properties (such as number of H-bond acceptors/donors, number of $OH/COOH/NH_2$ groups, number of acidic H or basic O^- in molecule) MZedDB generates a list of potential adducts and neutral loss fragments that are likely to be observed for each structure and calculates on the fly the accurate mass of every potential ionisation product which provides targets for searches based on accurate mass. Starting with a list of m/z signals MZedDB supports a range of manual or semi-automated (via R environment) annotation strategies based on either m/z mass or predicted elemental composition at a range of mass resolutions (12). The database can also be used to generate elemental compositions for m/z signals using the seven golden rules (13) (see Table 1), this process can be done manually on the Web interface or in batch mode via R (code available at http://maltese.dbs.aber.ac.uk:8888/hrmet/supp/rhrmet.html) (12).

A range of feature ranking methods can be used to identify m/z signals that significantly alter ($p \leq 0.001$) between sample classes (11, 14). Multivariate analyses such as principal components analysis (PCA) (15) or linear discriminant analysis (LDA) can be used in an exploratory manner to assess the shape of the model as well as to determine errors such as mislabelled or out-lying samples (16, 17). Univariate methods such as N-Way ANOVA and non-parametric Kruskal–Wallis tests, as well as Random Forest (RF) decision trees (11, 14, 18, 19), and evolutionary computation (20, 21) are used to identify significant differential metabolite variables which are of the most interest with regard to the biological system/question under study. This can be followed by correlation analyses such as the Pearson Coefficient to indicate related signals from the sample matrix (adducts, isotopes, neutral losses) (22). The selected metabolites can then undergo a simple calculation of the accurate mass differences between individual pairs of correlated signals indicating their likely relationships. This then allows any annotation suggestions to focus on, potentially, the correct ionisation products (12, 22). For example, a correlated signal with a mass difference of 1.0033 indicates the difference between ^{12}C and ^{13}C containing signals.

As several overlapping solutions predicting the presence of different metabolites can be possible, the most likely ions putatively identifying a specific metabolite can be confirmed by MS^n experiments (1, 3, 14, 23) to confirm if the tentative (database matched) identification is correct by comparison of the fragmentation patterns of the sample analytes and those of pure analytical

Table 1
The seven golden heuristic and chemical rules for the selection of accurate and correct elemental compositions

Rule	Description
Rule 1: Restriction of elements	Natural compounds contain restricted numbers of each element, thus by dividing the mass range by the element mass allows sensible ranges of atoms for that specific element to be predicted, i.e. C has a mass of 12 Da, data were collected over the mass range 1–1,000 Da 1,000/12 = 83, 83 C atoms is the maximum expected for a mass of 1,000 Da Heuristic filtering based upon information for the numbers of atom present for each element within compounds that are found commonly in the PubChem, Wiley, NIST02, and DNP databases, was used to reduce the predicted numbers of atoms further. Based upon database information, maximum element ratios can also be applied to heuristic filtering, e.g. for 47 C the maximum H is 150.
Rule 2: Lewis and Senior check	The LEWIS rule in simple terms demands that a molecule consisting of simple elements (C, H, N, O, especially) share electrons so that the *s*, *p*-valence shells are filled completely, i.e. the "octet" rule. However, the LEWIS rule excludes all nitroso compounds and so is combined with the, SENIOR rule that requires three essential conditions for the existence of an elemental composition: (a) The sum of valences or total number of odd number atoms is equal (b) The sum of valences is even to or greater than double the maximum valence (c) The sum of valences is even to or greater than double the total atom number − 1
Rule 3: Isotopic pattern filter	Natural compounds comprise monoisotopic and isotopic masses according to the natural average abundance of stable isotope abundances for each element. For MS instruments with low relative errors of 2–5% RSD and assuming high-quality data with a good signal-to-noise ratio and accurate detection of the M + 1 and M + 2 isotope ions, inclusion of the calculation for isotope ratio abundance permits the removal of the majority of incorrectly assigned elemental compositions. Of all seven golden rules this is the most important for the removal of incorrect elemental compositions.
Rule 4: H/C element ratio check	By including element ratio constraints to the heuristic filtering (especially for H/C), the calculated elemental compositions are further restricted to the most probable candidates. For most natural molecules, the H/C ratio is rarely greater than 3 or less than 0.125 and by applying a filtering range of 0.2 > 3.1 the majority of drug and natural compounds can be filtered for. However, in extreme cases such as fluorines, when the experimenter expects to find such compounds, the range needs to be extended for them to be fully accounted for.
Rule 5: Heteroatom ratio check	Many formulas, alkanes for one example, comprise no heteroatom. Cases of high ratios of heteroatom to carbon number are extremely rare, thus a simple exclusion of very high heteroatom ratio elemental compositions helps to further remove unlikely candidates.

(continued)

Table 1 (continued)

Rule	Description
Rule 6: Element probability check	Based upon the NIST02, Wiley, and DNP database searches and element combinations of N, O, P and N, O, S, with C and H, a high number of entries are found which have high element ratios. From this information specific thresholds for the numbers of atoms for each element can be accordingly defined.
Rule 7: TMS check	TMS derivatisation is commonly performed in GC-MS analyses in order to enhance volatility and permit the detection of otherwise undetectable compounds. To calculate elemental compositions of neutral masses, the replacement of acidic H^+ with TMS groups must be accounted for in order to calculate the non-derivatised molecules mass. The number of TMS groups is easily deduced via the calculation of isotopic abundances. The TMS check also mandates that for each Silicon there has to be three methyl groups.

Kind and Fiehn (13) developed an algorithm based upon seven heuristic and chemical rule-based filters for the accurate selection of the correct elemental formula from the hundreds that may be generated for any one given accurate mass. For liquid chromatography (LC) data, adducts must first be identified and removed, thus giving a list of neutral ions alone. Likewise, for gas chromatography (GC) data, products of derivatisation must be identified and the original neutral ion calculated. Elemental compositions are then generated for the accurate masses of each neutral ion. The algorithm performs at its best providing that the elemental compositions are based upon high resolution and mass accuracy data from instruments such as FT-ICR-MS and Thermo hybrid LTQ Orbitrap system (i.e. within 3 ppm mass accuracy and resolution of 100,000 >) for molecules which are purely resolved with either liquid chromatography, gas chromatography, or capillary electrophoresis. The seven golden rules are explained in the following table, when applied to the elemental compositions generated for 6,000 database entries, the seven golden rule algorithm selected the correct elemental composition as the top hit with an 80–99% probability rate. Adapted from ref. 13
Abbreviations: *DNP* Dictionary of Natural Products, *NIST02* National Institute of Standards and Technology 2002 MS library, *TMS* Trimethylsilyl

standards. Since FT-ICR-MS is a "trapping" instrument, multiple stages of analysis (MS^n) can be undertaken, such as fragmentation of an ion selected from a mixture, followed by further fragmentation of the product daughter fragment ions (as required for unambiguous confirmation of metabolite identification along with accurate mass measurement of the parent ion, and comparison to an analytical standard). The most commonly employed form of MS^n uses collision-induced dissociation (CID), although alternative methods that could potentially be applied to metabolite analysis include infrared multiphoton dissociation (IRMPD) and electron capture dissociation (ECD), although the latter is more commonly employed for the analysis of positively charged peptides (1, 3) since the target analyte must be doubly or higher charged. The method presented within this chapter will employ CID (24) as this is the most commonly used.

2. Materials

2.1. Harvest of Plant Material

1. Clean stainless steel scissors (sharp), forceps, and spatulas of appropriate size for sample material (see Note 1).
2. Liquid nitrogen, a 1–2 L Dilvac (Day-Impex, Colchester, Essex, UK) and long-arm forceps to retrieve tubes from the liquid nitrogen (see Note 2).
3. Pre-labelled (alcohol resistant marker pen) high-quality 2-mL polypropylene microcentrifuge tubes and/or 15- or 50-mL polypropylene falcon tubes (see Notes 3 and 4) (Greiner Bio One, Stonehouse, Gloucestershire, UK) depending upon volume of sample material.
4. Stainless steel 5-mm ball bearings (Retsch, Hunslet, Leeds, UK) cleaned in methanol and air-dried three times, placed in pre-labelled 2-mL microcentrifuge tubes (see Notes 3 and 4) (Greiner Bio One, UK).
5. Denver Instrument Balance—Summit SI-234 (Denver, Colorado, USA), or similar.
6. Appropriate freezer boxes suitable for long-term –80°C storage of samples.

2.2. Extraction for the Capture of Polar Metabolites and Chloroform Purification of Non-polar Metabolites

1. Ice and insulated ice box (see Note 5).
2. Liquid nitrogen, a 1–2 L Dilvac (Day-Impex, UK) and long-arm forceps (see Note 2).
3. Retsch MM200 ball mill and two 5 or 10 position microcentrifuge tube adapters (Retsch, UK).
4. Eppendorf Concentrator 5301 at 30°C and setting 1 (Eppendorf UK Ltd., Histon Cambridge, UK).
5. Pre-labelled (alcohol resistant marker pen) high-quality 2-mL polypropylene microcentrifuge tubes (Greiner Bio One, UK), two sets should be prepared for storage of the final extracts and one set for preparation of the extracts (see Note 3).
6. High-quality methanol (trace analysis grade), water (ultrapure), and chloroform (HPLC grade or better) (Mallinckrodt-J.T. Baker, Leadenhall Street, London, UK).
7. Prepare a mixture of 100 mL chloroform–250 mL methanol–100 mL water using a solvent washed (see Note 1) measuring cylinder and storage bottle fitted with a PTFE lined lid. Prepare and store at –20°C for 24 h minimum prior to extraction (see Note 6).
8. High-quality P1000 and P200 polypropylene pipette tips (Greiner Bio One, UK) (see Note 3).
9. Appropriate freezer boxes suitable for long-term, –80°C storage of samples.

2.3. Preparation of Metabolite Standards

1. High-quality methanol or isopropanol (trace analysis grade) and water (ultra-pure) (Mallinckrodt-J.T. Baker, UK).
2. Prepare 70% Aqueous methanol and 50% Aqueous isopropanol (see Note 1).
3. Substance P (Thermo Corp., DE) for calibration (see Note 7).
4. Acetaminophen ($[M+H]^+$ 152.0712, $[M-H]^-$ 150.0555) (Sigma-Aldrich Ltd., Gillingham, Dorset, UK).
5. Caffeine ($[M+H]^+$ 195.0882, $[M-H]^-$ 193.0726) (Sigma-Aldrich Ltd., UK).
6. Sulfaguanidine ($[M+H]^+$ 215.0603, $[M-H]^-$ 213.0446) (Sigma-Aldrich Ltd., UK).
7. Sulfamethoxine ($[M+H]^+$ 311.0814, $[M-H]^-$ 309.0658) (Sigma-Aldrich Ltd., UK).
8. Valine-Tyrosine-Valine ($[M+H]^+$ 380.2125, $[M-H]^-$ 378.2029) (Sigma-Aldrich Ltd., UK).
9. Terfenadine ($[M+H]^+$ 472.3216, $[M-H]^-$ 470.3059) (Sigma-Aldrich Ltd., UK).
10. Reserpine ($[M+H]^+$ 609.2812, $[M-H]^-$ 607.2656) (Sigma-Aldrich Ltd., UK) (see Note 7).
11. Erythromycine ($[M+H]^+$ 734.4691, $[M-H]^-$ 732.4534) (Sigma-Aldrich Ltd., UK).

2.4. Preparation of Samples and Addition of Internal Standards

1. 70% Aqueous methanol and 50% Aqueous isopropanol prepared from trace analysis grade solvents and ultra-pure water (Mallinckrodt-J.T. Baker, UK) (see Note 1).
2. Minisart RC4 single-use syringe filter non-sterile, regenerated cellulose membrane, polypropylene housing, pore size 0.20 μm (Sartorius, Goettingen, DE) (see Note 3).
3. Disposable polypropylene 1 mL syringes (Becton Dickinson, Oxford, UK) (see Note 3).
4. High quality P1000 and P200 polypropylene pipette tips (Greiner Bio One, UK) (see Note 3).

2.5. Instrumentation and Analysis

1. Standard multi-well plate for nanospray ESI or borosilicate glass mass spectrometry vials appropriate for the autosampler being employed for standard ESI.
2. TriVersa™-NanoMate chip technology (Advion Biosystems, NY, USA) coupled to an LTQ-FT™ ICR mass spectrometer (Thermo Corp, DE) (see Fig. 1).
3. Thermo Xcalibur version 2.0 (Thermo Corp, DE).

2.6. Data Processing and Statistical Analysis

1. MatLab R2008a (The Mathworks Inc., Natick, MA, USA).
2. R environment using the FIEMSpro metabolomics data analysis package (11, 23, 25) Web accessible (http://users.aber.ac.uk/jhd).

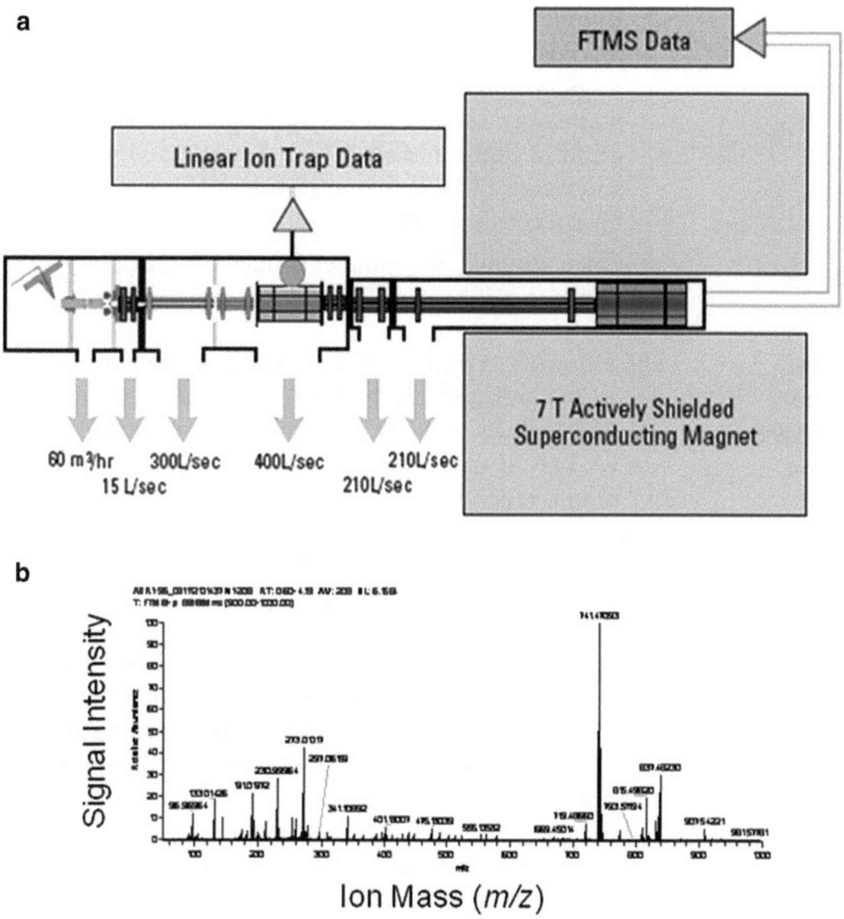

Fig. 1. FT-ICR-MS schematic and example FT-ICR-MS profile. (**a**) Diagram of the Thermo LTQ-FT-MS system (Reproduced with thanks to Thermo Fisher Scientific). (**b**) An example Nano-infusion FT-ICR-MS fingerprint of a polar extract taken from *Brachypodium distachyon* leaf tissue. The sample preparation and mass spectral acquisition was performed as presented in the methods within this chapter.

3. Determining the mathematical relationships between *m/z* and automated database searches are performed in R using the code described at http://maltese.dbs.aber.ac.uk:8888/hrmet/supp/rhrmet.html (12).

3. Methods

3.1. Harvest of Plant Material

1. Plant material should be rapidly excised using clean sharp scissors whilst maintaining that there are no soil particles coating the material and that contact is not made between the plant material and laboratory gloves (see Note 8).

2. Rapidly transfer the material (100 mg ±2 mg) with clean forceps into 2-mL microcentrifuge tubes each containing a single 5-mm stainless steel ball bearing (cleaned three times in methanol). Alternatively, for larger sample material directly grind in liquid nitrogen with a pestle and pre-cooled mortar and weigh the still-frozen powder (100 mg ±2 mg) into 2-mL microcentrifuge tubes.
3. Once weighed, the samples should again be plunged into liquid nitrogen prior to −80°C storage until extraction.

3.2. Extraction for the Capture of Polar Metabolites and Chloroform Purification of Non-polar Metabolites

The following extraction procedure was originally devised by Fiehn et al. (26) and updated by Lisec et al. (27). It was designed for GC-MS analyses and has been successfully applied to each of the META-PHOR target species of melon, broccoli, and rice (28, 29) but in our experience is equally applicable to direct infusion mass spectrometry with ESI for the analysis of polar (5) and non-polar metabolites (6) from the leaf material of *Arabidopsis thaliana* and *Brachypodium distachyon*. It is important to be well organised in advance of starting the procedure and to work quickly and precisely throughout using 1,000 and 200 μL pipettes (see Note 8). It must be taken into consideration that analysis of a single sample provides only a single metabolic snapshot without further information on biological variation or analytical errors. To estimate such variations, sufficient biological replicates and sufficient technical replicates must be prepared and analysed. If excess material is available then excess samples should also be prepared to allow for optimisation of reconstitution solvents and their final volume, as well as instrument conditions and to assess analytical and technical errors.

1. Samples should be removed from −80°C storage and flash frozen in liquid N_2, non-ground samples are homogenised using a Retsch MM200 ball mill set on a frequency of 30 Hz for 1 min, and placed on ice.
2. To each sample 1 mL of −20°C extraction solvent, chloroform–methanol–water (1:2.5:1), is added and the sample placed back on ice.
3. The samples are then mixed on a vortex and vigorously shaken in a cold room at 3°C for 15 min and returned back onto ice.
4. The samples are then centrifuged at 3°C and 14,500 × *g* for 3 min with a microcentrifuge, after which the supernatants are decanted to clean labelled 15-mL falcon tubes and kept on ice.
5. Repeat steps 2–4 on the same sample pellet, thus extracting each sample twice.
6. To 2 mL of the clean combined sample supernatants, 1 mL of ultra-pure water is added and the samples are then mixed with a vortex and centrifuged at 3°C and 14,500 × *g* for 3 min with a desktop centrifuge to aid solvent phase separation.

7. The polar phase is recovered (carefully avoiding the interphase) as 250–500 μL aliquots (depending upon sample concentration) into clean labelled 2-mL microcentrifuge tubes, approximately 200 μL of non-polar phase can also be recovered to a clean labelled 2-mL microcentrifuge tube.
8. The polar and non-polar samples are then dried via speed vacuum concentration in an Eppendorf Concentrator 5301, on setting 1, for 8 h and stored at −80°C prior to analysis. Alternatively, if the samples are for immediate analysis the extract can be directly injected into the mass spectrometry system (see Note 9).

3.3. Preparation of Metabolite Standards

For tuning the FT-ICR-MS across a suitable mass range for the analysis of polar phase plant extracts, a cocktail of analytical standards containing a final concentration of 100 μM of each standard (all of a minimum 99% purity) should be prepared. Standards should be weighed precisely on an accurate balance, when possible standards should be dissolved and diluted in 70% [v/v] aqueous methanol or 50% [v/v] aqueous isopropanol; on occasion standards may first require a pure non aqueous solvent to dissolve completely prior to dilution with aqueous solvents. Just prior to FT-ICR-MS tuning, further dilute the cocktail 1:10 with 70% [v/v] aqueous methanol or 50% [v/v] aqueous isopropanol (depending on the initial dissolvent).

The range of standards used should be appropriate to the mass range of metabolites present within the sample. The standards should also be of relevance to the plant biology of interest, i.e. if you wish to study glucosinolates then also use relevant glucosinolate standards within the calibration cocktails. This is of importance since, due to ESI suppression effects, pure compounds or compounds present in simple mixtures may respond differently to ESI than when present in a complex matrix such as a plant extract.

3.4. Preparation of Samples and Addition of Internal Standards

1. In our experience, lyophilised polar and non-polar samples are best reconstituted in 200 μL methanol (trace analysis grade)–water (ultra-pure) (70:30, [v/v]) for ESI applications.
2. Prior to analysis, reconstituted samples are sonicated for 15 min and either centrifuged at 0°C for 4 min at 14,000 × *g* (12) or may be filtered using Minisart RC4 syringe filters.
3. Prepare also an extraction blank in a clean 2-mL microcentrifuge tube which is also subjected to the above centrifugation or filtration steps. This sample permits the removal of mass signals which originate from plasticides within the pipette tips, microcentrifuge tubes, syringe, and filters (see Note 3).
4. The samples are then randomised and directly transferred into borosilicate glass mass spectrometry vials (200 μL) or multi-well plates (20 μL) (see Note 10) suitable for the auto-sampler

being employed. The remaining sample is stored in a liquid state at −80°C, and for long-term storage the vials are topped off with argon or nitrogen.

5. Prior to analysis, prepare also a representative sample pool containing an equal volume of every biological sample (~200 μL total volume) to serve as a quality control (QC). Aliquot 90 μL of QC into a clean 2-mL microcentrifuge tube and add 10 μL of the 100 μM calibration cocktail, this will provide an assessment of how the sample matrix effects the analytical standards FT-ICR-MS detection when compared to the cocktail of analytical standards in solution. The QC also provides a data quality check for the true experimental samples. The QC sample should be included after every tenth biological sample within the analytical run sequence.

3.5. Instrument Set up, Tuning and Calibration for FT-ICR-MS Sample Profiling and MS^n

For reasons of clarity, the described protocol focuses on the use of a single instrument, the Thermo-Finnigan LTQ fitted with a 7-Telsa FT-ICR mass analyser (Thermo-Finnigan, DE; Fig. 1), for sample profiling. If required, multiple MS/MS (MS^n) experiments are possible to follow up secondary ionisations of either the most abundant or predefined mass ions. Generally speaking, increases in mass resolution are concomitant with a proportional increase in data dimensionality, which in turn effects experimental design with regards to the numbers of replicates required to achieve statistical robustness (11, 30). A workflow from FT-ICR-MS analysis through to data processing, statistics, and metabolite assignments is available for reference (see Fig. 2).

1. Before starting the analytical run sequence, ensure that the LTQ instrument is fully operational according to the manufacturer's recommended instrumental conditions and performance (see Note 11). Also using a single representative sample (the QC being ideal), you must check that its concentration is optimal for FI-ESI-FT-ICR-MS analysis.
2. Place the extracted samples as described above into the autosampler (see Note 12). The tray holder is maintained at 5°C (31). An equivalent method for standard flow ESI is described by Beckmann et al. (25).
3. Typical nanospray conditions comprise 200 nL/min flow rate, 0.5 psi back pressure, and +1.6 kV (positive ion data) or −1.6 kV (negative ion data) electrospray voltage, controlled by Chipsoft software (Advion Biosystems, USA). Prior to starting a run sequence of polar plant extracts maintain that the nanospray is stable for at least 3 min. FT-ICR-MS parameters include an automatic control gain setting of 1×10^5 and a mass resolution of 100,000 (defined at m/z 400). Data is recorded for 5 min per replicate infusion using the Xcalibur software (Thermo Corp., DE) (12, 25).

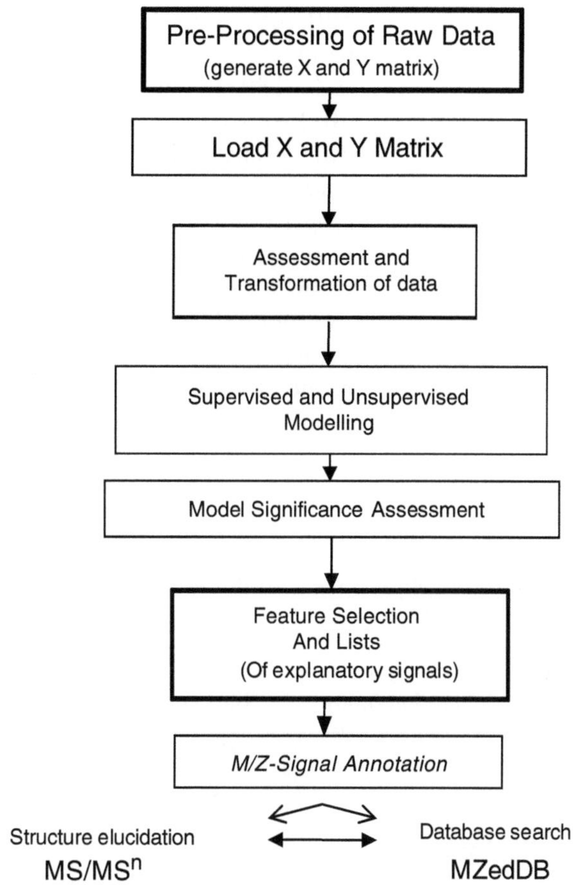

Fig. 2. Overall workflow for metabolic profiling using FT-ICR-MS. Overview of the major components of data analysis starting with raw-data conversion and first-pass data analysis, followed by data mining and finally annotation and database searches. Adapted from ref. 11.

4. To alleviate the loss of low mass ions when analysing wide mass ranges, ions are transferred from the linear ion trap to the ICR detector for full scans by segmenting the total *m/z* spectral range into an optimal number of smaller windows. This helps to minimise the loss of low mass ions due to time-of-flight effects. Mass resolution is fixed at 100,000 (defined for an ion at *m/z* 400) throughout. Automatic gain control (AGC) is set to correspond to the number of charges transferred from the front-stage ion trap to the ICR detector cell.
5. The data acquisition method set to acquire data in positive or negative ionisation mode is as follows: run time 5 min; one segment was used; number of scan events 17 (0.25 min per event), scan rate "normal" (1 scan/s), scan type "full," and data type centroid. The SIM window scan events are set as follows: scan event 1: positive polarity, mass range from *m/z* 50 to *m/z*

120; scan event 2: positive polarity, mass range from m/z 100 to m/z 200; scan event 3: positive polarity, mass range from m/z 180 to m/z 280; and so on until the mass range 50–1,400 m/z is covered (see Table 2). The number of events can be customised to meet the objective m/z range of the study. Each scan event is 0.25 min with the first scan event longer to incorporate a 0.75 min delay to allow the system time to normalise. The scan event acquisition time can be increased to allow acquisition of more scans per SIM window. For negative mode the same method is used only changing the polarity. Prior to any statistical analysis the data is log transformed to reduce the chance of high-intensity peaks dominating in the multivariate data analyses.

Table 2
FT-ICR-MS SIM window data acquisition method for polar plant leaf extracts

SIM window segment	Duration (min)	Start scan *(m/z)*	End scan *(m/z)*	Acquisition time (min)
1	1	50	120	1
2	0.25	100	200	1.25
3	0.25	180	280	1.5
4	0.25	260	360	1.75
5	0.25	340	440	2
6	0.25	420	520	2.25
7	0.25	500	600	2.5
8	0.25	580	680	2.75
9	0.25	660	760	3
10	0.25	740	840	3.25
11	0.25	820	920	3.5
12	0.25	900	1,000	3.75
13	0.25	980	1,080	4
14	0.25	1,060	1,160	4.25
15	0.25	1,140	1,240	4.5
16	0.25	1,220	1,320	4.75
17	0.25	1,300	1,400	5

In order for FT-ICR-MS to maintain high mass accuracy across a large mass range, especially with regard to metabolites of low m/z, SIM window methodologies are employed. The table presents clearly the recommended SIM window methodology for the analysis of polar extracts of plant leaf material

6. Run blank samples comprising extraction solvents, calibration cocktail, and QC samples (mix of all plant samples), interspersed at random into the run sequence, to monitor instrument performance and detect system peaks. The mass spectral response of the analytical standards within the calibration cocktail should also be compared to their response within the complex QC sample matrix to check for variation in the reported molecular weight (analyte m/z value) as well as to monitor ESI suppression effects and differential ionisation efficiencies.
7. Accurate mass measurements are performed in the FT-ICR-MS by the collection of 30 mass spectra and averaging the masses acquired over these scans. An initial scan window of 70 Da (50–120 m/z) is acquired and followed by scan windows of 100 Da with a 20-Da overlap (front and back) between windows across the mass range 100–1,400 m/z (see Table 2).
8. MS*n* data can be recorded throughout the profiling analysis for either the most abundant or predefined m/z targets, or alternatively, target m/z's selected by multivariate analysis of the profiling data can be analysed at a later stage. MS*n* is achieved via first isolating the target m/z and applying CID within the LTQ; collect 30 mass spectral scans and sum the data for each target analyte (25).
9. For XY-matrix generation, subsequent data mining and MZedDB searches, infusion data acquired in profile mode are obtained as processed mass spectra with associated peak lists (Xcalibur, Thermo Corp., DE) and exported as exact mass text files (see Note 13). Accurate mass alignment of all mass spectra, "peak-picking", integration, and centroiding of mass signals is performed in Matlab (11, 12, 25). Another data acquisition and XY-matrix generation strategy providing maximum m/z accuracy uses the stitching of transient files (i.e. scans recorded in the time domain), customised mass calibration using known m/z-ions for each SIM window and XY matrix processing in custom-written MATLAB software (31–33).

3.6. Data Analysis

1. Data within each biological XY matrix class are aligned and any peaks not represented in 70% of class replicates should be removed from the matrix.
2. Carry out all statistical tests in the R environment using the FIEMSpro metabolomics data analysis package (11) which is Web accessible (http://users.aber.ac.uk/jhd).
3. Perform explanatory feature selection using RF decision trees (11, 34, 35).
4. Perform signal correlation analysis by the Pearson correlation method on the explanatory m/z obtained by the feature selection methods such as RF, ANOVA, and non-parametric Kruskal–Wallis (11, 22).

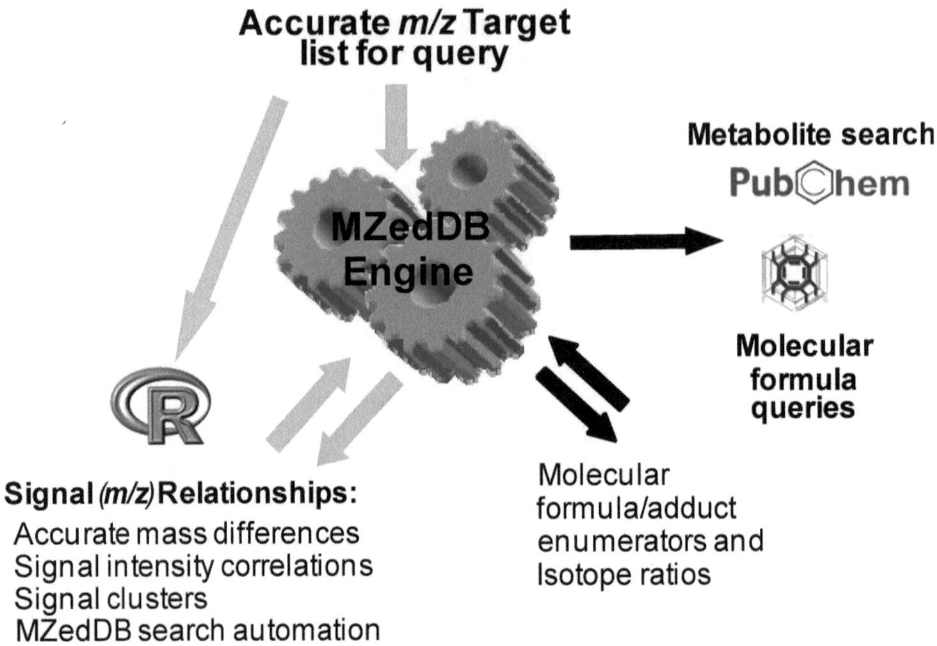

Fig. 3. MZedDB Web-resource workflow. MZedDB architecture for accurate *m/z* searches. *Grey arrows* represent MZedDB's general functionalities and *black arrows* indicate some common query pathways. Adapted from ref. 12.

5. Employ hierarchical cluster analysis based on the correlation coefficient to identify the set of clusters, which satisfy some setting, for example, signal correlation coefficient larger than 0.75.
6. Determine the mathematical relationships between *m/z* in R (see Fig. 3) using the code described at http://maltese.dbs.aber.ac.uk:8888/hrmet/supp/askMZedDBworkflow.r (12). This code searches for operator predetermined mass differences between measured accurate masses at an adjustable sensitivity (see Note 14).

4. Notes

1. Glassware such as bottles for the storage of extraction solvents and measuring cylinders for their preparation, as well as the scissors, forceps, and spatulas used to prepare and weigh samples and standards must be very clean. In our experience, washing and repeat washing clean glassware and metal ware with polar solvents such as methanol, ethanol, propan-2-ol, and acetonitrile, as well as non-polar solvents such as chloroform, prior to rinsing several times with HPLC grade water, oven drying, and capping with kitchen foil, helps to prevent sample contamination.

2. Liquid nitrogen requires careful handling. Please refer to your organisation's guidelines on safety for its use.
3. It is best to use high-quality polypropylene plastic ware (microcentrifuge and falcon tubes as well as pipette tips, syringes, and syringe filters) from a reputable supplier (e.g. Eppendorf, Greiner or Sarstedt), since this helps to reduce the range of plasticides that are introduced to the sample extracts. Alternatively, disposable borosilicate glass tubes may be used for extraction and borosilicate glass MS vials for storage and concentration of extracts. The mass signals of plasticides frequently mask the signals from metabolites of interest. As recommended in the methods an extraction blank should be prepared identically to the plant material samples, this sample can be used to account for mass signals introduced through sample preparation.
4. Do not label the lids of the 2-mL microcentrifuge tubes used for milling plant material since the lids can crack and transfer ink into the sample. If a lid does crack, provided that no plastic enters the sample material, remove the lid with scissors and replace with a lid removed from a clean microcentrifuge tube.
5. Perspex ice boxes are the best, but polystyrene can be used.
6. This solution can be stored at −20°C and used for up to 1 month after preparation.
7. Calibration is undertaken following procedures set out in the Thermo FT-MS handbook. However, please note that Reserpine and substance P do not calibrate into a low enough mass range for metabolite applications. To calibrate into a suitable low mass range, researchers should select appropriate low molecular weight standards, for a list of recommended compounds for online calibration of ESI-based instrumentation refer to Subheading 2.3. It must also be considered that deuterated internal standards could be added to samples or alternatively known metabolites within the samples can be used as lock-mass for off-line calibration (32, 33).
8. It is important to work on a sample-by-sample basis as rapidly and precisely as possible. When harvesting plant material and undertaking sample extraction, it is best to be well organised and to work quickly but precisely. The scissors, forceps, and spatulas used in the sample harvest procedure must be rinsed in HPLC grade water and dried between the collections of each sample. It is easy to underestimate the importance of this, but technical variance is frequently seen as being greater than the analytical variance of instruments such as FT-ICR-MS.
9. Polar samples may be too dilute to be amenable to the detection of minor metabolites of low concentration and may therefore require some form of concentration prior to injection.
10. For reduced contamination from multi-well plates buy pre-washed plates.

11. Re-calibrate the system if not performed within the previous 4 days before analysis.
12. The described method utilises an Advion Nanomate chip-based direct infusion nanospray ionisation system to introduce the sample. Chip nozzles block very easily, so ensure that samples are filtered or spun down, and are free of precipitates.
13. Check sample file size as an indicator that the sample ran correctly (i.e. spray current was stable throughout the run), e.g. if three replicate injections have file sizes of around 800 Mb and one has a file size of around 700 Mb, then something is not right with the fourth sample. Additionally, check occurrence of different levels of total ion count (TIC) for all SIM windows in the Xcalibur chromatogram view: a failed analysis shows TIC of near zero in combination with missing *m/z* signals in spectrum view especially at longer infusion times and should be removed for XY-matrix generation.
14. In theory, any mass difference can be searched for providing the operator knows the exact expected mass difference between the measured masses (see Fig. 3). This process is important in indicating possible isotope signals present in the matrix for which a prediction would not be wanted, and as an indication of ionisation products within the matrix.

Acknowledgements

JWA and RG would like to acknowledge the EU Frame work VI initiative for research funding and support as part of the plant metabolomics project META-PHOR (FOOD-CT-2006-036220). RG is also grateful to the UK BBSRC for financial support of the MCISB (Manchester Centre for Integrative Systems Biology). DP, JD, and MB would like to acknowledge research support received from Aberystwyth University and UK BBSRC grant BB/D006953/1; MB is further supported by a Research Councils UK Fellowship.

References

1. Brown, S.C., Kruppa, G., Dasseux, J.-L. (2005) Metabolomics applications of FT-ICR mass spectrometry. *Mass Spec. Rev.* **24**, 223–231.
2. Hughey, C.A., Rodgers, R.P., Marshall, A.G. (2002) Resolution of 11,000 compositionally distinct components in a single electrospray ionization Fourier transform ion cyclotron resonance mass spectrum of crude oil. *Anal. Chem.* **74**, 4145–4149.
3. Barrow, M.P., Burkitt, W.I., Derrick, P.J. (2005) Principles of Fourier transform ion cyclotron mass spectrometry and its application in structural biology. *The Analyst* **130**, 18–28.
4. Aharoni, A., De Vos, C.H.R., Verhoeven, H.A., Maliepaard, C.A., Kruppa, G., Bino, R., Goodenowe, D.B. (2002) Nontargeted Metabolome Analysis by Use of Fourier Transform Ion Cyclotron Mass Spectrometry. *Omics* **6**, 217–234.
5. Parker, D., Beckmann, M., Enot, D.P., Overy, D.P., Caracuel Rios, Z., Gilbert, M., Talbot, N., Draper, D. (2008) Rice blast infection of

Brachypodium distachyon as a model system to study dynamic host pathogen interactions. *Naure. Prot.* **3**, 435–445.

6. Allwood, J.W., Ellis, D.I., Heald, J.K., Goodacre, R., Mur, L.A.J. (2006) Metabolomic approaches reveal that phosphatidic and phosphatidyl glycerol phospholipids are major discriminatory non-polar metabolites in responses by *Brachypodium distachyon* to challenge by *Magnaporthe grisea*. *The Plant J* **46**, 351–368.
7. Koulman, A., Woffendin, G., Narayana, V.K., Welchman, H., Crone, C., Volmer, D.A. (2009) High-resolution extracted ion chromatography, a new tool for metabolomics and lipidomics using a second-generation orbitrap mass spectrometer. *Rapid Communications in Mass Spectr.* **23**, 1411 - 1418.
8. Hirai, M.Y., Yano, M., Goodenowe, D.B., Kanaya, S., Kimura, T., Awazuhara, M., Arita, M., Fujiwara, T., Saito, K. (2004) Integration of transcriptomics and metabolomics for understanding of global responses to nutritional stresses in *Arabidopsis thaliana*. *PNAS USA* **101**, 10205–10210.
9. Nakamura, Y., Kimura, A., Saga, H., Oikawa, A., Shinbo, Y., Kai, K., Sakurai, N., Suzuki, H., Kitayama, M., Shibata, D., Kanaya, S., Ohta, D. (2007) Differential metabolomics unravelling light/dark regulation of metabolic activities in *Arabidopsis* cell cultures. *Planta* **227**, 57–66.
10. Ohta, D., Shibata, D., Kanaya, S. (2007) Metabolic profiling using Fourier-transform ion-cyclotron-resonance mass spectrometry. *Anal. Bioanal. Chem.* **389**, 1469–1475.
11. Enot, D.P., Lin, W., Beckmann, M., Parker, D., Overy, D.P., Draper, J. (2008) Preprocessing, classification modelling and feature selection using flow injection electrospray mass spectrometry metabolite fingerprint data. *Nature Prot.* **3**, 446–470.
12. Draper, J., Enot, D.P., Parker, D., Beckmann, M., Snowdon, S., Lin, W., Zubair, H. (2009) Metabolite signal identification in accurate mass metabolomics data with MZedDB, an interactive m/z annotation tool utilising predicted ionisation behaviour 'rules'. *BMC Bioinformatics* **10**, 227.
13. Kind, T. and Fiehn, O. (2007) Seven Golden Rules for heuristic filtering of molecular formulas obtained by accurate mass spectrometry. *BMC Bioinformatics* **8**, 105.
14. Enot, D.P., Beckmann, M., Draper, J. (2007) Detecting a difference – assessing generalisability when modelling metabolome fingerprint data in longer term studies of genetically modified plants. *Metabolomics* **3**, 335–347.
15. Jolliffe (1986) *Principle Components Analysis*. Springer-Verlag, New York.
16. Goodacre, R. (2007) Metabolomics of a superorganism. *J. Nutrition* **137**, 259 S–266 S.
17. Goodacre, R. Vaidyanathan, S., Dunn, W.B., Harrigan, G.G., Kell, D.B. (2004) Metabolomics by numbers – acquiring and understanding global metabolite data. *Trends Biotech.* **22**, 245–252.
18. Enot, D.P., Beckmann, M., Overy, D., Draper, J. (2006) Predicting interpretability of metabolome models based on behavior, putative identity, and biological relevance of explanatory signals. *PNAS USA* **103,** 14865–14870.
19. Enot, D.P. and Draper, J. (2007) Statistical measures for validating plant genotype similarity assessments following multivariate analysis of metabolome fingerprint data. *Metabolomics* **3**, 349–355.
20. Goodacre, R., York, E.V., Heald, J.K., Scott, I.M. (2003) Chemometric discrimination of unfractionated plant extracts profiled by flow-injection electrospray mass spectrometry. *Phytochem.* **62**, 859–863.
21. Johnson, H.E., Broadhurst, D., Goodacre, R., Smith, A.R. (2003) Metabolic fingerprinting in salt-stressed tomatoes. *Phytochem.* **62**, 919–928.
22. Brown, M., Dunn, W.B., Dobson, P., Patel, Y., Winder, C.L., Francis-McIntyre, S., Begley, P., Carroll, K., Broadhurst, D., Tseng, A., Swainston, N., Spasic, I., Goodacre, R., Kell, D.B. (2009) Mass spectrometry tools and metabolite-specific databases for molecular identification in metabolomics. *The Analyst* **134**, 1322–1332.
23. Overy, D.P., Enot, D.P., Tailliart, K., Jenkins, H., Parker, D., Beckmann, M., Draper, J. (2008) Explanatory signal interpretation and metabolite identification strategies for nominal mass FIE-MS metabolite fingerprints. *Nature Prot.* **3**, 471–485.
24. Laskin, J. and Futrell, J.H. (2005) Activation of large ions in FT-ICR mass spectrometry. *Mass Spec. Rev.* **24**, 135–167.
25. Beckmann, M., Parker, D., Enot, D.P., Duval, E., Draper, J. (2008) High-throughput metabolome fingerprinting using Flow Injection Electrospray Mass Spectrometry. *Nature Prot.* **3**, 486–504.
26. Fiehn, O., Kopka, J., Dormann, P., Altmann, T., Trethewey, R.N., Willmitzer, L. (2000) Metabolite profiling for plant functional genomics. *Nat. Biotechnol.* **18**, 1157–1161.
27. Lisec, J., Schauer, N., Kopka, J., Willmitzer, L., Fernie, A.R. (2006) Gas chromatography mass spectrometry-based metabolite profiling in plants. *Nature Prot.* **1**, 387–396.
28. Biais, B. and Allwood, J.W., Deborde, C., Xu, Y., Maucourt, M., Beauvoit , B., Dunn, W.B.,

Jacob, D., Goodacre, R., Rolin, D., Moing, A. (2009) ^{1}H-NMR, GC-EI-TOF-MS, and data set correlation for fruit metabolomics, application to melon. *Anal. Chem.* **81**, 2884–2894.

29. Allwood, J.W. and Erban, A., de Koning, S., Dunn, W.B., Luedemann, A., Lommen, A., Kay, L., Löscher, R., Kopka, J., Goodacre, R. (2009) Inter-laboratory reproducibility of fast gas chromatography – electron impact – time of flight mass spectrometry (GC-EI-TOFMS) based plant metabolomics. *Metabolomics* **5**, 479–496.
30. Broadhurst, D.I. and Kell, D.B. (2006) Statistical strategies for avoiding false discoveries in metabolomics and related experiments. *Metabolomics* **2**, 171–196.
31. Taylor, N.S., Weber, R.J.M., Southam, A.D., Payne, T.G., Hrydziuszko, O., Arvanitis, T.N., Viant, M.R. (2009) A new approach to toxicity testing in *Daphnia magna*: application of high throughput FT-ICR mass spectrometry metabolomics. *Metabolomics* **5**, 44–58.
32. Southam, A.D., Payne, T.G., Cooper, H.J., Arvanitis, T.N., Viant, M.R. (2007) Dynamic Range and Mass Accuracy of Wide-Scan Direct Infusion Nanoelectrospray Fourier Transform Ion Cyclotron Resonance Mass Spectrometry-Based Metabolomics Increased by the Spectral Stitching Method. *Anal. Chem.* **79**, 4595–4602.
33. Payne, T.G., Southam, A.D., Arvanitis, T.N., Viant, M.R. (2009) A Signal Filtering Method for Improved Quantification and Noise Discrimination in Fourier Transform Ion Cyclotron Resonance Mass Spectrometry-Based Metabolomics Data. *JASMS* **20,** 1087–1095.
34. Beckmann, M., Enot, D.P., Overy, D.P., Draper, J. (2007) Representation, comparison, and interpretation of metabolome fingerprint data for total composition analysis and quality trait investigation in potato cultivars. *J. Ag. Food Chem.* **55**, 3444–3451.
35. Breitling, R., Pitt, A.R., Barrett, M.P. (2006) Precision mapping of the metabolome. *Trends Biotech.* **24**, 543–548.

Chapter 12

Combined NMR and Flow Injection ESI-MS for Brassicaceae Metabolomics

John M. Baker, Jane L. Ward, and Michael H. Beale

Abstract

High-throughput screening of large collections of plants, whether in the context of gene function analysis, quality trait selection, or metabolic engineering requires robust and rapid methodologies that provide maximum information with minimum sample pre-fractionation. Here, we present a protocol for high-throughput plant metabolomic analysis developed for Arabidopsis and generally applicable to plant green tissue, including other Brassicaceae. The methodology uses combined, flow injection electrospray mass spectrometry (FI-ESI-MS) and nuclear magnetic resonance (NMR) spectroscopy analysis. The protocol covers all steps of the process including sample extraction, data acquisition, data processing, and multivariate statistical analysis.

Key words: Metabolomics, NMR spectroscopy, Flow injection electrospray mass spectrometry, Multivariate analysis

1. Introduction

The comparison of metabolite composition of biological systems (known as metabolomics) is now a mature field and has been applied to a range of problems in plant and crop science. These include determination of individual gene function (1), analysis of natural variation (2), quality trait localisation (3), investigating the effects of stress (4) and pathogen or pest attack (5), and the assessment of substantial equivalence of genetically modified varieties (6). A wide variety of analytical techniques have been employed in metabolomics, and each has its own advantages and drawbacks. Data collection is typically carried out on large samples sets and is thus a key feature of the subsequent analysis is the use of multivariate statistical techniques such as principal components analysis (PCA). These are used to cluster samples, reveal trajectories, and identify the important metabolite signals that change between samples (7).

Nigel W. Hardy and Robert D. Hall (eds.), *Plant Metabolomics: Methods and Protocols*, Methods in Molecular Biology, vol. 860, DOI 10.1007/978-1-61779-594-7_12, © Springer Science+Business Media, LLC 2012

The analytical techniques used to collect metabolomic data can be, broadly, split into two categories—those which separate the components of the crude solvent extracts prior to detection (detection is usually by mass spectrometry (MS)) and those which directly analyse crude, unfractionated mixtures. Techniques such as high pressure liquid chromatography, HPLC(–MS), ultra performance liquid chromatography UPLC(–MS), gas chromatography GC(–MS), and capillary electrophoresis CE(–MS) separate the plant extracts and benefit from improved resolution of metabolites and are ideal for analysing targeted compound groups. However, the behaviour of chromatographic systems can change over time, posing significant (but not insurmountable) challenges when used in high throughput, where data quality and analysis rely on the use of multivariate statistics. Direct techniques for the analysis of unfractionated solvent extracts are inherently more stable and hence ideally suited to high-throughput metabolomics applications. Here, we present a protocol which combines two of these methods; Flow Injection Electro-Spray Ionisation Mass Spectrometry (FI-ESI-MS) and Nuclear Magnetic Resonance Spectroscopy (NMR).

NMR-based plant metabolomics is a well-established technique (2, 8, 9) and has the advantage of detecting a wide range of metabolites in an inherently quantitative and unbiased manner; it is also extremely robust and allows for accurate interpretation of signals against those of metabolite standards. However, NMR is perhaps less sensitive than other analytical methods and can suffer from problems with signal overlap, particularly in the carbohydrate region of the spectra. FI-ESI-MS is also well established (10, 11) and benefits from greater sensitivity than NMR and suffers less from signal overlap. It is selective in nature, but is complementary to NMR in that it responds well to many compounds which are only present as small peaks in the NMR spectrum. However, the assignment of signals is more ambiguous for FI-ESI-MS, particularly when the data are collected at nominal mass.

By analysing an identical plant extract with a combination of these two techniques we can utilise the advantages of both methods, and the assignment of unknown metabolites from both the NMR and ESI spectra is made much easier when both molecular weight and chemical shift data are available. In broad terms for plant extracts made with a polar solvent, NMR analysis gives quantitative information on the major primary metabolites, while the ESI spectra also contain semi-quantitative data that also includes secondary metabolites. In the Brassicaceae, this includes the glucosinolates and flavonoid glycosides.

Whichever analytical technique is being employed it is first necessary to extract the metabolites from the plant tissue prior to analysis. For metabolomics, it is desirable that this step extracts a broad range of metabolites and that it is reliable, robust, and suitable for the high throughput of samples that typifies a metabolomics experiment. This protocol describes a method for producing polar extracts of freeze-dried green plant tissue which has proved to be highly reliable over several years of operation at a throughput of some 2,000 samples/month. The method utilises an extraction solvent comprising 20% methanol in water. Other solvent mixtures have been used for plant NMR metabolomics (e.g. 9, 12). Some, containing buffers, are not compatible with ESI-MS. The methanol–water mixture used here produces stable extracts and performs well in electrospray MS.

2. Materials

1. Eppendorf type polypropylene tubes, 1.5 ml (Eppendorf UK, Cambridge, UK).
2. [^{1}H]-NMR extraction solvent, prepared in advance, in sufficient volume to process the whole sample set, and comprising (v/v) 80% deuterium oxide (D_2O, 99.9%D); 20% deuteromethanol (CD_3OD, 99.8%D), and 0.05% (w/v) sodium deuterotrimethysilylpropionate (d_4-TSP).
3. Clean, dry, 5-mm thin wall NMR tubes.
4. ESI-MS dilution/flow solvent, prepared in advance comprising (v/v) 80% water (polished to 18.2 MΩ) and 20% methanol (HPLC grade) which has been filtered through a 4.5-μm nylon filter.
5. 2-ml HPLC autosampler vials with caps, pre-fitted with silicone-PTFE septa, (Chromacol, Welwyn, UK).
6. Modern NMR spectrometer, with 5 mm ^{1}H probe and autosampler. We use a Bruker Avance with a 5-mm inverse, (SEI) probe operating at a frequency of 600 MHz (Bruker Biospin, Coventry, UK). The instrument should run contemporary software, ideally with batch processing and bucketing facility (we utilise Bruker's Topspin 1.3 and Amix—(Analysis of Mixtures, Bruker Biospin, Germany)).
7. Modern electrospray mass spectrometer and HPLC system; we use an Esquire3000 (an ion trap instrument) (Bruker Daltonics, Coventry, UK) with an Agilent 1100 HPLC system. The two are connected by a 2-μm in-line filter (4 mm diameter, Grace Davison, Carnforth, UK). For more comprehensive data collection, a spectrometer that is capable of switched positive and negative ionisation is preferred. Again a batch processing and bucketing facility is ideal (we utilise Bruker Daltonics Data Analysis 3.2 and Amix).
8. SIMCA-P multivariate statistical software (Umetrics, Umea, Sweden) or other similar software packages such as Pirouette (Infometrix, Bothell, WA, USA), Genstat (VSN International, Hemel Hempstead, UK), or Spotfire (Spotfire Inc., Mass. USA).

3. Methods

3.1. Metabolite Extraction and Sample Preparation

1. From each biological replicate of freeze-dried green tissue (see Notes 1 and 2), weigh three replicate 15 mg (±0.03 mg) samples into separate, labelled 1.5-ml Eppendorf tubes. Randomise the biological and technical replicates across the experiment (see Note 3).
2. Add 1 ml of the NMR extraction solvent (see above) and close the tubes.
3. Vortex-mix the contents of the tubes, until the green tissue is completely dis-aggregated (usually approximately 30 s) (see Note 4).
4. Heat the tubes at 50 (±1)°C for exactly 10 min. This is easily accomplished by use of a polystyrene raft and a pre-heated water bath. The tubes should be positioned so that all their contents are below the water level of the bath.
5. Immediately after removal from the water bath transfer the tubes to a micro-centrifuge and spin at full speed for 5 min.
6. From each tube transfer 850 μL of the supernatant to a clean labelled 1.5-ml Eppendorf tube.
7. Heat-shock the solutions (see Note 5) at 90 (±2)°C for 2 min, using a pre-heated water bath as before.
8. Immediately after removing the raft from the water bath, place the tubes in a refrigerator (4°C) and leave at this temperature for 30 min.
9. Remove samples from the cold and micro-centrifuge at full speed for 5 min.
10. Transfer 600 μL of supernatant to a clean, dry 5-mm thin wall NMR tube and cap ready for analysis (see Note 6).
11. Transfer a further 50 μL of the supernatant to a clean labelled HPLC autosampler vial.
12. To the HPLC autosampler vial add 950 μL of ESI-MS dilution solvent (see above) (see Note 7).

3.2. NMR Data Collection

1. Load NMR tubes into the NMR auto-sampler rack.
2. Ensure that the NMR probe temperature is stable at 300 K.
3. Enter the sample details into the automation program's sample list, taking care to accurately enter the appropriate sample label. Select the sample lock solvent as D_2O then select a suitable pulse sequence and number of scans (see Note 8).
4. Start the automation sequence. The NMR software should then automatically load each sample into the NMR magnet,

find the D_2O signal and lock onto it, optimise the intensity of this signal (via an automated shimming routine) (see Note 9), set the receiver gain and then collect the NMR data. At the end of the data collection, the NMR automation routine automatically processes the data before proceeding to the next sample (see Note 10).

5. Once data have been collected and assessed for quality (see Note 11), NMR samples are removed from the NMR tubes and transferred to screw cap glass vials. These vials are stored in a refrigerator in case future analyses are required.

3.3. Flow Injection ESI-MS Data Collection

1. The HPLC and ESI-MS should be configured with the out flow from the autosampler connected directly to the mass spectrometer vial a 2-μm in-line filter. One of the solvent reservoirs should be filled with enough ESI-MS flow solvent (see above) to run all of the samples (1 ml of solvent per sample is usually a good guide). Fresh flow solvent should be prepared regularly.
2. An HPLC method should be setup with a flow rate of 0.1 ml/min of 100% flow solvent with a runtime, after injection, sufficient to allow the entire injected sample to have flowed into the mass spectrometer plus at least 1 min (see Note 12).
3. The MS method should be set up with conditions which produce mass spectra with good signal to noise ratios (see Note 13). The spectrometer's divert valve should be set to send the flow, from the HPLC, to the source for all but the first and last few seconds of each sample's run (see Note 14). To reduce data size, the method should only save the mass spectral data for the period when the analyte is flowing into the spectrometer (see Note 12).
4. Load ESI samples into the HPLC auto-sampler. Set the injection volume for each sample to 100 μl, enter the details for each sample into the sample list and start the run (see Note 15).

3.4. Data Processing, Databasing, and Spectral Bucketing of the NMR Data

Illustrations of typical NMR spectra of Arabidopsis green tissue, generated by this protocol, are available in refs. (2) and (8). Prior to analysis of the data in statistical packages, some further processing is required. The first stage of this process is removal of noise from the spectra and its inclusion in the Bruker NMR spectrometer's database (SBase). The second is the reduction of the spectra to a "bucket table". The rationale for this step is to ensure a high comparability of the data sets and to reduce the complexity in the data from many different spectra of 128 k data-points to a matrix consisting of ~1 k data-points. This "bucketing" process also negates alignment problems that can sometimes arise from minor chemical shift differences in some signals due small variation in pH of samples. We carry out this process using Amix software; other methods are available.

1. Using the "Prepare Data" tool in Amix save each of the spectra into the spectra base (SBase) (see Note 16) using the following parameters. The noise level should be calculated from the noise region (δ –0.5 to –0.6) using a noise factor of 10 (see Note 17) and all of the negative peaks should be removed. At this stage, no exclusion regions should be used. Each spectrum should be saved as the sample's name (see Note 18).
2. Using the "buckets, statistics" tool in Amix, create a new bucket table of simple rectangular buckets from the data in the SBase using the following parameters (see Notes 19–22). The data range to be bucketed should be δ 9.5–0.5 ppm, the bucket width δ 0.01, all positive peaks should be bucketed and scaled to reference region of δ 0.05 to –0.05, two regions should be excluded (δ 4.875 to 4.705: HOD and δ 3.335 to 3.275: CD_2HOD).
3. While it is possible to perform PCA and other multivariate statistical analyses, within Amix, directly on the bucket table, it is often easier to export the data as a comma separated value (CSV) file for use in other packages.
4. Open the bucket table CSV file in a spreadsheet such as Excel in order to add extra rows of annotation to assist in future data analysis (e.g. line, treatment, timepoint) and save ready for multivariate analysis.

3.5. Data Processing and Spectral Bucketing of the ESI-MS Data

The ESI-MS data takes the form of a broad one-peak "ion chromatogram" (see Note 23). The data-points of the "chromatogram" alternate between positive and negative ionisation modes, each data point being the average of 25 scans as shown in Fig. 1. It is necessary to separate the positive and negative traces and generate the corresponding average mass spectra over the whole "ion chromatogram" (Fig. 2). These data are then exported as ASCII files containing the retention time of the peak and the mass spectra as mass intensity pair lists. These spectra must then be combined into bucket tables in order to be interpreted using multivariate techniques. The conversion into bucket tables also acts to reduce the effect of the small variability (ca m/z 0.1) in the reported masses of the ions. The mass intensity pair lists are generated in a batch process using the program Bruker Daltonics Data Analysis; the bucket tables are generated in Amix (see Note 24).

1. Open all of the "chromatograms" to be processed in the Bruker Daltonics Data Analysis program.
2. For each of the chromatograms generate the negative mode base peak chromatogram and from this generate the average negative mode mass spectrum for the entire "chromatogram".
3. For each of the samples export the mass spectrum as an ASCII file (3 Note 25).

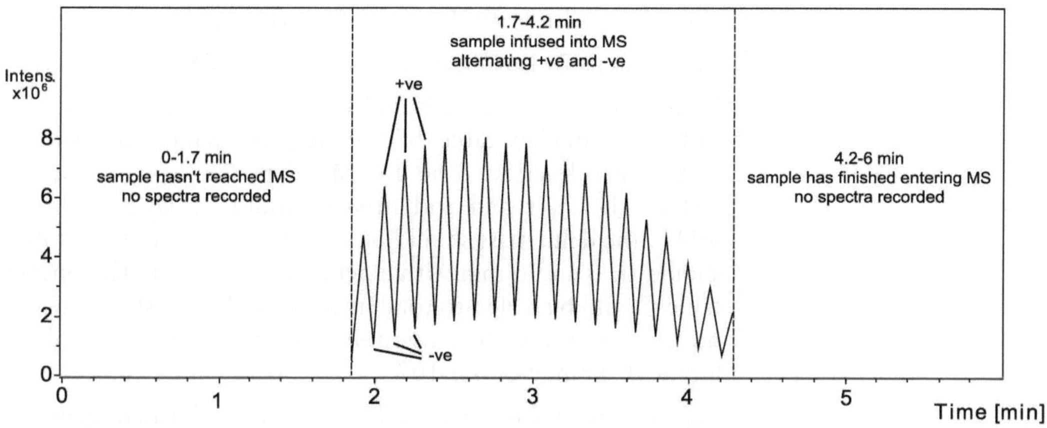

Fig. 1. Total ion current versus time trace for the direct infusion of an Arabidopsis extract (Columbia ecotype) into the mass spectrometer.

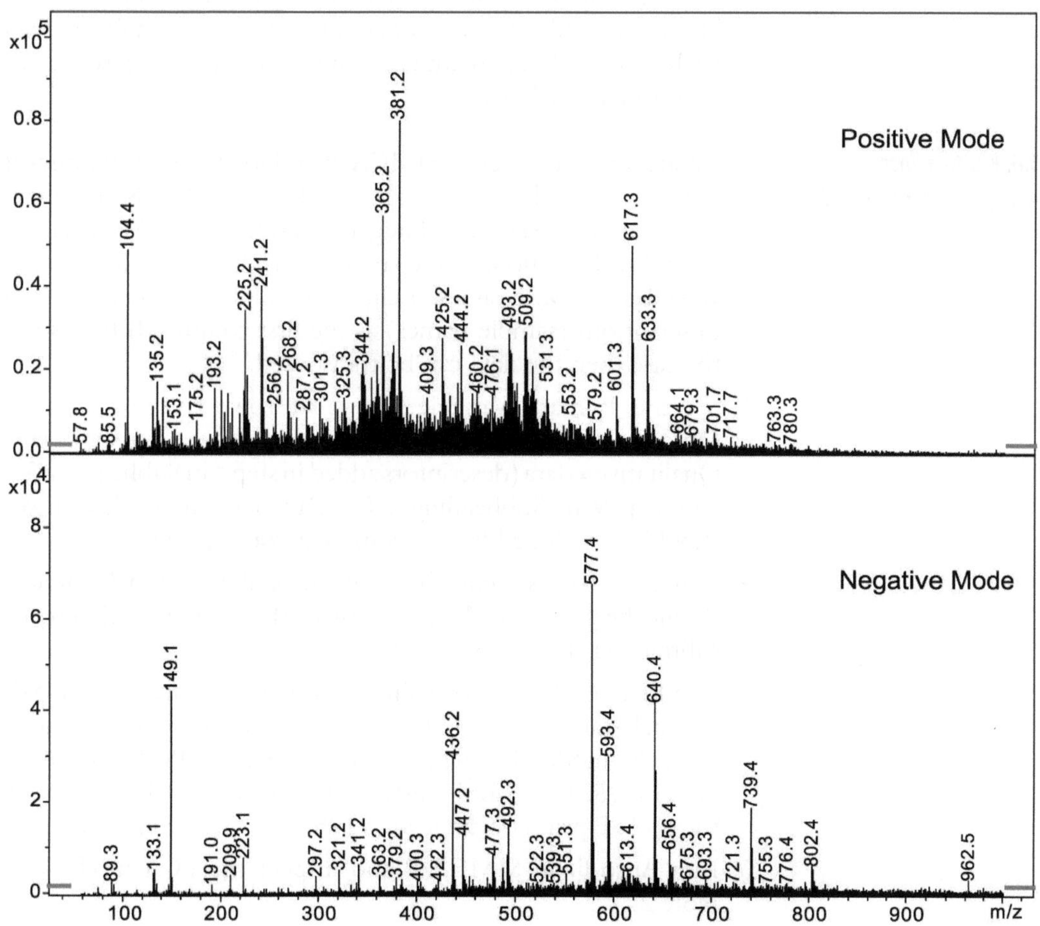

Fig. 2. Positive (*upper panel*) and negative (*lower panel*) FI-ESI-MS spectra of an Arabidopsis extract (Columbia ecotype).

4. Copy the negative mode ASCII files to a suitable directory.
5. Repeat steps 2–4 using positive mode ESI-MS data.
6. Using the "buckets, statistics" tool in Amix, generate a new "LC-MS" bucket table of the negative data using simple rectangular buckets. The LC-MS data file type should be ASCII, the start and end masses should be *m/z* 50.5 and 999.5, the delta mass should be *m/z* 1, the start and end times should be 0 and 6 min with a delta time of 6 min. The bucket table should be scaled to total intensity and there should be no noise removal therefore the noise level should be set to absolute with a range of 0 to10^{20}.
7. Repeat step 6 using the positive mode ESI-MS data to generate a separate data set.
8. As with the NMR data, the ESI-MS data can be analysed using Amix, but it is often easier to export the bucket table as a CSV file for use in other packages.
9. Open the bucket table CSV files in a spreadsheet such as Excel in order to add extra rows of annotation to assist in future data analysis (e.g. line, treatment, timepoint) and save ready for multivariate analysis.

3.6. Multivariate Analysis (See Note 26)

1. Create a new project in SIMCA-P and load one of the bucket tables generated in Subheadings 3.4 or 3.5. The NMR, positive mode FI-ESI-MS and negative mode FI-ESI-MS data sets should each be modelled separately. The data table should have variables (i.e. *m/z* values or chemical shifts) as the first row and observations (sample names) as the first column. If this is not the case then the data can be transposed.
2. Set the Primary Variable IDs (first row) and the Primary Observation IDs (first column) likewise assign any Qualitative x data (descriptors added in step 4 in Subheading 3.4 and step 9 in Subheading 3.5). Obviously these descriptors should be excluded from any models constructed.
3. Using the workset edit function, the scaling for each variable should be set to "ctr". This centres the data around zero by subtracting the average.
4. The FI-ESI-MS data contain peaks that result from the NMR internal standard (d_4-TSP). These peaks should be excluded from the model. For negative mode FI-ESI-MS they occur at *m/z* 149, 321, 493, 665, and 837. For positive mode FI-ESI-MS the d_4-TSP peaks occur at *m/z* 195, 367, 539, 711, 883.
5. Run Auto-fit in SIMCA-P and inspect the PCA model (see Note 27)

6. The PCA scores plot can now be analysed. Plots of various components should be analysed for different clustering patterns. Principal Component (PC) 1 versus PC2 should always be examined as these must represent the largest variance in the data set.
7. By colour coding the variables according to your descriptor information (line, treatment, etc.) it is easier to see trends in the data set.
8. For each scores plot, the two corresponding loadings plots (see Note 28) should be generated to describe the metabolites responsible for differences in clustering.
9. By comparison of NMR loadings plots with a library of NMR spectra from authenticated pure compounds, the positive and negative peaks in loadings plots can be assigned in terms of metabolites changing between the clusters (see Note 29).
10. FI-ESI-MS peaks can be tentatively assigned based on their mass (see Note 30). Ideally, there will be a very large degree of agreement between the assignments from the NMR and FI-ESI-MS data (see Note 31).
11. If there is no clear clustering, a discriminant analysis (e.g. Partial Least Squares-Discriminant Analysis PLS-DA and/or Orthogonal Signal Correction, OSC) can be performed. This involves assigning classes to the data set prior to modelling and using this information to "force" differences in the data set. Corresponding scores and loadings plots can be examined as described above (see Note 32).
12. In the process of generating loadings plots, information is gathered on those data points that are responsible for the differences. These give *clues* only. Examination of the original data should also be explored in order to confirm metabolite assignments and changes. All regions of the loadings plot should be examined, as in many cases changes in the intensity of very small peaks can be more significant than smaller changes in the very large peaks in the spectrum.

4. Notes

1. Careful sampling and recording of the plant material is of utmost importance. The metabolome is highly dynamic and all sampling should be carried out at the same time in the photoperiodic cycle and the tissue harvested into liquid nitrogen to arrest metabolism. Pooling of plants or plant parts is best carried out at the point of harvest. Tissue should then be stored at –80°C prior to processing.

2. Labels of samples should reflect the biology and contain identifiers such as line, plot/tray number, treatment, and biological and technical replicates. Careful consideration of this here makes the later data processing and in particular the multivariate analysis much easier.
3. The inclusion of analytical replicates and tracking samples, randomised across the experimental array can help to quality assure the whole experiment. At the data analysis stage, the technical replicates should cluster together as should the tracking samples. These samples help to assess the reproducibility of the extraction process.
4. Any material not suspended in the solvent at this stage will lead to a higher variability in the extraction process.
5. We have found the heat shock step to be helpful. Even though the solvent is 20% methanol, hydrolytic enzymes such as α-amylase remain active. The result of this can be change in the carbohydrate profile in the extract with time. This becomes evident from analysis of the NMR spectra technical replicates which should be randomised across the sample array. We have demonstrated that the 90°C, 2 min heat shock eliminates this problem and the NMR spectra remain stable. The problem is much less pronounced in freeze-dried green tissue than in other materials such as grain, but it is wise to incorporate this heat shock into all metabolomic extraction protocols that include aqueous solvents.
6. The stability of the samples should be assessed prior to undertaking a large experiment. This is achieved by collecting an NMR spectrum of a freshly prepared sample and comparing it to a spectrum collected several days later. This important step ensures that the samples will remain stable during the time spent in the autosampler prior to data collection. For Arabidopsis green tissue, prepared by the described method, samples remain stable for many days.
7. Diluting the samples to 5% in protonated solvent is sufficient to ensure that the major mass-spectrometric peak for most compounds (i.e. those with 18 or fewer exchangeable hydrogens) is non-deuterated.
8. We collect our spectra with a simple pre-sat pulse sequence with a 90°C pulse and pre-saturation during the 5 s relaxation delay. Other more complex water suppression sequences are available, but are unnecessary unless the samples contain more H_2O than those used here. The relaxation delay should be long enough to allow complete relaxation of the samples between scans. Each FID was collected with 128 k data-points covering a sweep width of 14,368 Hz (24 ppm). On our equipment, 128 scans were sufficient to give good signal to noise; for other systems more or fewer scans may be needed.

9. We have found that recording our spectra with the sample spinning at 20 Hz and using an automated deuterium gradient shimming system to optimise shims z-z^5 gives extremely reliable and narrow (≤1.2 Hz) line widths. Other systems may require a different approach. Whatever approach to shimming is employed, consistency in sample depth and the amount of sample in each NMR tube will improve the reliability of the process. It is important that line widths are uniform for later bucketing and PCA steps.
10. Each spectrum is automatically Fourier transformed after zero filling to 128 k data points and the application of an exponential window function with a line broadening of 0.5 Hz. Spectra are also phase corrected and baseline corrected (2nd order polynomial) and referenced to d_4-TSP (at δ 0.00) in automation.
11. NMR data are assessed for quality in three ways. Firstly the line shape of the d_4-TSP peak is automatically checked by measuring the peak width at half height. This should be less than 1.2 Hz (after application of the window function), any larger than this and the spectrum should be re-recorded. The second check is a visual one of overlaid batches of spectra to ensure that there are no gross abnormalities with any of the spectra, that there are no significant peak-shifts and that the automatic phasing has been carried out adequately. Finally, analytical replicate spectra of the same tissue should be overlaid to ensure that the reproducibility is good. It should be noted however that relatively few samples actually produce poor quality data if care has been taken with the sample preparation. In a large experiment, it is wise to quality assess the data as the experiment is in progress so that any samples which may need to be re-run can be done so quickly and before the sample is removed from the NMR auto-sampler. To avoid confusion, if samples are re-run, they should be done so using the same file name as the original data set, thus over-writing the poor spectrum.
12. The time taken for each "slug" of sample to flow from the HPLC into the mass spectrometer and give a signal will vary depending on such factors as the diameter and length of tubing connecting the two and any dead volumes (e.g. auto-sampler or inline filter). In the case of our instrument, each sample's run lasts for 6 min, it is 1.7 min after injection until any signal can be detected and the last of the sample has entered the instrument by 4.2 min. The remaining 3.5 min in which solvent is flowing along the sample path serves to wash it between samples (see Fig. 1).
13. The optimum mass spectral parameters will be highly dependent on instrument model. However, for our instrument (a Bruker Esquire 3000) good results can be obtained using the following parameters. Samples were introduced into the spectrometer with

a nebuliser pressure of 20 psi with dry gas of 6 L/min at 350°C. Mass spectra were recorded using Bruker's "smart tune" facility, with a target *m/z* of 300 and trap drive and stability values of 100%. Positive and negative mode ion chromatograms were collected from the same sample by alternating between polarities every 25 scans. The spectra were collected under ion charge control conditions over a mass range of *m/z* 50–1,000, with a maximum accumulation time of 40 ms, in normal scan mode with wide optimisation and ICC targets of 10,000 and 40,000 for negative and positive modes respectively.

14. We have noticed that if a fault is going to occur, it will tend to happen at the very beginning or end of a sample's run. Often, in these cases, the pump will continue to run while the other processes stop. If the divert valve is set to waste for the beginning and end of a run, then the chances of the spectrometer being damaged, by having solvent pumped into it when it has switched off, are reduced.

15. The control software saves each data set with a filename derived from the sample name entry (see Note 2).

16. The SBase is database of spectra which have been reduced in size by the removal of noise and much of the metadata (i.e. FIDs and imaginary spectra). Using Amix it is much easier to search this SBase for specific spectra than it would be in normal NMR processing software. Part of the reason for this is that each spectrum can be renamed when loaded into the SBase and this new name can reflect the nature of the sample. If a thoughtful sample naming regime has been employed (see Note 2), then comparison of different subsets of spectra from very large data sets is relatively simple. Obviously, great care should be taken when performing any data entry steps as errors in sample naming or label transposition can cause immense confusion when analysing the data.

17. The removal of noise from NMR spectra has been shown to improve the quality of PCA results (13). However, in spectra with excessive baseline distortions, care should be taken when using this method of noise removal.

18. Amix contains a facility to load large numbers of spectra into a SBase automatically. Briefly, the first spectrum should be loaded into the SBase as described. Using Amix's batch processing tool, the rest of the spectra, can be processed using the first spectrum as a template. This method retrieves the contents of the "title" text file associated with each spectrum and uses it as the filename for each processed spectrum; it is therefore important to ensure that the title for each spectrum is accurate and in a format suitable for use as a filename.

19. In data sets where changes in chemical shift are a problem (possibly due to pH effects), increasing the bucket width (δ 0.04 is often employed) can help prevent peaks from shifting between buckets.
20. As the internal standard (d_4-TSP) was added to the initial solvent mixture added to each tissue sample, scaling to this peak (δ 0.00) renders all spectra in the experiment comparable.
21. In data sets where there is variable sample tissue (e.g. moisture or starch content), it can sometimes be useful to scale to total intensity rather than scaling to the reference region.
22. The exclusion regions used here effectively remove the residual water and methanol peaks on our instrument. For instruments of different field strengths, different values may be more useful.
23. Because it is a familiar concept and because it is how the data processing software treats it, the word "ion chromatogram" has been used to describe the total ion current versus time trace; obviously, as there is no separation of components in the sample, it is not a true chromatogram.
24. The process described here works well when using Bruker Daltonics Data Analysis and Amix. Other systems will require different processing protocols, but the method of treating the data as a one peak chromatogram should work on most systems.
25. This process can be automated using a suitable processing script.
26. The series of steps suggested for multivariate analysis are not exhaustive. The approach to statistical data analysis depends on the nature of the biological questions being asked. In many cases, a series of data models need to be constructed. In all cases, findings from multivariate analysis can be checked by inspection of the original NMR or ESI-MS data sets.
27. For a good model R^2 and Q^2 (measures of explained and predicted variance in PCA analysis) values should be as near to 1 as possible. When the generation of additional components causes Q^2 to decrease, no further components should be generated.
28. Loadings plots describe the differences in chemical shift intensities or m/z intensities responsible for the separation of clusters in the PCA scores plots. The loadings plots can be represented as 2D scatter plots or as line plots. The line plot format is especially useful as this resembles the initial NMR or mass spectrum but with peaks in both the positive and negative directions. Peaks which are positive in the loadings plot of a given component represent signals which are more intense in those samples which have a high score for that component. Peaks which are negative in the loadings plot of a given component represent signals which are more intense in those samples which have a low score for that component.

29. A database of NMR spectra collected on the same spectrometer under the same conditions (solvent, temperature, pulse program) of authenticated natural compounds needs to be constructed and bucketed in AMIX under the same conditions as the samples. Once this is in place, it is relatively straightforward to compare spectra of standards against the loadings plot.
30. In the negative mode FI-ESI-MS, peaks tend to be present as the $[M-H]^-$ ion. Therefore kaempferol dirhamnoside (MW 578), the major flavonoid of Arabidopsis, gives a signal at *m/z* 577. Inherently, negatively charged molecules such as glucosinolates tend to appear as the $[M]^-$ ion; therefore, glucoraphanin (MW 436 without counter ion) gives a signal at *m/z* 436. In the positive mode, FI-ESI-MS peaks tend to be present as either $[M+H]^+$ or as a pair of peaks $[M+Na]^+$ and $[M+K]^+$. For example, sucrose (MW 342) gives a pair of peaks at *m/z* 365 and 381. Inherently, positively charged molecules will tend to appear as $[M]^+$; therefore, choline (MW 104 without counter ion) should give a signal at *m/z* 104.
31. It is impossible to be certain of the identity of a FI-ESI-MS peak if it is based solely on mass. Even when molecular formulae are available, from accurate mass instruments, there is no way of being sure which isomer is present. One example of this is the peak at *m/z* 593 in the negative mode spectrum of Arabidopsis (Fig. 2). Accurate mass analysis of this peak should give a molecular formula of $C_{27}H_{30}O_{15}$. When the database KNApSAcK (14) is searched for this formula, over 50 different compounds are returned. Even when the search is restricted to only those compounds which have been previously been found in Arabidopsis, there are still multiple possibilities. In fact, in Arabidopsis seeds, this signal tends to be a quercetin dirhamnoside and in leaves it tends to be a kaempferol rhamnoside glucoside (15). It is important that there should always be secondary confirmation of the identity of the metabolite (e.g. MS^n or correlation with the NMR data).
32. When using discriminant analysis, the data should be split into a training and validation set to test the robustness of the statistical model. In the scores plot, samples in the training set and in the validation set would be expected to cluster together.

Acknowledgements

This work has been funded by the EU Framework VI programme META-PHOR (FOOD-CT-2006-036220) and the UK Biotech nology and Biological Sciences Research Council (BBSRC).

References

1. Fukushima, A., Kusano, M., Nakamichi, N., Kobayashi, M., Hayashi, N., Sakakibara, H., Mizuno, T., Saito, K. (2009) Impact of clock-associated *Arabidopsis* pseudo-response regulators in metabolic coordination. *P. Natl. Acad. Sci. USA* **106**, 7251–7256.
2. Ward J.L., Harris C., Lewis J., Beale M.H. (2003) Assessment of ^{1}H-NMR spectroscopy and multivariate analysis as a technique for metabolite fingerprinting of *Arabidopsis thaliana. Phytochem.* **62**, 949–957.
3. Fu, J., Keurentjes, J.J.B., Bouwmeester, H., America, T., Verstappen, F.W.A., Ward, J.L., Beale, M.H., de Vos, R.C.H., Dijkstra, M., Scheltema, R.A., Johannes, F., Koornneef, M.,Vreugdenhil,D., Breitling R. and Jansen R.C. (2009) System-wide molecular evidence for phenotypic buffering in Arabidopsis. *Nature Genet.* **41**, 166–167.
4. Carmo-Silva, A.E., Keys, A.J., Beale, M.H., Ward, J.L., Baker, J.M., Hawkins, N.D., Arrabaca, M.C., and Parry, M.A.J. (2009) Drought stress increases the production of 5-hydroxynorvaline in two C-4 grasses. *Phytochem.* **70**, 664–671.
5. Parker, D, Beckmann, M, Zubair, H, Enot, D.P., Caracuel-Rios, Z, Overy, DP.,Snowdon, S., Talbot, N.J and Draper, J. (2009) Metabolomic analysis reveals a common pattern of metabolic re-programming during invasion of three host plant species by *Magnaporthe grisea. Plant J* **59**, 723–737.
6. Baker, J.M., Hawkins, N.D., Ward, J.L., Lovegrove, A., Napier, J.A., Shewry, P.R. and Beale, M.H. (2006) A metabolomic study of substantial equivalence of field-grown genetically modified wheat. *Plant Biotech. J.* **4**, 381–392.
7. Lindon, J.C, Holmes, E. and Nicholson, J.K. (2001) Pattern recognition methods and applications in biomedical magnetic resonance. *Prog. Nucl. Magn. Res.* **39**, 1–40.
8. Ward J.L., Baker, J.M., Beale M.H. (2007) Recent applications of NMR spectroscopy in plant metabolomics. *FEBS J.* **274**, 1126–1131.
9. Le Gall G, Colquhoun IJ, Davis AL, Collins GJ, Verhoeyen ME (2003)Metabolite profiling of tomato using ^{1}H NMR spectroscopy as a tool to detect potential unintended effects following a genetic modification. *J. Agric. Food Chem.* **51**, 2447–2456.
10. Beckmann, M., Parker, D., Enot, D.P., Duval, E. (2008) High-throughput, non targeted metabolite fingerprinting using nominal mass flow injection electrospray mass spectrometry. *Nat. Protoc.* **3**, 486–504.
11. Aharoni, A., de Vos, R., Verhoeven, H., Maliepaard, C., Kruppa, G., Bino R and Goodenowe, D (2002) Non-Targeted Metabolic Profiling Using Fourier Transform ion cyclotron Mass Spectrometry (FTMS). *OMICS: A Journal of Integrative Biol* **6**, 217–234.
12. Deborde C, Maucourt M, Baldet P, Bernillon S, Biais B, Talon G, Ferrand C, Jacob D, Ferry-Dumazet H, de Daruvar A, Rolin D, Moing A (2009) Proton NMR quantitative profiling for quality assessment of greenhouse-grown tomato fruit *Metabolomics* **5**, 183–198.
13. Halouska, S., and Powers, R. (2006) Negative impact of noise on the principal component analysis of NMR data *J.Magn. Reson.* **176**, 88–95.
14. Shinbo, Y., Nakamura, Y.,Altaf-Ul-Amin, M., Asahi, H., Kurokawi, K., Arita, M., Saito, K., Ohta, D., Shibata, D., Kanaya, S. (2006) KNApSAcK: a comprehensive species metabolite database. *Biotechnol. Agr. Forest.* **57**, 166–181.
15. Veit M. and Pauli, G.F. (1999) Major flavonoids from *Arabidopsis thaliana* leaves. *Journal of Natural Products* **62**, 1301–1303.

Chapter 13

ICP-MS and LC-ICP-MS for Analysis of Trace Element Content and Speciation in Cereal Grains

D.P. Persson, T.H. Hansen, K.H. Laursen, S. Husted, and J.K. Schjoerring

Abstract

Trace elements are unevenly distributed and speciated throughout the cereal grain. The germ and the outer layers of the grain have the highest concentrations of trace elements. A large fraction of the trace elements is therefore lost during the milling process. The bioavailability of the remaining trace elements is very low. This is usually ascribed to the formation of poorly soluble complexes with the phosphorus storage compound phytic acid. Hence, analysis of the total concentration of trace elements in grain tissues must be combined with a speciation analysis in order to assess their contribution to human nutrition. This chapter deals with the fractionation of anatomically very different cereal tissues. Procedures for microscaling of digestion procedures are outlined together with requirements for the use of certified reference materials in elemental profiling of grain tissue fractions. Methods for extraction and analysis of complexes containing trace elements in the grain tissue fractions are described. Finally, the chapter concludes with criteria for choice of chromatographic methods and setting of ICP-MS instrument parameters.

Key words: Aleurone layer, Cereal grain, Chromatography, Endosperm, ICP-MS, Iron, Micronutrients, Microscaled digestion, Polyatomic interference, SEC-ICP-MS, Size exclusion, Speciation, Trace elements, Zinc

1. Introduction

The majority of the world's population is dependent on cereal-based foods for their survival. Cereal products are an important source of not only carbohydrates but also trace elements. A large proportion of the trace elements in cereal grains are present in poorly soluble forms which are largely biounavailable (1). Trace element deficiencies, especially iron and zinc, are frequent in populations where cereal grains constitute the major food source (2). There are two main reasons for these malnutrition problems: Firstly, trace elements are

Nigel W. Hardy and Robert D. Hall (eds.), *Plant Metabolomics: Methods and Protocols*, Methods in Molecular Biology, vol. 860, DOI 10.1007/978-1-61779-594-7_13, © Springer Science+Business Media, LLC 2012

predominantly located in the bran layers (pericarp and aleurone) and in the germ of the grain. These parts are usually removed by milling, mainly to prevent the grain from becoming rancid (3). Secondly, a major part of the trace elements present in the grain forms poorly soluble complexes with phytic acid which is the major phosphorus storage compound in grains. Phytic acid is a negatively charged molecule with a high affinity for cationic micronutrients such as Fe and Zn (4).

Several methods have been developed for analysis of the total concentration of trace elements and different bioligands in the cereal grain (5, 6). However, only a few methods deal with analysis of intact trace element complexes, defined by IUPAC as a speciation analysis (7). Gentle separation by size exclusion chromatography (SEC) combined with highly sensitive elemental analysis by inductively coupled plasma-mass spectrometry (ICP-MS) constitutes an attractive tool for both quantitative and qualitative speciation analysis. However, speciation analysis cannot stand alone but has to be accompanied by quantification of the total concentration of the trace elements of interest in order to enable assessment of extraction efficiencies and contamination risks (8). Major challenges in the quantification of total trace element concentrations include limitations of sample material, efficiencies of digestion procedures, minimization of contamination and stability of instrumental performance. Insufficient quantity of sample material may be a problem when grain tissue fractions are analyzed individually. During sample preparation and digestion, precautions must be taken to minimize contamination while at the same time ensuring efficient digestion of the material available. The instrumental performance must be carefully monitored, both in relation to changes in sensitivity but also regarding precision and accuracy of the analysis. The latter is often evaluated by the use of relevant certified reference materials (CRMs).

A flow diagram of the major steps in the analysis of the concentration of trace elements and their speciation in cereal grain tissues is shown in Fig. 1. As the first step, the cereal grain is separated into four main fractions: awns (usually present in grains of barley and oats, but not in rice and wheat), bran (including pericarp, testa, and aleurone), germ, and endosperm (9). The masses of all fractions are recorded precisely, whereupon they are digested by the use of macro- or microscaled microwave-assisted methods (10). For comparison, whole grains are also digested. In the digests, the concentrations of the trace elements of interest are determined using ICP-MS. Results are usually only accepted if the accuracy is >90% compared to the CRM. The quantity of trace elements in each of the four grain fractions is calculated by multiplication of element concentrations and grain fraction weights. Finally, the summarized values are compared with the total grain content in order to obtain a mass balance for the individual trace elements.

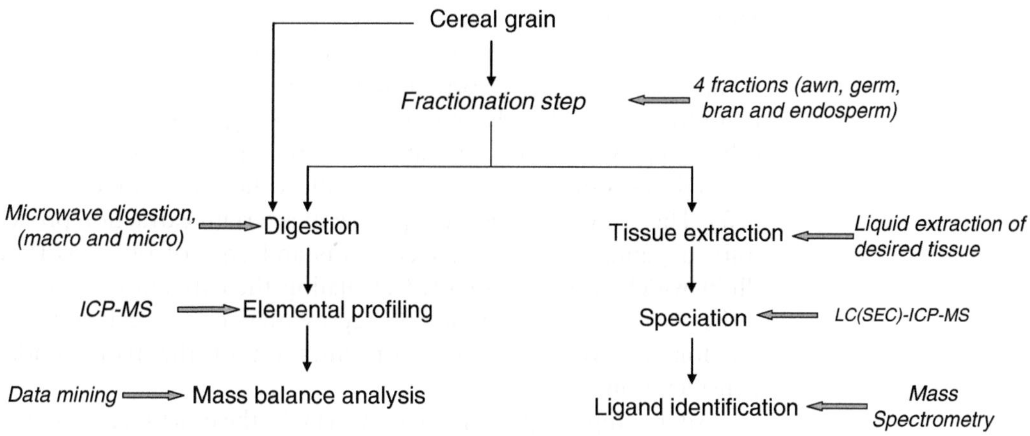

Fig. 1. Schematic flow diagram showing the methodological and analytical steps of elemental profiling and speciation analysis of cereal grain tissues.

When trace element concentrations of the grain fractions are obtained, the speciation analysis can be performed. The right hand side of Fig. 1 shows an experimental flow chart for speciation analysis of plant tissue. The major challenge of any speciation analysis is to evaluate whether the produced results reflect the naturally occurring species or are artifacts. Species artifacts may occur during extraction (e.g., via oxidization), and/or during chromatographic analysis (e.g., via ligand exchange). In order to obtain information on the type of bioligand, phosphorus and sulfur are analyzed along with the trace elements. Phosphorus is important due to its prevalence in phytic acid, and sulfur is an important constituent of metal binding proteins and peptides (4, 6, 11).

ICP-MS analysis of sulfur is challenging since the major isotope (^{32}S) cannot be analyzed due to interference from $^{16}O_2$ which has the same nominal mass. The standard procedure in this case is to analyze the second most abundant S-isotope (^{34}S), but this implies a large decrease in sensitivity, since ^{34}S only constitutes 4.3% of the total naturally occurring sulfur. The resulting loss of sensitivity can be overcome by addition of a reaction gas mixture to the octopole, in this case oxygen and helium. Using O_2 as a reaction gas promotes the formation of $^{48}SO^+$ as the polyatomic product ion of the ^{32}S and ^{16}O isotopes ($32+16=48$), thus enabling analysis of the major sulfur isotope (12). The addition of O_2 also affects the other elements analyzed, and therefore, this type of method requires careful monitoring of the elements of interest. The ion intensities of these elements should be checked by comparison of values obtained in both standard (no gas mode) and oxygen mode (11).

As the final step, results from the speciation analyses are matched with the total concentration of each element in order to evaluate the efficiencies of extraction and the amount of species relative to the total concentration (speciation recovery). Extraction efficiencies, hence also speciation recoveries, are rarely close to

100%. One reason for the lower recoveries is that, typically, only water-soluble species are extracted. Elements may be present in complexes which are poorly soluble in water and may, moreover, be fixed in the cell walls or attached to cell organelles. In addition, other factors such as the stability of the species, the choice of extraction solution and its pH value affect the extraction efficiency (13). The critical limit for acceptance of a certain extraction procedure depends on the target elements and type of tissue as they differ widely in extractability. Calculating the extraction efficiency is the only way to determine how representative the speciation data are for the total elemental concentration of the tissue under consideration.

After completing the speciation analysis, the exact identity of the metal binding ligands can be pursued by the use of e.g., ion exchange or reverse-phase chromatography on collected SEC-peaks (2nd dimension chromatography) coupled to various mass spectrometry techniques, such as ESI-MS, MALDI-TOF-MS, or Ion Traps.

2. Materials

2.1. Grain Samples

The starting material is samples of whole grains. Based on rice, a minimum of 300 mg dry matter, corresponding to 10–15 seeds, is required for each sample in order to obtain a sufficient quantity of each of the grain tissue fractions. In order to minimize contamination it is essential to use ultraclean water, acid-washed vials, and ultrapure chemicals.

2.2. Sample Fractionation

1. Analytical grade quartz sand (SiO_2), 40–150 mesh.
2. 7% nitric acid (HNO_3) prepared from 70% HNO_3 and Milli-Q water (Milli-Q Plus, Millipore Corporation, Bedford, MA, USA).
3. Milling device, e.g., Retsch MM301 ball mill (Retsch, Haan, Germany) equipped with an adapter rack for microcentrifuge tubes.
4. 2.5-mL Eppendorf tubes with round bottom.
5. Drying oven at 60°C or Freeze Dryer.
6. Scalpel.
7. Microbalance.

2.3. Sample Digestion for Mass Balance Analysis

1. Microwave oven (e.g., the Multiwave 3000, Anton Paar GmbH, Graz, Austria).
2. For microscaled digestion (1–20 mg dry matter/sample), a 64MG5 rotor (Anton Paar GmbH, Graz, Austria) with capacity for 64 samples is used. This rotor accommodates 5-mL digestion bombs, e.g., 5-mL glass digestion vials equipped with lip seals and screw caps capable of withstanding pressures up to max. 20 bar (see Note 1).

3. For macroscaled digestion (>250 mg dry matter per sample), a 16HF100 rotor (Anton Paar GmbH, Graz, Austria), with capacity for 16 samples is used. This rotor accommodates 100-mL digestion bombs, e.g., 100 mL Teflon liners inserted into ceramic vessels and closed with vessel jackets (max 70 bar, 240°C).
4. A certified reference material (CRM), representative for the matrix and elements of interest, e.g., NIST 1567a durum wheat grain (National Institute of Standards and Technology, Gaithersburg, MD, USA) (see Note 2).
5. 30% H_2O_2 and 15% H_2O_2 (prepared from 30% H_2O_2 and Milli-Q water).
6. 70% HNO_3 and 5% HNO_3 (prepared from 70% HNO_3 and Milli-Q water).
7. Milli-Q water.
8. 70 mL HD polyethylene vials (Capitol Vial, Fulton Ville, NY, USA).

2.4. Sample Extraction

1. Mortar and a pestle, acid washed in 7% nitric acid (HNO_3).
2. 7 and 10% HNO_3 prepared from 70% HNO_3 and Milli-Q water.
3. Quartz sand (SiO_2).
4. Ice.
5. Inert gas (N_2 or Ar).
6. Tris HCl buffer solution, 50 mM with pH 7.5, prepared from Trizma hydrochloride and Trizma base.
7. Ultrasonication bath, such as the Branson 2510 (Branson Ultrasonics, Danbury, USA).
8. Ion exchange column packed with chelating resin Chelex-100, Sodium form.
9. Ultrafilters with 50-kDa cutoff (Microcon YM-50; Millipore Corporation, Bedford, MA, USA).
10. Pipettes.
11. Centrifuge capable of yielding a relative centrifugal force of $16{,}000 \times g$.

2.5. Direct ICP Analysis and Online Size Exclusion Chromatography

1. An ICP-MS equipped with an octopole reaction cell, such as the Agilent 7500ce (Agilent Technologies, Manchester, UK).
2. The ICP-MS should also be equipped with a mass flow controller capable of handling an octopole gas flow rate of 0.5 mL/min.
3. The ICP-MS should be equipped with an auto sampler for direct injection.
4. Inorganic standards for ICP-MS calibration (e.g., P/N 4400-ICP-MSCS, P/N4400-132565A and P/N4400-132565B, CPI International, Amsterdam, Holland).

5. A perfluroalkoxy (PFA) microflow nebulizer for nebulization of liquid samples.
6. All HPLC connections should be tubings with 0.17 mm id and made of polyether ether ketone (PEEK).
7. An HPLC, such as the Agilent 1100 Series (Agilent Technologies, Manchester, UK) equipped with a Diode Array Detector (DAD) for hyphenated HPLC-ICP-MS.
8. Size exclusion chromatography column, such as the Superdex 75 10/300 GL (Glass, 10×300 mm, 13 μm cross-linked agarose/dextran, Amersham Biosciences, Uppsala, Sweden), with an optimum separation range between 7,000 and 70,000 Da (see Note 3).
9. Calibration kit for the SEC column (Amersham Biosciences, Uppsala, Sweden) (see Note 4).
10. Tris HCl buffer solution, 50 mM with pH 7.5, prepared from Trizma hydrochloride and Trizma base.
11. Ultrasonication bath such as Branson 2510 (Branson Ultrasonics, Danbury, USA).
12. Gas cylinder containing 10% oxygen gas mixed in helium, connected to the reaction cell of the ICP-MS.
13. Buffer solution: 50 mM Tris HCl buffer solution and 5 mM EDTA (Ethylened iaminetetraacetic acid Disodium salt Dihydrate, >99% pure).
14. Wash solution: 1.75% HNO_3 and 0.2% HF, prepared from 70% HNO_3, 40% hydrogen fluoride, and Milli-Q water.
15. 2% HNO_3, prepared from 70% HNO_3 and Milli-Q water.
16. Solution containing 100 ppb sulfur in 50 mM Tris HCl buffer solution (for example prepared based on CPI standard P/N4400-132565A, CPI International, Amsterdam, Holland and 50 mM Tris HCl buffer solution). If the recommended CPI standard is used, the tune solution will also contain other elements, including the micro nutrients Zn, Fe, Mn, Cu, and Ni in the 1–5 μg/L range.
17. Internal standard solution: for example Erbium, typically in the 50 μg/L range (Agilent Technologies, Manchester, UK).
18. Solution containing pepsin (1 mg/mL), NaCl (0.5 M), acetic acid (0.1 M) and Milli-Q water. Prepared from pepsin (Pepsin A; EC 3.4.23.1; Sigma-Aldrich Chemie GmbH, Steinheim, Germany), NaCl, acetic acid, and Milli-Q water.
19. 0.1% TFA, prepared from 100% Trifluoroacetic acid and Milli-Q water.
20. HPLC needle wash solution consisting of 50% EtOH and Milli-Q water.

2.6. Software for Analyzing Data

1. Software for chromatographic data acquisition, quantification and processing, such as the MassHunter v. B-01-01 (Agilent Technologies, Manchester, UK).
2. Visualization software (e.g., SigmaPlot 11.0 Systat Software Inc., USA).

3. Methods

3.1. Fractionation of Cereal Grains

A cereal grain consists of the following four main components: (1) an outer layer consisting of awns fused with the grain pericarp (this layer is usually absent in rice and wheat grains, but present in barley and oats), (2) the bran layers (including pericarp, testa, and aleurone), (3) the germ (also termed embryo; includes the scutellum), and (4) the endosperm (9). The fractionation method described below is developed with the aim of separating and collecting these four main fractions prior to trace element analysis. At least four replicates should be included for statistical purposes.

1. To minimize surface contamination, wash the cereal grain of choice three times in Milli-Q water. The amount of starting material for each sample should be around 300 mg in order to obtain a sufficient quantity of each tissue fraction.
2. Dry the grains in an oven at 60°C or in a freeze dryer overnight. To ensure that the grain batch is totally dry, weigh the batch over 2 h intervals. When weights are stable over time, the drying process is complete.
3. If present, gently peel off the outer layer of awns by use of a scalpel (fraction 1).
4. Gently loosen and remove the germ using the tip of a scalpel (fraction 2).
5. To separate the bran and the endosperm from each other, a polishing process has to be performed. This can be done by high-speed shaking in a ball mill (Retsch MM301) at 30 Hz using an adapter rack for microcentrifuge tubes (see Note 5).
6. Prepare a batch of ultrapure acid-washed quartz sand by shaking the sand in 7% HNO_3 three times. After the third decantation, the sand is washed three times with Milli-Q water or until the pH is neutral in the suspension. Thereafter, dry the sand in an oven at 60°C.
7. A predefined and exact weight, 250–300 mg, of the acid-washed sand is used for polishing. Save the mixture of sand and abraded material from the grain as fraction 3 (bran layers).

8. Transfer the remains of the grain to a new microcentrifuge tube and wash three times with Milli-Q water to remove surface dust. Dry afterward. Save this fraction as fraction 4 (endosperm).

3.2. Mass Balance Analysis

The dry matter mass of the individual grain fractions are quantified gravimetrically and should together match the weight of the whole grain. When trace element concentrations have been determined for each fraction, multiplying with the corresponding dry weights and summing up the results for all fractions should produce a cumulated value which is equal to the content of the whole grain. The mass balance analysis can be performed in micro- or macroscale, depending on sample quantity.

3.2.1. Macroscaled Digestion

A rotor with the capacity for 16 samples, designed for digestion of samples with dry matter mass between 125 and 300 mg, is used for macroscaled digestion. Include at least one CRM and one true blank for each duty cycle; end up with at least seven replicate CRM samples and blank samples for later validation purposes (see Subheading 3.2.3). The digestion of whole grain samples, the bran layer including sand from the polishing procedure (fraction 3) and the endosperm (fraction 4) is performed using the following procedure:

1. Bran layers + sand (fraction 3) and the endosperm (fraction 4) and whole grain samples are suspended in 5 mL of 70% HNO_3 and 5 mL 15% H_2O_2 in 100-mL vessels.
2. The 100-mL digestion bombs are closed with vessel jackets and screw caps and subsequently microwaved as follows: 10 min ramping to the max temperature of 210°C; keep this temperature for 36 min and then cool for 30 min. The pressure in the bombs must be kept below 40 bar and the energy input to the microwave generator below 1,400 W.
3. Samples are transferred to 70-mL HD polyethylene vials and diluted with Milli-Q water to give a volume of 50 mL, resulting in 7% HNO_3.
4. Directly before analysis by ICP-MS, the samples are diluted 1:1 with Milli-Q water, giving a final HNO_3 concentration of 3.5% (see Note 6).

3.2.2. Microscaled Digestion

For fractions with a dry matter mass between 1 and 20 mg, a macroscaled microwave digestion cannot be performed. Instead, a rotor with the capacity for 64 samples, but less volume per sample, is used (10). Include at least three CRMs and three true blanks in each digestion cycle, in order to be able to monitor fluctuations between individual digestion cycles. For statistical purposes, always end up with at least seven CRMs and seven blanks in the complete sample set to be analyzed no matter how small the total number of samples may be.

1. Acid-wash a number of glass vials (5 mL) in 5% HNO_3 and let them air-dry (see Note 7).
2. Fractions 1 and 2 are weighed and transferred to the glass vials and suspended in 250 μL 70% HNO_3 and 125 μL 30% H_2O_2 if the sample quantity is below 10 mg. For samples above 10 mg, use double the amount of the chemicals.
3. At least seven true blanks and seven CRM samples are required for data validation. Use the same quantity of CRM (NIST 1567a) as used for the samples to be analyzed, in this case 1–20 mg.
4. The vials are closed with lip-seals and screw caps, forming the digestion bomb.
5. Microwave the digestion bombs for 100 min (see Note 8) using the following program: 10 min ramping to max. temperature 140°C; keep this temperature for 80 min and then cool for 10 min. Place the samples in a freezer for 30 min before releasing the pressure inside the bomb.
6. Before analysis, the samples must be diluted to match the requirements of the analytical method of choice, e.g., ICP-MS or inductively coupled plasma-optical emission spectroscopy (ICP-OES) (see Note 9).

3.2.3. ICP-MS Analysis

ICP-MS is performed with external calibration covering the elements of interest in concentrations comparable to the samples. For quality control of the sample preparation procedure, an internal standard such as Erbium (Er) can be included in each sample digestion, spiked to the acid used. This element is not present at detectable levels in normal grain samples and is therefore a good choice of internal standard.

In the ICP-MS from Agilent, a built-in sample injector is available that includes 89 samples (5 mL) and 3 large samples (100 mL). One of the large samples is used as wash sample (1.75% HNO_3/0.2% HF). The wash is included after each sample to ensure that contamination does not build up in the system. One of the other large samples is usually a CRM sample which is analyzed for every ten samples in order to ensure that there is no sensitivity loss throughout the run series.

1. Tune the machine as described by the manufacturer.
2. Build a method including the elements of interest and include the mass 76. This m/z of 76 refers to the $^{38}Ar^{38}Ar$ interference which is a very stable and reliable signal and hence an ideal way to monitor instrumental drift throughout the entire analysis (see Note 10). Also, include the m/z value of the selected internal standard (for example ^{166}Er).

3. Perform an external calibration of the elements of interest, covering a linear range of at least five orders of magnitude (>10 calibration points).
4. After analyzing the calibration standards, include two extra wash cycles to ensure that the system is fully decontaminated.
5. From now on analyze the CRM (NIST 1567a) for every ten samples. This is used to evaluate and correct for possible drift throughout the run series.
6. Thereafter, analyze the seven replicate NIST 1567a CRM samples (see Subheading 3.2.1). These samples must contain the elements of interest and have the same matrix as the grain samples (see Note 2). Use the CRMs for validation and only accept elements that are determined with accuracy above 90% (see Note 11).
7. Run two extra wash cycles followed by the seven blanks. These are used for estimating the limit of detection (LOD) and the limit of quantification (LOQ), represented by 3σ (three times the standard deviation) and 10σ, respectively.
8. Analyze at least three replicates of whole grain and subfractions 1–4.
9. When the analysis is complete, check the calibration curves (exclude nonlinear points). Perhaps the calibration curve should be split into two sections, depending on the linear range at the time of analysis, using the software. Quantify the elements of interest.
10. Open MassHunter and import the results and save both counts per second (cps) and concentration (ppm).
11. Open the Excel files and calculate the drift using the counts/s at *m/z* 76, corresponding to the $^{38}Ar^{38}Ar$ peak (drift should be less than 10% throughout the whole analytical period).
12. Check that accuracy of the seven CRMs is >90% of the certified values and that the relative standard deviation (RSD) is below 10%. Calculate the LOD and LOQ from the standard deviation between the seven true blanks and compare with a typical sample. Ensure that the concentration of Er does not vary more than 10% between samples.
13. Perform the mass balance calculations:

$$X_{\text{grain}} = \left(\sum_{n=1}^{n=4} X_n w_n\right) / w_{\text{grain}} \quad (1)$$

where X_n denotes the concentration of a given element in grain fraction n and w_n the corresponding dry matter mass; X_{grain} and w_{grain} denote the corresponding values for the whole grain.

3.3. SEC-ICP-MS Analysis of Cereal Grain Tissue Fractions

SEC-ICP-MS is a hyphenated technique, which means that a chromatographic separation step is directly coupled to a detector, in this case an ICP-MS. The ICP-MS allows for high sensitivity detection of trace elements and for this reason any contamination has to be kept at a minimum level in all steps from sample preparation to injection. In addition, instrumental plasma-based interferences as well as matrix-based interferences may bias the analytical result. Almost all such interferences can be efficiently reduced or even eliminated. However, if results are not checked carefully, interferences may be overlooked (see Note 12).

3.3.1. Extraction of Tissue Fractions

The extraction of elemental species from plant material is challenging, especially because the identity and quantity of the target species to a large extent are unknown. In order to maintain the integrity of the species, the pH must be kept stable, oxidation must be avoided, and ligand exchange must be minimized (see Note 13). After analysis, the efficiency of the extraction for any given element should be calculated as the percentage of the total concentration.

1. Acid wash a 1 L bottle overnight in 10% HNO_3. Also, acid-wash one mortar and one pestle for each sample to be extracted.
2. Degas the extraction solution, in this case 50 mM Tris HCl buffer solution, pH 7.5, for 30 min at room temperature (see Note 14). Degassing can easily be performed in an ultrasonic bath, for example a Branson 2510. Make sure that the degassed solution is free from metal contamination by running it through a Chelex-100 column.
3. Weigh 10–50 mg of tissue material and put it in the mortar together with 600–800 mg of acid-washed sand (see step 6 in Subheading 3.1) and 2 mL of degassed Tris HCl buffer solution.
4. Perform the extraction on ice and under a flow of inert gas (N_2 or Ar) in order to prevent oxidation of chemical species (see Note 13).
5. Make sure that all solid material is finely crushed so that it becomes a slurry. Wait for 15 min and stir it up again. Repeat four times while keeping it on ice, resulting in a 1-h extraction procedure.
6. Centrifuge each sample in a 2-mL Eppendorf vial at 16,000 × *g* for 10 min at 4°C.
7. Transfer the supernatant to an ultrafilter vial with a 50-kDa cutoff using a clean pipette.
8. Keep cold on ice and analyze within 6 h; otherwise, store the sample, preferably at −80°C.

3.3.2. Preparing the Size Exclusion Column

Size exclusion chromatography (SEC) is generally considered to be a gentle separation technique when it comes to maintaining the integrity of species. SEC separates compounds by size, which means that large compounds elute faster than smaller compounds. Species and chemical compounds with similar molecular weights will therefore elute at the same position in the chromatogram. For exact identification of a certain metal binding ligand, further separation may be achieved using, e.g., ion exchange, ion pairing or reverse-phase chromatography on the collected SEC fractions (2nd dimension chromatography).

Free cations in the samples are frequently bound to the stationary phase of the column. Such unspecific binding may induce ligand exchange in subsequent analyses which will bias the results and hinder reproducibility. Therefore, regeneration and equilibration of the column is a very critical step. We have observed that repetitive injection of an EDTA solution efficiently can rinse the column and ensure reproducible analytical conditions (see Note 15).

1. Make sure that the column is perfectly clean both with regards to proteins and metal cations. Equilibrate it with degassed 50 mM Tris HCl buffer solution, at a flow rate of 1 mL/min.
2. Protein contaminants can be removed from the column using a solution consisting of pepsin (1 mg/mL), NaCl (0.5 M) and acetic acid (0.1 M). Inject 100 μL of the pepsin solution and leave overnight at room temperature or 1 h at 37°C. After the enzymatic treatment, wash the column with 25 mL of 0.1% TFA at a flow rate of 0.5 mL/min. Immediately hereafter rinse the column with 25 mL of Milli-Q water followed by 50 mL of mobile phase (in this case a 50 mM Tris HCl buffer solution), at a flow rate of 0.5 mL/min.
3. Minor metal contaminations can be removed with 0.1% TFA. Wash the column with 25 mL of 0.1% TFA at a flow rate of 0.5 mL/min. Immediately hereafter rinse the column in the same way as after the enzymatic treatment.
4. Major metal contaminations can be removed with repeated injections of a 5 mM EDTA/50 mM Tris HCl buffer solution (see Note 15).
5. Calibrate the column, using the calibration kit.
6. Set the flow at 1 mL/min and the runtime to 25 min. Make sure that the pressure is not exceeding the recommended limit (18 bar for the Superdex 75 SEC column).
7. Connect the end tubing from the column to the nebulizer of the ICP-MS via an open T-piece. The T-piece ensures that the flow of liquid into the nebulizer is appropriate so that the spray chamber is not overloaded. Check that the mobile phase is running through the tubing into the nebulizer and that the waste liquid is discarded from the back of the spray chamber.

8. Note the background signals of the elements of interest and save it. During the course of analysis, make sure that the background signals stay approximately the same. If not, the column should be cleaned and/or the mobile phase changed.

3.3.3. ICP-MS Settings Using Oxygen as Reaction Gas

The settings for the SEC-ICP-MS are important as they determine both the sensitivity and the avoidance of interferences. Use of a reaction gas of 10% O_2 in 90% He promotes the formation of $^{48}SO^+$ as the product ion of ^{32}S and oxygen (32 + 16 = 48). This increases the sensitivity at least five times (^{34}S in no gas mode vs. $^{48}SO^+$ in oxygen mode (11, 12)). However, it is important to note that the addition of oxygen to the octopole will decrease the ion transmission and consequently lower the analytical sensitivity of ions for which the bias is not by-passed by oxygen addition (see Note 16). The elements of interest must therefore be carefully monitored during tuning of the instrument.

1. Tune the ICP-MS in standard mode. Save the tune file.
2. Find the settings for maximum oxide formation but with as little decrease in sensitivity of other analytes as possible. Start with the settings usually used in reaction mode. Thereafter ensure that the kinetic energy discrimination is neutral by having the same voltage at the exit of the octopole as at the entrance of the quadrupole. This setting allows the formed sulfur oxides to reach the detector. Tune with a buffer solution containing 100 μg/L sulfur in 50 mM Tris HCl buffer solution together with the elements of primary interest. Note the conditions where maximum sensitivity on $^{48}SO^+$ and minimum loss of analyte signals are obtained. Typically, micronutrients are tuned in the 1–5 μg/L range. Make a ramp flow from 0 to 1 mL/min and note where the highest response is obtained (ion intensity).
3. Make a new tune file with the tuned settings. We usually work with the following settings: Oxygen flow: 0.5 mL/min (= 50% with a microflow-controller). OctBias: −16 V. QPBias: −16 V. Cell exit: −36 V. QP focus: −15 V.
4. Tune again, this time manually with a blank solution (=mobile phase; 50 mM Tris HCl buffer solution) and with a buffer solution of 100 μg/L S in 50 mM Tris. Note the sensitivity, since it can be useful when comparing results from different days.
5. Double check the ion intensity of the elements of interest other than sulfur. Compare identical injections in standard mode with injections in oxygen mode.

3.3.4. Analysis

1. Create a method in the software. Choose the elements of interest. If possible, choose at least two isotopes of each element. When using a flow rate of 1 mL/min, the runtime should be 25 min.
2. Check the backgrounds in the tune-window.

3. Inject the sample. Use the needle wash facility on your HPLC of choice, if possible. Use a solution of 50% EtOH in Milli-Q water as needle wash solution.
4. After analysis, inject five times 20–100 μL buffer solution consisting of 5 mM EDTA in 50 mM Tris HCl buffer solution, with a 1 min delay between injections. If the flow rate is lower than 1 mL/min, choose a 3-min delay. The cleaning procedure can be followed "online" to ensure that all background signals return to their original level. The efficiency of the procedure can also be evaluated by running the cleaning process two times in a row (see Note 15).

3.3.5. Calibration

1. Disconnect the column and connect the HPLC tubings directly to the ICP-MS.
2. Use a run time of 3 min.
3. Inject at least three blank samples between the calibration standards.
4. Calibrate by injecting an identical volume of the calibration solutions, starting with the lowest concentration (see Note 17).
5. Integrate the calibration peaks and make sure that the peak areas cover the ranges of the analyzed samples.
6. Create a linear regression with concentration and peak area. Insert the peak area of the element of choice from the sample.
7. Calculate how much was recovered from the column and how much was speciated out of the total concentration.

3.4. Identification of Metal Binding Ligands

Complete identification of the metal binding ligands can be achieved by use of additional chromatography and mass spectrometry. Usually 2nd dimension chromatography is performed on collected SEC-peaks, in hyphenation to a mass spectrometer. The 2nd dimension chromatography may be ion exchange, ion pairing, reverse or normal phase, depending on the element species and on the compatibility with the mass spectrometer of choice. The most frequently used techniques are electrospray ionization mass spectrometry (ESI-MS) and matrix-assisted laser desorption ionization (MALDI), but there are a lot of additional methods and techniques to choose from. The final identification of metal binding ligands, and thus of the entire metal complex, requires its own descriptions which are outside the scope of this chapter.

4. Notes

1. Micro-digestion can also be performed using a rotor with capacity for 16 samples, following the vial-in-vial procedure as described in Jakobsen et al. (14).

2. CRMs are offered by many different suppliers but most are only certified for a limited number of elements. Always use CRMs which closely match the matrix and include the elements of interest in concentrations matching the samples.
3. Size exclusion columns are available in different size separation ranges; for low (0.7–7 kDa), medium (7–70 kDa), and high (70–200 kDa) molecular weight compounds.
4. Calibration kits are available for the different size exclusion columns. Usually UV detection is used to identify the calibration compounds.
5. The use of Eppendorf tubes with round bottoms ensures a uniform polishing outcome for all grains. Please notice that the required milling time needed to ensure a complete removal of the bran layers varies among cereal species and genotypes within the same species. A preliminary study is therefore required to estimate the optimal milling time. This can be done by analyzing the elemental concentration of grains that have been polished for, e.g., 60, 70, 80, 90, and 100 s. A negative concentration gradient should be observed with increasing time of polishing until the elemental concentrations reach a steady state when the bran layers have been completely removed. For rice, approximately 80 s of milling will in most cases be optimal, while for barley longer milling times must be expected due to a thicker bran layer.
6. It is our experience that 7% HNO_3, but not 3.5%, ensures stability of the samples until analysis if prior storage is needed.
7. These vials can be reused, if acid washed. It is important to rinse the screw caps immediately after uncapping so that the acid does not destroy them.
8. To ensure a total digestion of nonmilled samples, the digestion program has to be longer than for milled samples. Alternatively, the nonmilled samples can be predigested overnight.
9. It is our experience that the ICP-MS runs very stably with 3.5% HNO_3, but not with stronger acidities. By contrast, a higher acid content is possible with ICP-OES, which may help to overcome the inherently lower sensitivity of this analytical technique compared to ICP-MS.
10. It should be noted that the signals from ^{76}Se and ^{76}Ge may cause isobaric overlap, thereby interfering with the $^{38}Ar^{38}Ar$ signal if these analytes are present in high concentrations. However, in cereal grain samples the contributions from these isotopes are usually marginal relative to the $^{38}Ar^{38}Ar$ signal. Hence, in such cases the $^{38}Ar^{38}Ar$ signal may be used throughout the whole sample set.
11. Standard additions may be used if no CRM is available for the element of interest. By adding different amounts of a standard

solution to the unknown samples the concentration of elements of interest can be calculated by extrapolation. However, as the method of standard addition assumes a linear response between measured signal and analyte concentration, care must be taken.

12. Some interferences may mimic the anticipated signals of other elements. In one study of barley whole grains, using oxygen mode, we observed that two major isotopes of molybdenum (^{95}Mo and ^{98}Mo; 15.92 and 24.13% of total Mo, respectively) created molybdenum oxides at the masses 111 and 114, thus resembling two major isotopes of cadmium (^{111}Cd and ^{114}Cd; 12.80 and 28.73% of total, respectively). Calculation of the isotope ratios was therefore necessary in order to elucidate the true identity of the signal.
13. Prevention of oxidation is absolutely necessary during analysis of metal-binding sulfur-rich compounds. When harvesting fresh tissue, put samples directly into liquid nitrogen or immediately start lyophilization.
14. Most biological metal ion species are stable at neutral pH, matching the pH of the cytosol of the living plant cell. Using SEC coupled to ICP-MS, it is very important to carefully consider which mobile phase to use. The mobile phase must ensure that compounds and complexes remain intact and are recovered from the column, but should neither decrease the analytical sensitivity nor create polyatomic interferences with the elements of primary interest. It is our experience that a 50 mM ammonium acetate buffer works equally well as the Tris HCl buffer solution.
15. In a previous article by Persson et al. (15) several wash procedures were tested. The fastest and most efficient one was repetitive injections of 5 mM EDTA dissolved in the mobile phase, in this case 50 mM ammonium acetate at a pH of 7.5. Similar results were obtained in tests with 50 mM Tris HCl buffer solution. In between every analytical run, the EDTA-solution was injected and the areas of the eluting contaminant peaks were recorded online by ICP-MS (see Fig. 2). Optimization of the procedure showed that by injecting 20 μL 5 mM EDTA-solution repeatedly five times with a delay of 3 min between injections, the level of Cd and Cu contamination could be reduced by approximately 85% per injection. The procedure provided chromatographic results with a high reproducibility. The standard deviation of the integrated peak areas of fraction 1, 2, and 3 was 3, 2, and 4%, respectively, for the repetitions (see Fig. 2). Zn, Fe, Mn, and Ni are also removed from the column in a similar manner. As concentrations and extraction efficiencies differ, the EDTA concentration and number of injections must be adjusted to fully rinse the sample of choice.

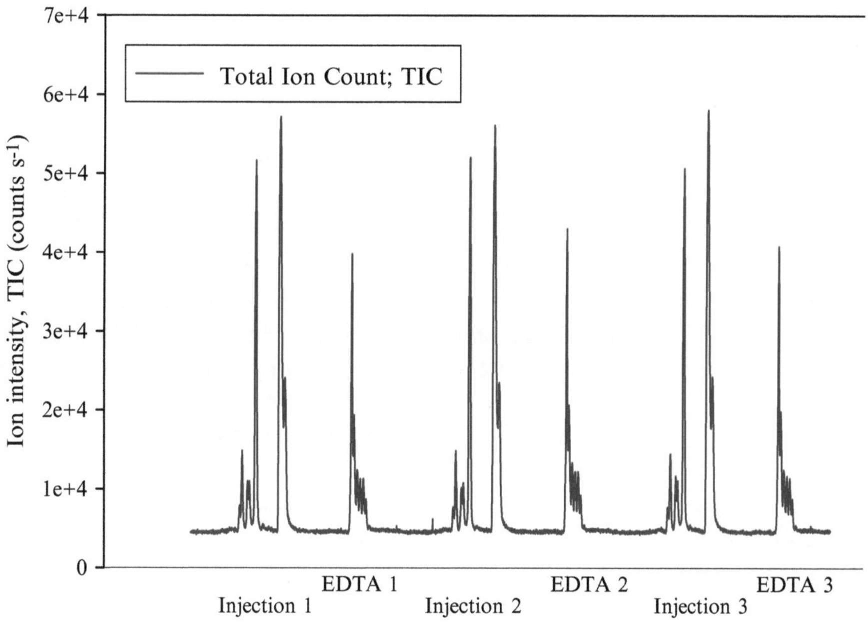

Fig. 2. Speciation chromatogram from a barley grain sample showing the total ion count. Injections 1–3 show the sample injections and EDTA 1–3 show the online cleaning procedure.

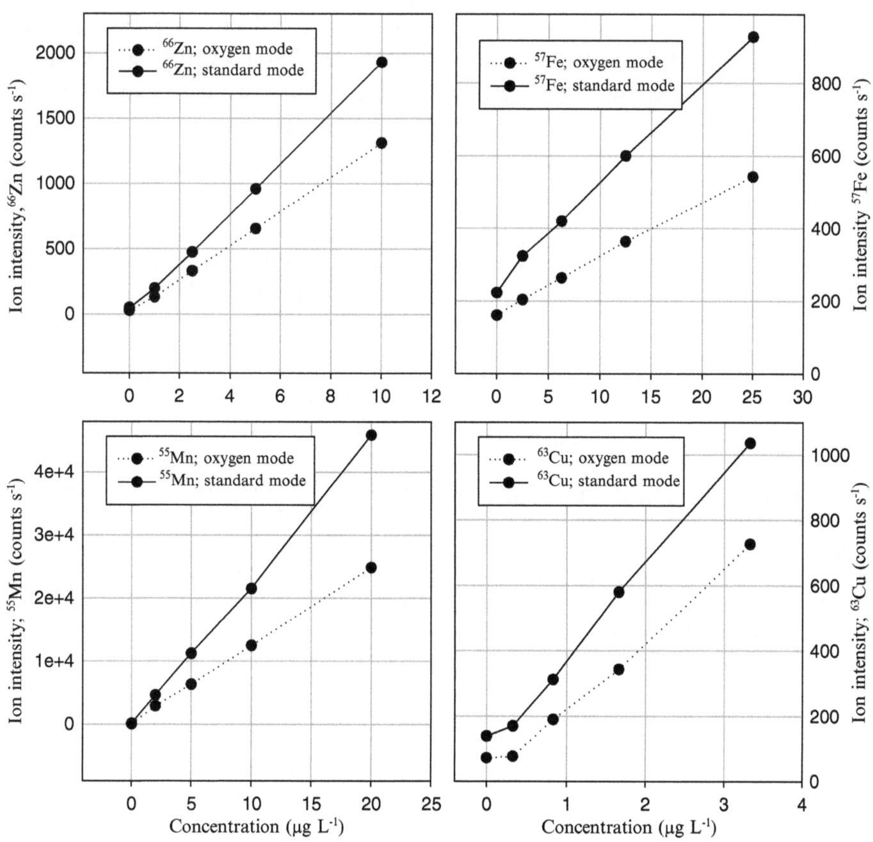

Fig. 3. The response factors for Zn, Fe, Mn, and Cu in standard and oxygen mode.

16. Oxygen addition generally decreases sensitivity of elements. The nonoxide signals of Zn, Mn, Fe, and Cu decreases with approximately 10%, compared to standard mode (see Fig. 3). Recently, it was found that also Fe can be monitored as its oxide product ion; $^{72}FeO^{+}$, which lowered LOD 20-fold compared to ^{57}Fe in standard mode (11).
17. Sometimes peak shape can be problematic when injecting calibration standards in flow injection. Addition of 1 mM EDTA to the standards may improve peak shape significantly. Remember to subtract a blank sample of the EDTA since it usually contains trace amounts of most metals.

Acknowledgements

Financial support from the EU-FP6 project META-PHOR (FOOD-CT-2006-03622), the EU-FP6 project PHIME (FOOD-CT-2006-016253), The Danish Research Council for Technology and Production Sciences (project 23-04-0082 and 10-100087), and The Danish Ministry of Food, Agriculture and Fisheries (via the OrgTrace project (project number 3304-FOJO-05-45-01) coordinated by the International Centre for Research in Organic Food Systems, ICROFS) is gratefully acknowledged.

References

1. Lönnerdal, B. (2002) Phytic acid – trace element (Zn, Cu, Mn) interactions. *Int. J. Food Sci. Tech.* **37**,727–39.
2. Welch R. M. and Graham R. D. (1999) A new paradigm for world agriculture: meeting human needs Productive, sustainable, nutritious. *Field Crops Res.* **60**, 1–10.
3. Wikipedia (2010) Online: http://en.wikipedia.org/wiki/white_rice
4. Ockenden, I., Dorsch, J. A., Reid, M. M., Lin, L., Grant, L. K., Raboy V., and Lott, J. N. A. (2004) Characterization of the phosphorus, inositol phosphate and cations in the grain tissues of four barley (*Hordeum vulgare* L.) low phytic acid genotypes. *Plant Sci.* **167**, 1131–42.
5. Talamond, P., Doulbeau, S., Rochette, I., Guyot, J.-P., and Treche S. (2000) Anion-exchange high-performance liquid chromatography with conductivity detection for the analysis of phytic acid in food. *J. Chromatogr. A.* **871**, 7–12.
6. Peroza, E. A. and Freisinger, E. (2007) Metal ion binding properties of *Triticium aestivum* E_c-1 metallothionein: evidence supporting two separate metal thiolate clusters. *J. Biol. Inorg. Chem.* **12**, 377–91.
7. Templeton, D. M., Ariese F., Cornelis, R., Danielsson L-G., Muntau, H., van Leewen H. P. and Lobinski, R. (2000) Guidelines for terms related to Chemical Speciation and fractionation of elements. Definitions, structural aspects, and methodological approaches; IUPAC recommendations. *Pure Appl. Chem.* **72**, 8, 1453–1470.
8. Francesconi, K. A. and Sperling, M. (2005) Speciation analysis with HPLC–mass spectrometry: time to take stock. *The Analyst.* **130**, 998–1001.
9. Encyclopædia Britannica (2010) Online: http://www.britannica.com/EBchecked/topic/502259/rice
10. Hansen, T. H., Laursen K. H., Persson, D.P., Pedas P., Husted, S. and Schjoerring J. K. (2009) Micro-scaled high-throughput digestion of plant tissue samples for multi-elemental analysis. *Plant Methods.* **5**, 1–11.
11. Persson, D. P., Hansen, T. H., Laursen, K. H., Schjoerring, J. K., and Husted, S. (2009) Simultaneous zinc, iron, sulphur and phosphorus speciation analysis of the barley grain tissues using SEC-ICP-MS and IP-ICP-MS. *Metallomics.* **1**, 418–426.

12. Hann, S., Koellensperger, G., Obinger, C., Furtmüller, P.G., and Stingeder, G. (2004) SEC-ICP-DRCMS and SEC-ICP-SFMS for determination of metal-sulfur ratios in metalloproteins. *J. Anal. At. Spectrom.* **19**, 74–79.
13. Nischwitz, V., Michalke, B., and Kettrup, A. (2003) Optimisation of extraction procedures for metallothionein-isoforms and superoxide dismutase from liver samples using spiking experiments. *The Analyst* **128**, 109–115.
14. Jakobsen, M. K., Poulsen, L. R., Schulz, A., Fleurat-Lessard, P., Møller, A., Husted, S., Schiøtt, M., Amtmann, A., and Palmgren, M. G. (2005) Pollen development and fertilization in *Arabidopsis* is dependent on the MALE GAMETOGENESIS IMPAIRED ANTHERS gene encoding a Type V P-type ATPase. *Genes Dev.* **19**, 2757–2769.
15. Persson, D. P., Hansen, T. H., Holm, P. E., Schjoerring, J. K., Hansen, H. C. B., Nielsen, J., Cakmak, I., and Husted, S. (2006) Multi-elemental speciation analysis of barley genotypes differing in tolerance to cadmium toxicity using SEC-ICP-MS and ESI-TOF-MS. *J. Anal. At. Spectrom.* **21**, 996–1005.

Chapter 14

The Use of Genomics and Metabolomics Methods to Quantify Fungal Endosymbionts and Alkaloids in Grasses

Susanne Rasmussen, Geoffrey A. Lane, Wade Mace, Anthony J. Parsons, Karl Fraser, and Hong Xue

Abstract

The association of plants with endosymbiotic micro-organisms poses a particular challenge to metabolomics studies. The presence of endosymbionts can alter metabolic profiles of plant tissues by introducing non-plant metabolites such as fungal specific alkaloids, and by metabolic interactions between the two organisms. An accurate quantification of the endosymbiont and its metabolites is therefore critical for studies of interactions between the two symbionts and the environment.

Here, we describe methods that allow the quantification of the ryegrass *Neotyphodium lolii* fungal endosymbiont and major alkaloids in its host plant *Lolium perenne*. Fungal concentrations were quantified in total genomic DNA (*g*DNA) isolated from infected plant tissues by quantitative PCR (qPCR) using primers specific for chitinase A from *N. lolii*. To quantify the fungal alkaloids, we describe LC-MS based methods which provide coverage of a wide range of alkaloids of the indolediterpene and ergot alkaloid classes, together with peramine.

Key words: *Neotyphodium lolii*, *Lolium perenne*, Endosymbiosis, Quantitative PCR, Chitinase A, Indolediterpenes, Ergot alkaloids, Peramine

1. Introduction

Temperate grasses (subfamily Poëideae) are often associated with clavicipitaceous fungi (genus *Epichloë*), which live mainly in the apoplastic spaces of above ground plant parts (1). These endosymbionts usually cause no visible symptoms of infection and can, especially in agricultural ecosystems, contribute considerably to improved plant productivity (2). A range of anti-herbivorous alkaloids are produced by *Epichloë* fungi, depending on species and strain (3). The common strains of the asexual *Neotyphodium lolii*

Nigel W. Hardy and Robert D. Hall (eds.), *Plant Metabolomics: Methods and Protocols*, Methods in Molecular Biology, vol. 860, DOI 10.1007/978-1-61779-594-7_14, © Springer Science+Business Media, LLC 2012

produce the alkaloids peramine, lolitrem B, and ergovaline, together with a range of minor indolediterpenes (4) and ergot alkaloids (5) in their natural host, *Lolium perenne*.

Studies of endophyte hyphae and alkaloid distribution within host plants have shown that concentrations of hyphae and individual alkaloids follow distinct spatial and temporal gradients, and that the ratios between hyphal abundance and that of alkaloids of different classes differ in each tissue (6). A controlled environmental study has also shown that both endophyte hyphal (assessed by quantitative PCR; qPCR) and alkaloid concentrations are strongly affected by nutrient supply (e.g. nitrogen) and host plant metabolic composition (7). Conversely, studies on the impact of endophyte infection on the metabolic profiles of the symbiotum (combined plant and fungal metabolome analysis) have revealed that these are tissue specific (8) and depend also to a great extent on environmental factors and plant and fungal genetics (9). It is therefore of critical importance to relate a given metabolic profile of these and other endosymbiotic associations to actual abundances of the endosymbionts, to specific tissues, and to the environmental conditions the symbiotum was subjected to.

A diverse range of bioactive alkaloids with differing polarities are present in the symbiotum at widely differing concentrations. With standard LC-UV and LC-fluorescence methods only a few major compounds can be measured (6). Recently developed LC-MS-based analytical methods (10) provide improved quantification limits and selectivity, and facilitate the measurement of an extended range of alkaloids. Mass spectrometry using a Linear Ion Trap also provides high quality MS^2 spectral information for peak identification, confirmation, and quantification. With this methodology, a more comprehensive view is afforded of the tissue distribution and effects of environment on the broad alkaloid profile.

2. Materials

2.1. Analysis of Endophyte Abundance

1. Genomic DNA (*g*DNA) extraction: DNeasy® Plant Mini Kit (Qiagen).
2. 96–100% ethanol.
3. Eppendorf Thermo mixer (Eppendorf).
4. NanoDrop® ND-100 Spectrophotometer (NanoDrop Technologies).
5. Primers for PCR: *N. lolii* chitinase A (forward primer: aagtccaggctcgaattgtg, reverse primer: ttgaggtagcggttgttcttc, amplicon size: 353 bp).
6. Plasmids for qPCR calibration: TOPO vectors and One Shot® *E. coli* cells (Invitrogen).

7. Luria-Bertani (LB; 1% (w/v) tryptone, 0.5% (w/v) yeast extract, 1.0% (w/v) NaCl) medium, LB agar (1.5% agar; Invitrogen) plates.
8. Ampicillin or kanamycin (Sigma).
9. X-gal (5-bromo-4-chloro-3-indolyl-b-D-galactopyranoside; Invitrogen).
10. QIAprep® Miniprep (Qiagen) kit.
11. Quantitative PCR: Primers, *g*DNA.
12. MilliQ® water.
13. iQ SYBR Green Supermix (Bio-Rad Laboratories Pty. Ltd.).
14. MyiQTM cycler (Bio-Rad).
15. 200 μl PCR strip tubes or 96-well plates (Axygen).

2.2. Analysis of Alkaloids

1. Instrument: The procedure assumes the use of a Thermo LTQ linear ion trap mass spectrometer equipped with an HPLC system using a Jasco X-LC-3080DG degasser, two Jasco X-LC 3185PU high pressure LC pumps, a Jasco X-LC3180MX high pressure mixer and a HTS-Combi-PAL auto sampler, but should be adaptable to other LC-MS/MS instrumentation.
2. Acetonitrile (Baker Analyzed HPLC Solvent, J.T. Baker).
3. Water (MilliQ®) (40: 60 v/v) containing 0.1% acetic acid (solvent A).
4. Acetonitrile containing 0.1% acetic acid (Analar, BDH) (solvent B).
5. Luna C18 column (150 × 2.0 mm; Phenomenex).
6. Tuning standard: 1 μg/ml paxilline (Sigma) in isopropanol–water (50:50 v/v).
7. 5 mM ammonium acetate (Analar, BDH) in water (MilliQ®) (solvent C).
8. Acetonitrile (solvent D).
9. Gemini C18 column (150 × 2.0 mm; Phenomenex).
10. Tuning standard: 1 μg/ml agroclavine (Sigma) in methanol–water (50:50 v/v).

3. Methods

Both endophyte abundance and endophyte alkaloid concentrations in infected plant tissues are strongly affected by a wide range of factors. These include tissue localization, environmental parameters such as nutrients, temperature, light, and re-growth periods, and genetic parameters such as endophytic strain and plant genotype/cultivar.

It is therefore advised to carefully plan and design experimental conditions for studies on the grass–endophyte symbiosis.

The selection of an appropriate gene sequence is critical for successful and meaningful endophyte quantification data. We recommend using the chitinase A gene from *N. lolii* for quantification of *N. lolii* strains; however, the quantification of other *Neotyphodium* or *Epichloë* species requires the identification of a gene from those species. These genes can be identified from in-house resources or public databases. Check that the selected sequence is specific for the fungal species to be analysed and that it represents a single copy gene by Southern blot analysis.

The LC-MS/MS procedures described here provide for the relative quantification of a wide range of the alkaloids produced by common toxic strains of *N. lolii* in *L. perenne*. Similar LC-MS/MS methodology can be applied to the analysis of the alkaloids of other classes of endophyte strains in other hosts with differing alkaloid profiles, but this requires prior exploratory LC-MS/MS analyses to define the appropriate MS^2 selection and filter ions, and chromatographic time windows for their detection.

3.1. Get Ready for Quantitative PCR

1. Transfer approx. 10 mg freeze-dried and finely ground plant tissue powder into a 2-ml Eppendorf tube (see Note 1). Isolate *g*DNA using DNeasy® Plant Mini Kit following the manufacturer's instructions (see Note 2).
2. Measure *g*DNA concentration using a NanoDrop® spectrophotometer. Blank the instrument, place 2 μl of AE buffer (from DNeasy® Kit) on the sensor; the reading should be less than 0.5 ng/ml. Wipe the sensor and place 2 μl of *g*DNA solution on it; the instrument automatically calculates the DNA concentration. Check the quality of your sample DNA (the ratio of absorbance at 260/280 nm should be approx. 1.8), and repeat this step three times for each sample with fresh 2 μl aliquots (see Note 3). The mean value of the three measurements is used to adjust the *g*DNA concentration to 0.5 ng/μl by dilution with AE buffer for subsequent qPCR.
3. To increase PCR efficiency, design primers to a sequence region within the selected gene which does not form strong secondary structures using http://www.bioinfo.rpi.edu/applications/mfold/dna/form1.cgi.
4. Once the target sequence is selected use Primer Express 3.0 software (Applied Biosystems) to design primers suitable for qPCR. Major criteria for qPCR primers are: 20–25 bases long, a predicted melting temperature of 60 ± 1°C, a guanine–cytosine (GC) content between 50 and 60%, and a maximum 3′ complementarity of 3.00 (for additional recommendations see the BioRad iCycler iQ handbook).

5. Perform a BLASTN search against GenBank to ensure that primers are unique to the gene of interest (see Note 4).
6. Prepare stock solutions of desalted primers (as obtained from manufacturer) with sterilised MilliQ® water to a final concentration of 100 μM (10 × working concentration) and store aliquots at −20°C.
7. To prepare plasmid DNA, perform standard PCR with the above primers, clone the PCR product (directly use the PCR reaction mix) into TOPO 2.1 vectors, and transform OneShot® *E. coli* cells with the vector following the manufacturer's instructions.
8. Plate transformed cultures in three dilutions (10 μl, 50 μl, 100 μl) onto pre-warmed LB agar plates containing 100 μg/ml ampicillin or 50 μg/ml kanamycin and 40 μl of 40 mg/ml X-gal, and incubate at 37°C overnight.
9. Transfer three individual white or light blue colonies with a sterilised tip into three individual 15-ml tubes containing 5 ml LB medium with 100 μg/ml ampicillin or 50 μg/ml kanamycin, and incubate overnight at 37°C in a shaking incubator at 200 rpm.
10. Isolate plasmid DNA from the transformed culture using the QiaPrep Miniprep® Kit following the manufacturer's instructions. Sequence the isolated plasmid DNA to ensure correct insert sequence using standard sequencing procedures.
11. Accurately determine the plasmid DNA concentration using the NanoDrop® spectrophotometer and dilute to a stock concentration of 2×10^8/μl. Prepare a set of serial tenfold dilutions from 2×10^5 to 2×10 copies by transferring 100 μl of the previous dilution into a new 1.5-ml Eppendorf tube containing 900 μl AE (from DNeasy® kit) buffer (see Note 5).

3.2. Quantitative PCR

1. For general considerations see Note 6. Set PCR reactions up in strip tubes or 96-well plates, each set-up should include *g*DNA to be tested (three technical replicates), serial dilutions of plasmid DNA containing the template of interest (see Subheading 3.1 step 4), and one negative control (autoclaved MilliQ® water).
2. Prepare a master mix containing per reaction 12.5 μl 2 × SYBR Green reagent, 0.75 μl forward primer (10 μM), 0.75 μl reverse primer (10 μM), and 1 μl autoclaved MilliQ® water in a 2-ml Eppendorf tube. Invert the tube several times to mix. Transfer 15 μl of the master mix into each tube or well, add 10 μl sample *g*DNA (containing 5 ng DNA, see Subheading 3.1 step 1), plasmid DNA (standards, see Subheading 3.1 step 4), or water (negative control). Mix the samples by vortexing for 3 × 1 s, followed by a brief spin (up to 2,500 ×*g*) in a centrifuge

to collect all reagents at the bottom of the well and to remove air bubbles.

3. Perform qPCR using the following thermocycle programme: 5 min at 95°C to activate polymerase, followed by 40 cycles of 20 s at 95°C, 30 s at 60°C, and 30 s at 72°C. Monitor fluorescence during the annealing step at 60°C.

4. Check that the negative control has not generated any signals which would indicate primer-dimer formation or master mix contamination. Conduct a dissociation analysis after the final amplification cycle following the manufacturer's instructions and examine the melt curve profiles—they should show only one sharp peak. The presence of several peaks is an indication of unspecific amplification products and primers need to be re-designed. Display the PCR amplification plots in linear view and use baseline subtraction to adjust baseline settings.

5. Examine the shape of the logarithmic PCR amplification plots to identify abnormal plots. Figure 1 shows a correct amplification plot with a linear baseline region, an exponential curve for the amplification phase, followed by a plateau, and parallel slopes. Incorrect amplification reactions must be discarded from the data set. Set the threshold level in the exponential region of the amplification where slopes are parallel and above the background noise of the baseline. Threshold levels should be held constant for all samples amplified with the same primers.

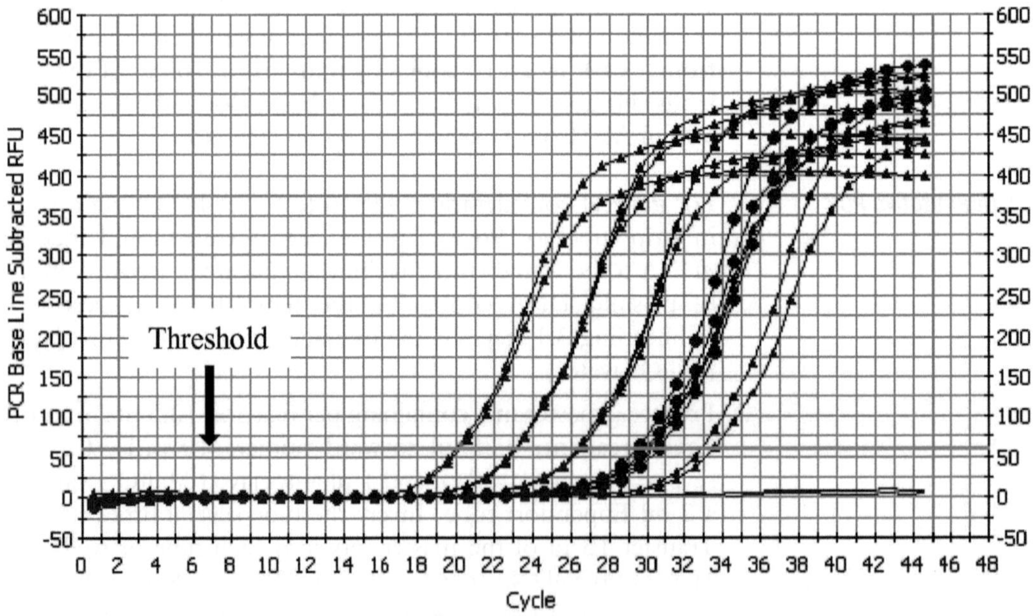

Fig. 1. PCR amplification plots of plasmid DNA dilution standards (*triangles*), gDNA test samples (*circles*), and negative control (*no symbol*).

6. Check that the slope of the standard curve is between −3.2 and −3.5 and the correlation coefficient is larger than 0.99. Delete all sample data points outside of the standard curve. Check that the C_t (cycle threshold) values for the three technical replicates are within 0.5 C_t of each other and delete outliers.
7. The copy number of template DNA in the samples is calculated by the MyiQTM cycler programme based on the plasmid standard curve and expressed as copy number per 5 ng *g*DNA.

3.3. Indolediterpenoid Analysis

1. Extract a 50 mg freeze-dried and ground sample with 1 ml isopropanol (Analar, BDH) by rotating in a 2 ml vial for 1 h. Centrifuge (8,000 × *g*, 5 min), and transfer supernatant to a glass 2-ml HPLC vial. Suitable controls should be prepared in a similar manner (see Note 7).
2. Prepare the elution solvents and tuning standard solution (see Subheading 2.2 item 2).
3. Set the MS to operate in positive ESI mode with the capillary at 275°C, the probe voltage at 5 kV and with N_2 as carrier gas (see Note 8). Optimise the MS instrument tuning parameters while infusing the paxilline tuning standard solution.
4. Perform HPLC with a flow rate of 0.2 ml/min with the column oven set at 25°C. Inject 5 μl of the sample extract. Apply a linear gradient from 80% A:20% B to 50% A:50% B over 20 min, then a further linear gradient to 100% B over another 20 min, and hold at 100% B for 10 min before returning to 80% A:20% B over 5 min, with a hold for re-equilibration of 5 min before injecting the next sample.
5. Collect data in both full scan and selective reaction monitoring mode. Set up four chromatographic segments with targeted MS^2 events as shown in Table 1 (see Notes 9–11). Set the isolation width for MS^2 precursor ions to ±1 amu (see Note 12).
6. Process the data using Xcalibur software for quantification.

3.4. Ergot Alkaloids and Peramine Analysis

1. Prepare extracts and control samples as described in Subheading 3.3 step 1, but with 1 ml isopropanol: water (1:1 v/v) as extraction solvent (see Note 7).
2. Prepare the elution solvents and tuning standard solution (see Subheading 2.2 item 3).
3. Set the MS to operate in positive ESI method with the capillary at 275°C, the probe voltage at 5 kV and N_2 as carrier gas. Optimise the MS instrument tuning parameters while infusing the agroclavine tuning standard solution.
4. Perform HPLC with a flow rate of 0.2 ml/min with the column oven set at 25°C. Inject 15 μl of the sample extract. Apply a linear gradient from 95% C:5% D to 50% C:50% D over 38 min,

Table 1
Indolediterpenoid analysis by LC-MS selected reaction monitoring the following: chromatogram segments, analyte, selected ions, and retention times

Segment	1	Time	0–13.5 min
Analyte	**MS1 precursor ion (m/z)**	**MS2 filter ions (m/z)**	**Retention time (min)**
lolitrem N	620.4	562.4	9
lolitriol[a]	620.4	562.4	10.8
Segment	**2**	**Time**	**13.5–22.0 min**
Analyte	**MS1 precursor ion**	**MS2 filter ions**	**Retention time**
Paspaline-B[a]	436.3	420.3	15.4
Paxilline[a]	436.3	420.2	16.2
Terpendole E[a]	438.3	422.2	16.7
Lolitrem K	602.3	544.4	16.8
Lolitrem M	604.3	546.4	15.5
Lollicine	604.3	546.4	18.6
Lolitrem J	662.4	604.4	15
Segment	**3**	**Time**	**22.0–28.6 min**
Analyte	**MS1 precursor ion**	**MS2 filter ions**	**Retention time**
13-desoxypaxilline[a]	420.3	402.2, 405.2	22.8
Lolitrem A	702.4	644.3	25.7
Segment	**4**	**Time**	**28.6–40.0 min**
Analyte	**MS1 precursor ion**	**MS2 filter ions**	**Retention time**
Paspalline	422.3	130.2, 406.3	33.9
Terpendole C[a]	520.3	504.3	28.4
Lolitrem B[a]	686.4	628.3	29.8
Lolitrem E[a]	688.4	630.3	31.3

[a]Identified relative to authentic standard by co-chromatography and mass spectrum match

then a further linear gradient to 100% D over another 7 min and hold at 100% D for 10 min before returning to 95% C:5% B over 3 min, with a hold for re-equilibration of 8 min before injecting the next sample. Divert the column flow for the first 6 min and last 20 min of the run.

Table 2
Ergot alkaloid and peramine analysis by LC-MS selected reaction monitoring the following: chromatogram segments, analyte, selected ions, and retention times

Segment	**1**	**Time**	**0–21.3 min**
Analyte	**MS1 precursor ion (m/z)**	**MS2 filter ions (m/z)**	**Retention time (min)**
Peramine[a]	248.1	206	16.4
Chanoclavine[a]	257.2	226.1	20.1
Lysergic acid[a]	269.2	223.2	14.9
Isolysergic acid	269.2	223.2	16.9
Lysergylalanine	340.3	208.2, 223.2	18.5
Isolysergylalanine	340.3	208.2, 223.2	20.8
Segment	**2**	**Time**	**21.3–32.2 min**
Analyte	**MS1 precursor ion**	**MS2 filter ions**	**Retention time**
Agroclavine[a]	239.2	183.1	29.5
Elymoclavine[a]	255.2	224.1	22.4
Lysergol[a]	255.2	240.2	22.7
Isosetoclavine	255.2	237.1	26.3
Setoclavine[a]	255.2	237.1	28.8
Ergine	268.3	223.2	22.5
Erginine[a]	268.3	223.2	26.8
Segment	**3**	**Time**	**32.2–44 min**
Analyte	**MS1 precursor ion**	**MS2 filter ions**	**Retention time**
Dehydroergovaline	532.3	208.2, 223.2, 268.2, 320.2	34.4
Dehydroergovalinine	532.3	514.2	41.0
Ergovaline[a]	534.3	208.2, 223.2, 268.2, 320.2	35.6
Ergovalinine[a]	534.3	516.2	42.0

[a]Identified relative to authentic standard by co-chromatography and mass spectrum match

5. Collect data in both full scan and selective reaction monitoring mode, averaging three micro scans for every point collected. Set up three chromatographic segments each with targeted MS^2 events as shown in Table 2 (see Notes 9–11). Set the isolation width for MS^2 precursor ions to ±1 amu (see Notes 12 and 13).
6. Process the data using Xcalibur software for quantification.

4. Notes

1. For accurate qPCR results it is important to avoid any DNA degradation to ensure that plant tissues remain frozen during storage and grinding. PCR amplification and LC-MS analysis are very sensitive methods; to avoid cross contamination of samples use a clean, autoclaved mortar and pestle for each sample. To improve extraction efficiency, grind the tissue samples to a very fine powder in liquid nitrogen.
2. The quality of the isolated *g*DNA is critical for accurate qPCR quantification. Contaminations such as cell walls, proteins, polysaccharides, detergents, alcohol, and high salt concentrations may affect qPCR efficiency and accuracy. It is therefore important to follow the manufacturer's protocol for the DNeasy Mini spin kit (Qiagen). Particularly important: Do not mix buffer AP1 with RNAse A before use. Remove any clumps of tissue by pipetting and vortexing or using a sterilised micropestle to ensure complete lysis of cells—insufficient lysis will result in low *g*DNA yields. It is important to dispense buffer AP3/E directly onto the lysate and to mix immediately. Buffers AW and AP3/E are supplied as concentrates—before using the kit for the first time, add the appropriate amount of ethanol as indicated on the bottles to obtain working solutions. When removing the DNeasy Mini spin column from the collection tube the column must not come into contact with the flow through, as this will result in carryover of ethanol.
3. The sensor of the NanoDrop instrument must be washed with distilled water and wiped clean with a soft Kimwipes® tissue between measurements. To ensure accurate reading, DNA must be dispensed carefully onto the sensor without any air bubbles. Each DNA sample should be quantified three times with a difference of less than 2 ng/μl between measurements.
4. To ensure efficiency, specificity, sensitivity, and absence of primer-dimers, at least three pairs of primers should be designed per gene and tested by preliminary qPCR using each individual primer pair on the same sample set and including samples in which target DNA should be absent (*g*DNA isolated from endophyte free samples). The primer pair with the lowest C_t and without any non-specific products and primer-dimers should be chosen for the qPCR experiments.
5. Accurate plasmid DNA standard dilutions are critical for the analysis of qPCR data, as test *g*DNA samples are calibrated against the plasmid DNA standard curves. Accurate pipetting, regular calibration of pipettes, and the use of pipettes appropriate for the

sample volume are prerequisites. To avoid sample contamination, use dedicated pipette sets, change pipette tips for each dilution step, and dispense the AE buffer first into each Eppendorf tube. Each dilution of plasmid DNA should be divided into small aliquots and stored at −20°C until use.

6. General and very important considerations for qPCR are high quality *g*DNA, accurate DNA quantification, accurate dilution of sample DNA, accurate pipetting, use of filter barrier pipette tips, and no drawings on PCR tubes, plates, or lids. Change gloves regularly and maintain clean and dedicated areas for DNA preparation, PCR set-up, PCR amplification, and PCR product analysis.
7. To analyse alkaloids, a suitable control would be a sample known to contain appreciable levels of the indolediterpenes (3.3) or ergot alkaloids (3.4). A composite control can be prepared by combining aliquots from each extracted sample.
8. Indolediterpenes can also be detected in positive APCI mode which provides higher sensitivity for the higher MWt compounds, but lower sensitivity for the lower MWt compounds. Note that source fragmentation can occur in APCI mode, and these source fragments may be useful selection ions for MS/MS analysis.
9. A full scan MS1 (e.g. range 180–800 *m/z*) can also be included in each segment to provide untargeted analysis data.
10. Chromatographic segment times shown in Table 1 (indolediterpenoids) and Table 2 (ergot alkaloids and peramine) are the values used for collection of the data shown in Figs. 2 and 3. Pre-equilibration of the column with several blank runs is advised to ensure stable segment windows. Some adjustment is likely to be required for application on another instrument and between batches of runs. This can be carried out with a control sample (see Note 7).
11. Peak assignments shown in Tables 1 and 2 are based on authentic standards or published data (4, 5). Paxilline, lysergol, and agroclavine were obtained from commercial suppliers (Sigma). We gratefully acknowledge the assistance of the following in providing authentic standards: Miroslav Flieger and Vladimir Kren (Academy of Sciences of the Czech Republic, Prague, Czech Republic; clavines and lysergyl compounds); Satoshi Ōmura (Kitasato University, Tokyo, Japan; terpendoles), Forrest Smith (Auburn University, USA; ergovaline and ergovalinine); Barry Scott (Massey University, Palmerston North, New Zealand; simple indolediterpenes); Brian Tapper (AgResearch, Palmerston North, New Zealand; peramine); and Sarah Finch and Chris Miles (AgResearch, Hamilton, New Zealand; indolediterpenes).

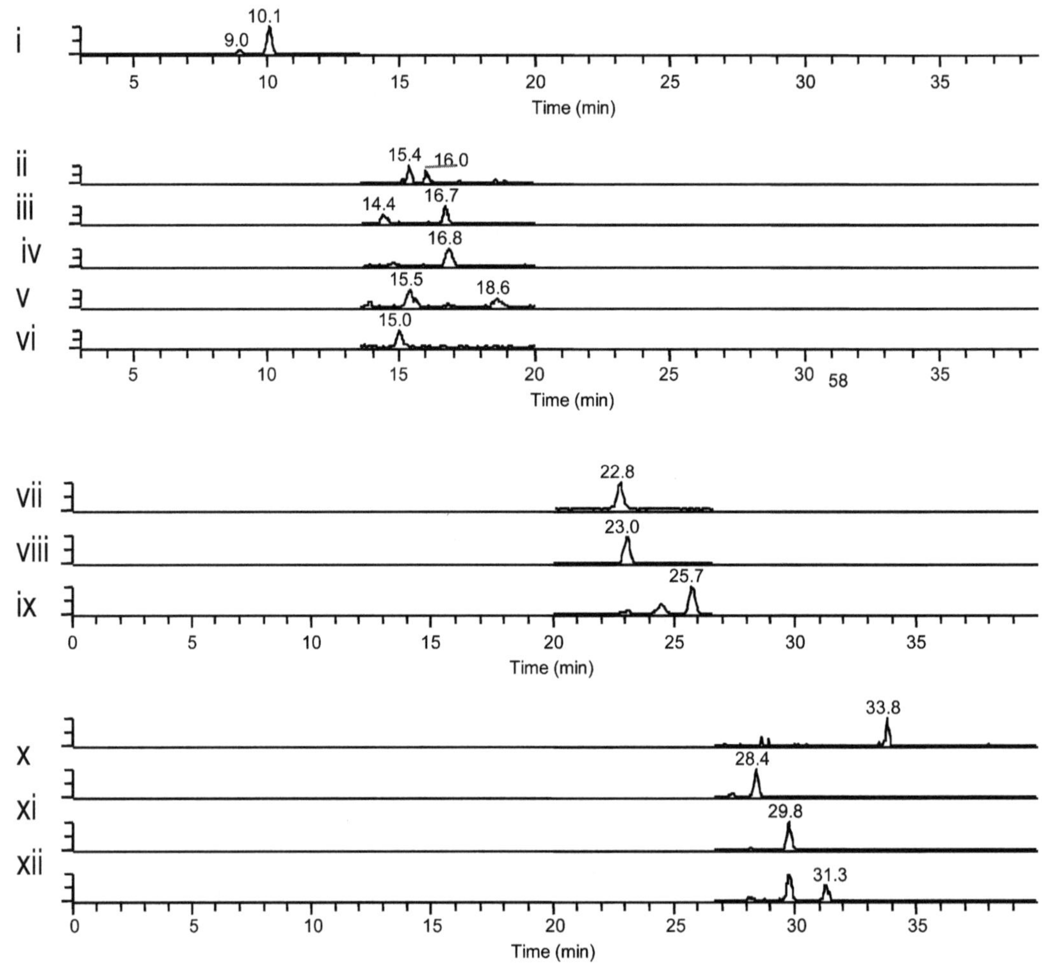

Fig. 2. Extracted ion chromatograms from LC-MS analysis by selective reaction monitoring of indolediterpenoids in an extract of perennial ryegrass (*L. perenne)* infected with an *N. lolii* endophyte strain. The traces show signals for MS^2 filter ions from fragmentation of selected MS^1 ions: (i) 620.4 > 562.4; (ii) 436.3 > 420.3; (iii) 438.3 > 422.2; (iv) 602.3 > 544.4; (v) 604.3 > 546.4; (vi) 662.4 > 604.4; (vii) 420.3 > 402.2, 405.2; (viii) 534.3 > 518.3; (ix) 702.4 > 644.3; (x) 422.3 > 130.2, 406.3; (xi) 520.3 > 504.3; (xii) 686.4 > 628.3; (xiii) 688.4 > 630.3. Assigned peaks in the chromatograms are listed in Table 1.

12. LC-MS analysis reveals further complexity in the profile of indolediterpenoids and ergot alkaloids. Some peaks evident in Fig. 2 (putative indolediterpenoids) and Fig. 3 (putative clavines) remain unassigned. Additional candidate indolediterpenoids were detected with the following selective reactions: segment 1: 454.3 > 438.3; 602.3 > 544.3; 618.3 > 560.3, 636.3 > 578.3; segment 2: 702.4 > 642.3; segment 3: 644.4 > 586.4, 684.4 > 626.3; 700.4 > 642.3; 704.4 > 646.3; 744.4 > 686.4; 760.4 > 702.3; segment 4: 518.3 > 398.2, 434.3 > 416.3; 684.4 > 626.3; 688.4 > 630.3; 704.4 > 646.3. Additional candidate clavines were detected with the following selective reaction: segment 1: 241.2 > 210.2.

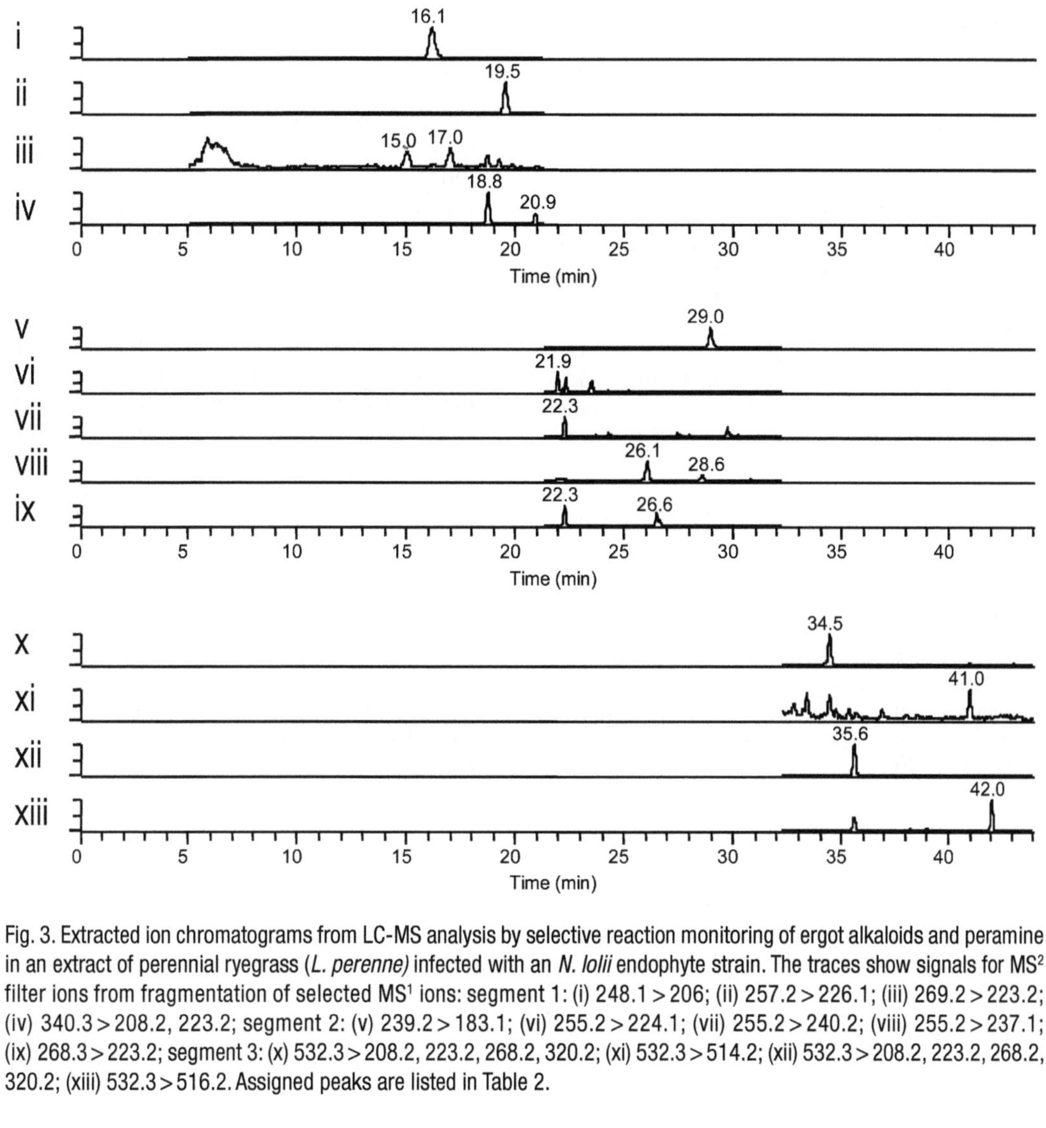

Fig. 3. Extracted ion chromatograms from LC-MS analysis by selective reaction monitoring of ergot alkaloids and peramine in an extract of perennial ryegrass (*L. perenne)* infected with an *N. lolii* endophyte strain. The traces show signals for MS^2 filter ions from fragmentation of selected MS^1 ions: segment 1: (i) 248.1 > 206; (ii) 257.2 > 226.1; (iii) 269.2 > 223.2; (iv) 340.3 > 208.2, 223.2; segment 2: (v) 239.2 > 183.1; (vi) 255.2 > 224.1; (vii) 255.2 > 240.2; (viii) 255.2 > 237.1; (ix) 268.3 > 223.2; segment 3: (x) 532.3 > 208.2, 223.2, 268.2, 320.2; (xi) 532.3 > 514.2; (xii) 532.3 > 208.2, 223.2, 268.2, 320.2; (xiii) 532.3 > 516.2. Assigned peaks are listed in Table 2.

13. LC-MS/MS quantification of ergopeptides is only semi-quantitative because of the variable degree of epimerisation of ergopeptides at C8 which can take place during the handling and storage of plant material, and during extraction and the storage of extracts. While UV and fluorescence detector responses to the epimers is likely to be similar, this is not the case for collision-induced fragmentation in the ion-trap, where epimers show widely differing product ions ratios, as reflected in the suggested MS^2 filter ions for epimer pairs such as ergovaline and ergovalinine (see Table 2).

References

1. Leuchtmann, A. (1992) Systematics, distribution, and host specificity of grass endophytes. *Nat. Toxins* **1**, 150–162.
2. Schardl, C., Leuchtmann, L.A., and Spiering, M.J. (2004) Symbiosis of grasses with seedborne fungal endophytes. *Ann. Rev. Plant Biol.* **55**, 315–340.
3. Bush, L.P., Wilkinson, H.H., and Schardl, C.L. (1997) Bioprotective alkaloids of grass-fungal endophyte symbiosis. *Plant Physiol.* **114**, 1–7.
4. Gatenby, W.A., Munday-Finch, S.C., Wilkins, A.L., and Miles, C.O. (1996). Terpendole M, a novel indole-diterpenoid isolated from *Lolium perenne* infected with the endophytic fungus *Neotyphodium lolii*. *J. Agric. Food Chem.*, **47**, 1092–1097.
5. Panaccione, D.G., Tapper, B.A., Lane, G.A., Davies, E. and Fraser. K. (2003) Biochemical outcome of blocking the ergot alkaloid pathway of a grass endophyte. *J. Agric. Food Chem.* **51**, 6429–6437.
6. Spiering, M.J., Lane, G.A., Christensen, M.J., and Schmid, J. (2005) Distribution of the fungal endophyte *Neotyphodium lolii* is not a major determinant of the distribution of fungal alkaloids in *Lolium perenne* plants. *Phytochemistry* **66**, 195–202.
7. Rasmussen, S., Parsons, A.J., Bassett, S., Christensen, M.J., Hume, D.E., Johnson, L.J., Johnson, R.D., Simpson, W.R., Stacke, C., Voisey, C.,R., Xue, H., and Newman, J.A. (2007) High nitrogen supply and carbohydrate content reduce fungal endophyte and alkaloid concentration in *Lolium perenne*. *New Phytol.* **173**, 787–797.
8. Cao, M., Koulman, A., Johnson, L.J., Lane, G.A., and Rasmussen, S. (2008) Advanced data-mining strategies for the analysis of direct-infusion ion trap mass spectrometry data from the association of perennial ryegrass with its endophytic fungus, *Neotyphodium lolii*. *Plant Physiol.* **146**, 1501–1514.
9. Rasmussen, S., Parsons, A.J., Fraser, K., Xue, H. and Newman, J.A. (2008) Metabolic profiles of *Lolium perenne* are differentially affected by nitrogen supply, carbohydrate content, and fungal endophyte infection. *Plant Physiol.* **146**, 1440–1453.
10. Koulman A, Lane GA, Christensen MJ, Fraser K, Tapper BA. (2006). Peramine and other fungal alkaloids are exuded in the guttation fluid of endophyte-infected grasses. Phytochemistry **68**, 355–360.

Part III

Data Analysis

Chapter 15

Data (Pre-)processing of Nominal and Accurate Mass LC-MS or GC-MS Data Using MetAlign

Arjen Lommen

Abstract

This paper gives a step-by-step account of how to install, set up, and run MetAlign software, which can be downloaded freely (http://www.metalign.wur.nl/UK/Download+and+publications). The software is used for accurate mass and nominal mass data coming from different kinds of GC-MS and LC-MS platforms. The algorithms are beyond the scope of this paper and were published separately.

Key words: GC-MS, LC-MS, Alignment, Preprocessing, MetAlign, Accurate mass, Nominal mass

1. Introduction

MetAlign (1) is a software package, which is used for the (pre-) processing of nominal and accurate mass GC-MS as well as LC-MS data from different manufacturers (2–11). Data derived from a separation technology combined with a mass spectrometer are becoming more and more informative due to improvements in separation technology and MS technology. A critical aspect for the future is the size of the data and the extraction of relevant information from the raw data. The time and money needed to perform the analysis of raw data is rapidly becoming a major bottleneck in metabolomics-based research. Preprocessing as described here is the process of deriving peak-picked (as in individual mass peaks) data from raw data. Alignment as described here is the process of creating a format of data derived from multiple preprocessed datasets in such a way that peaks can be compared on a peak to peak basis using, for instance, multivariate statistics. Details on the algorithms behind MetAlign were given in a previous paper (11).

Nigel W. Hardy and Robert D. Hall (eds.), *Plant Metabolomics: Methods and Protocols*, Methods in Molecular Biology, vol. 860, DOI 10.1007/978-1-61779-594-7_15, © Springer Science+Business Media, LLC 2012

This chapter has been written to help potential users to set up MetAlign for their applications.

2. Materials

2.1. Installation Requirements for MetAlign

MetAlign can be downloaded free of charge at http://www.metalign.wur.nl/UK/Download+and+publications/ (1). The installation requirements are the following:

(a) Windows XP, Windows NT, or Windows 2000 as the operating system on your PC.

(b) At least 1 GB of internal memory (SDRAM or better). It is recommended to exit any other memory consuming programs during execution.

(c) Free disk space of 80 GB is recommended to ensure that no disk space problems arise during alignment.

(d) The MetAlign program should be run at a screen resolution of 1,024 × 768 with small fonts or at higher resolution with large fonts.

(e) To install and run MetAlign you must have administrator rights.

2.2. Acquiring Data for MetAlign Processing

A detailed account of how to plan the sequence of your experiments is given in ref. (8) and in the documentation supplied with the download, i.e., "experimental_design_and_checks.ppt." It is advised to take small aliquots of all your samples and make a mixed sample as a control reference sample. Briefly, a sequence of triplicate samples would look like this: 5× mix sample—all first replica's randomized—1× mix sample—all second replica's randomized—1× mix sample—all third replica's randomized—1× mix sample. Before or after this sequence, additional references or blank controls may be run. Also in the case of accurate mass experiments, it could be advantageous to spike all of the samples with one or two deuterated reference compounds as a check on the precision of the measured accurate mass.

3. Methods

3.1. Installing MetAlign

After unzipping the MetAlign download, the software needs to be installed. This is done by double-clicking setup.exe in your MetAlign folder and clicking on the button "Complete installation of metAlign." For this to succeed, you need to have administrator rights on your PC. (To uninstall, use the button "Uninstall metAlign.")

3.2. Starting MetAlign

MetAlign is started by double-clicking on ms.exe in your MetAlign folder. To run MetAlign you need to have administrator rights. The screen in Fig. 1 should appear. In principle, the numbers given to describe buttons and parameters should be followed in ascending order. The interface is divided into three parts, namely, A, B, and C (see Fig. 1 for description). Standard parameters which are not system dependent are given in Table 1.

3.3. Configuring MetAlign

Clicking the button "1A. Program configuration" starts the config. exe subprogram as shown in Fig. 2. Start by defining where to find data and where to put data. In the box "Definition of Folders" this can be done by clicking the "Browse" buttons (see Note 1). If you want to load the settings and files from a previous session you should use the top "Browse" button in "Start from a Previous Metalign Session" box and follow Note 2.

Next use the "Data Format and Function Selection" to define "INPUT" and "OUTPUT FORMAT" as follows.

3.3.1. Masslynx Format

Masslynx format is accessed in line through Dbridge.exe (12). The Masslynx version on the MetAlign PC should be the same as or newer than that used for the MS machine. If Masslynx was installed

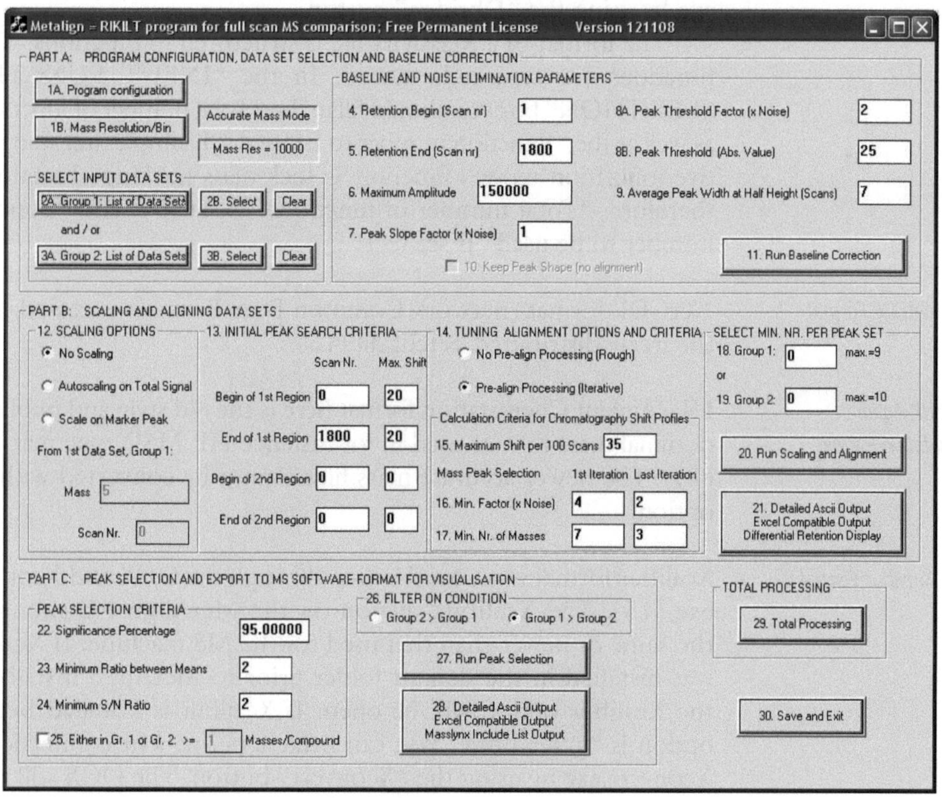

Fig. 1. The MetAlign interface (after double-clicking ms.exe).

Table 1
Standard parameter settings for the MetAlign interface. All other parameters are more system dependent and should be established by the user

	GCMS	LCMS
"7. Peak Slope Factor (× Noise)"	1	1
"8A. Peak Threshold Factor (× Noise)"	2	2
"15. Maximum Shift per 100 Scans"	15–50[a]	15–50
First "16. Min. Factor (×Noise)"	3	3
First "17. Min. Nr. of Masses"	8	4
Last "16. Min. Factor (×Noise)"	2	2
Last "17.Min. Nr. of Masses"	3	3

[a]If broad peaks occur due to saturation 50 would be the best choice

in the default folder prior to MetAlign installation the Masslynx option will be open. If Masslynx is installed but the option is "grayed out" you can make a permanent connection to Dbridge.exe by using the "Dbridge" button.

The format of a Masslynx file is structured in functions. These functions are separately stored. In the "INPUT FUNCTION SELECTION" box you should fill in the "Total number of functions" as well as the "Function number to be used." Example: full scan positive ionization mode = function 1; lock mass tracking = function 2; therefore, "Total number of functions" is set to 2 and "Function number to be used" is set to 1.

3.3.2. netCDF Format

NetCDF format (network Common Data Form) is accessed using the freely distributed netcdf.dll (13).

3.3.3. HP/Agilent Chemstation Format

HP/Agilent Chemstation format here is the old style and published nominal mass format used in for instance HP-MSD type machines (14). The newer accurate mass files cannot be converted with this option.

3.3.4. Xcalibur Format

Xcalibur format is accessed in line through the OCX and Xconvert.exe (15). The Xcalibur version on the MetAlign PC should be the same or newer than that used for the MS machine. If Xcalibur was installed in the default folder prior to MetAlign installation the Xcalibur option will be open. If Xcalibur is installed but the option is "grayed out" you can make a permanent connection to Xconvert.exe by using the "Xconvert" button. The OCX will automatically be found and registered.

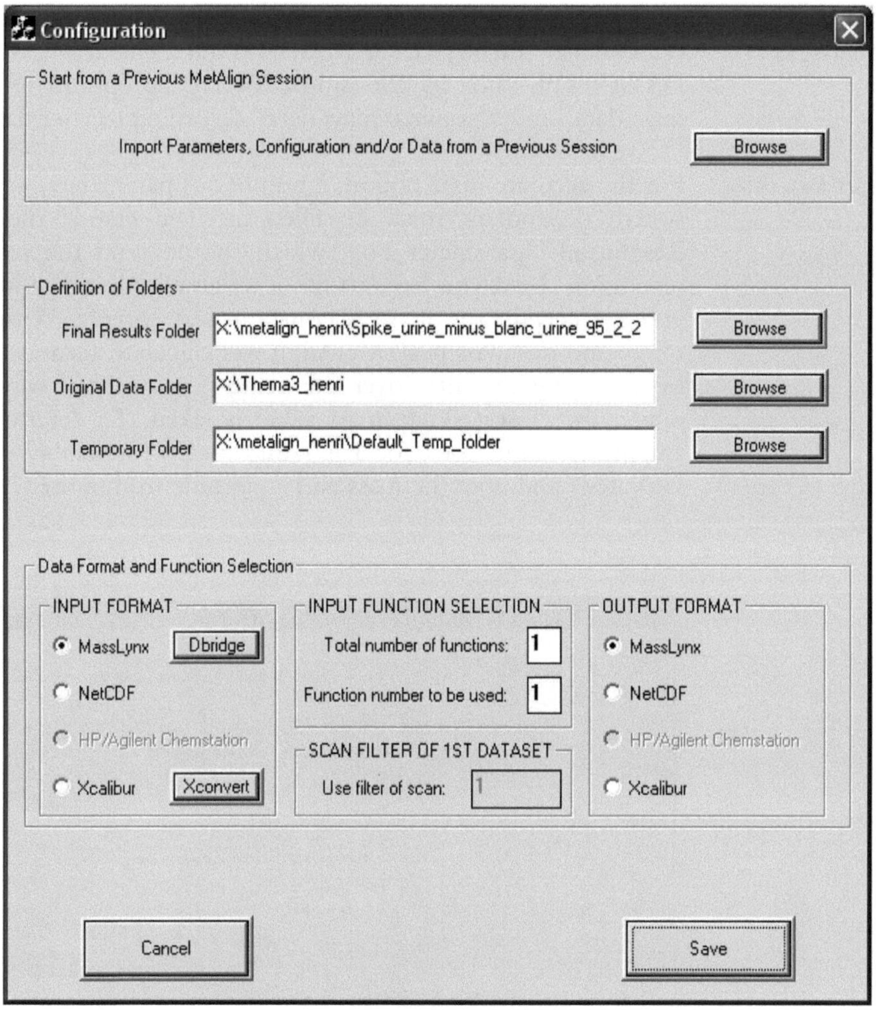

Fig. 2. The configuration interface (after "1A Program configuration").

The format of an Xcalibur file is one continuous array of scans containing all activities done in the time order. Thus, for example, positive ionization mode and (signal triggered) MS^n may be irregularly interleaved in the file. Xcalibur recognizes these modes by the Xcalibur "scan filters," which are tags. Xconvert does not make use of scan filters and therefore can not separate the modes used in the netCDF format. MetAlign therefore reads Xcalibur files using the provided OCX and selects on a user-defined scan filter. This scan filter should be defined in the "SCAN FILTER OF 1ST DATASET" box by entering an example scan from the first dataset (to be defined through button 2A) into the parameter box "Use filter of scan:" The OCX does not support writing to Xcalibur format; Xconvert.exe is utilized in-line to convert netCDF to Xcalibur format.

3.4. Defining Accurate and Nominal Mass

Clicking the button "1B. Mass resolution/bin" starts the "Accurate or Nominal?" menu (see Fig. 3). Start in this menu in the "SELECT DATA TYPE" box by choosing between the options "Accurate mass data" and "Nominal mass data" according to your data type.

3.4.1. Accurate Mass Data

For the accurate mass option, a number of parameters, which are system dependent, must be filled in. The first is the "Mass Resolution:" parameter box, which should hold the real mass resolution. Next you have to fill in an amplitude range in which you are certain the mass is constant and correct. Within this range and per mass peak MetAlign will calculate accurate masses by averaging the mass over the peak; if no value is within this range, the closest single mass value is taken. To determine the amplitude range, look for a few high peaks (preferably detector saturated) and note the mass and amplitude from noise level over

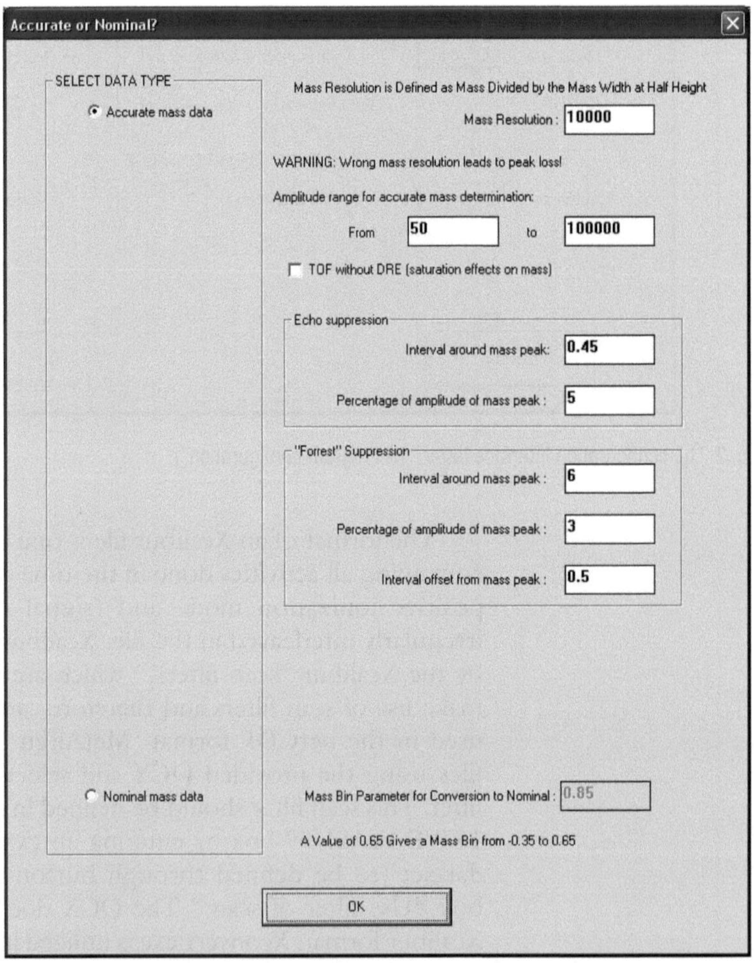

Fig. 3. The "Accurate or Nominal?" interface (after "1B. Mass Resolution/Bin").

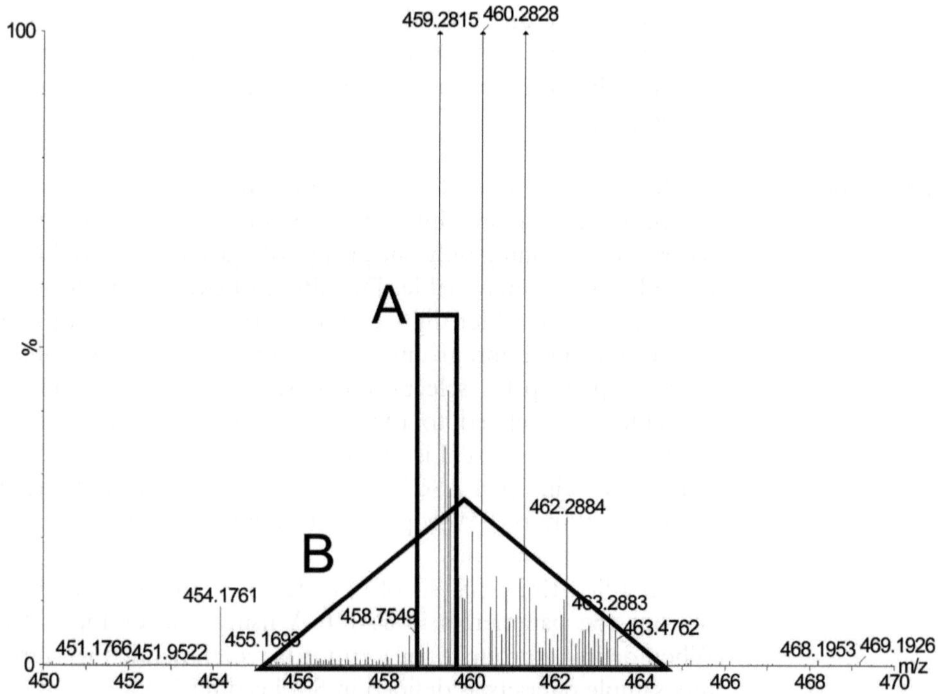

Fig. 4. Example of mass filters applied to arbitrary mass 459.2815. Mass peaks within rectangle A ("Echo suppression") and triangle B ("Forest suppression") are eliminated. Half the width of A is determined by parameter box "Interval around mass peak:" (in Dalton). The height of A is a percentage of the amplitude of the mass peak and is filled in the parameter box "Percentage of amplitude of mass peak:" Half the width of triangle B is determined by parameter box "Interval around mass peak:" (in Dalton). The height of B is a percentage of the amplitude of the mass peak and is filled in the parameter box "Percentage of amplitude of mass peak:" The triangle is placed at an offset from mass 459.2815, which is defined by parameter box "Interval offset from mass peak:" (in Dalton).

the maximum and again to the noise level. This should give you the desired information. If no saturation occurs and no deviation at the highest amplitudes you can fill in a maximum for the range, that is higher than any amplitude observed.

The check box "TOF without DRE (saturation effects on mass)" should be flagged if the MS per definition is amplitude dependent as in for instance a QTOF old style without Dynamic Range Extension. MetAlign will need to compensate for this extreme behavior.

As a last step, two filters should be set to eliminate artifact mass peaks as shown in Fig. 4.

For each mass peak in an entire dataset, filters ("Echo suppression" = rectangle A and "Forest suppression" = triangle B) are constructed to eliminate artifacts. The way to set the parameters for the filters is explained in the figure legend of Fig. 4.

3.4.2. Nominal Mass Data

This option uses nominal mass data directly if available or converts data to nominal mass using a mass bin, which should be defined in the parameter box "Mass Bin Parameter for Conversion to

Nominal." A value of 0.85 means that all mass peaks between for example 199.85 and 200.85 are rounded off to 200; if two mass peaks within the bin are present within the same scan they are added together.

3.5. Selecting Datasets

In the box "SELECT INPUT DATA SETS" two groups of data can be defined. In principle you need only define one group of data to proceed. Defining only one group will leave PART C of MetAlign grayed-out and unavailable. Definition of two groups is needed if you want to use MetAlign PART C for selection of differences between group 1 and group 2. The buttons "2B. Select" and "3B. Select" open up file selection as described in Note 1. The mask available is correlated to the format choice in Subheading 3.3. Buttons "2A. Group1: List of Data Sets" and "2B. Group2: List of Data Sets" will open ASCII text files with the selected files using Microsoft Windows Wordpad.exe (see Note 3). The "Clear" buttons clear the selections.

In a first time analysis of new data it is recommended to first try out the parameters in PART A using one example dataset. When defining group 1 for a run it is recommended to start with mix sample datasets as defined in Subheading 2.2.

3.6. Setting Up the Baseline Correction

In the box "BASELINE AND NOISE ELIMINATION PARAMETERS" several parameters have to be set, which are used for noise estimation, smoothing, peak finding, and dealing with saturation.

3.6.1. Importance of the Beginning and End of the Chromatogram

Parameter "4. Retention Begin (Scan nr)" and "5. Retention End (Scan nr)" are important parameters for the definition of noise in the dataset. Noise components come from chemical background and the detector. Chemical noise is mass and concentration dependent and is seen as a changing baseline. To be able to estimate noise, parameter 5 is especially important and should correlate to a position at the end of the chromatogram, where a maximum of chemical noise is expected (see Fig. 5) (see also Note 4). Local noise (as a function of mass and time) is estimated for all datasets.

Simultaneously these parameters will also cut out this part of the chromatogram for further processing.

3.6.2. Dealing with Saturation Artifacts

In metabolomics experiments, overloading of compounds often occurs in an effort to increase dynamic range. Due to high concentrations of compounds, saturation of the detector may occur. This is noticed as flattening or disfiguring of tops of broad peaks. Flattened or disfigured tops have badly defined maxima; multiple maxima may occur due to enhanced noise on the tops of saturated masses. MetAlign deals with this by creating a unique artificial top if an amplitude is higher than the user-defined value in parameter box "6. Maximum Amplitude." An example of saturation is given in Fig. 6. Normally a value of ca. 70% of a saturated mass peak

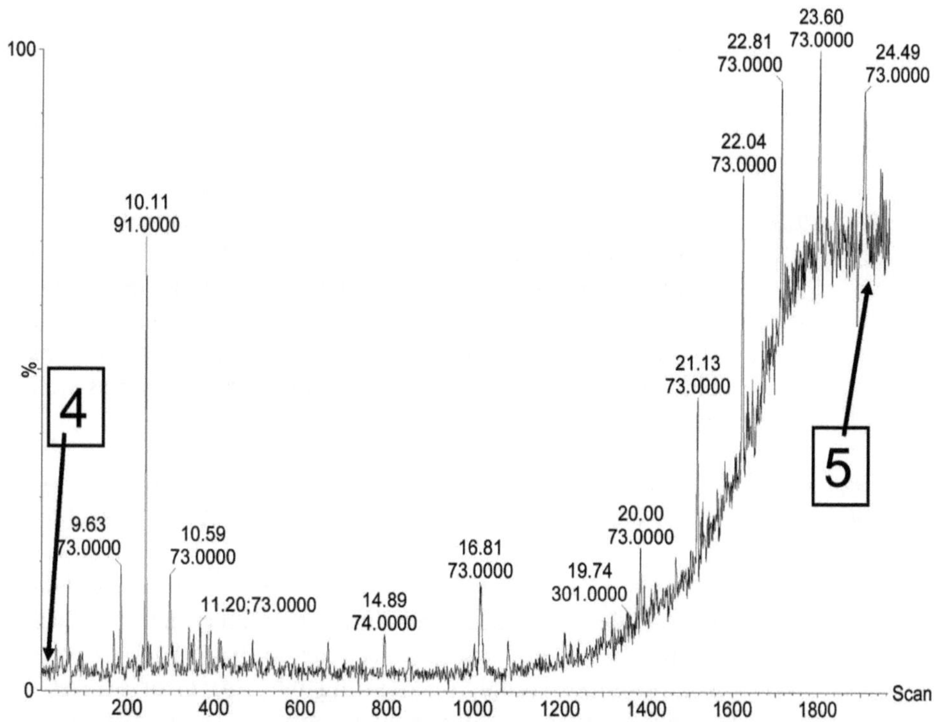

Fig. 5. Example (mass 208 from a GC-MS dataset) of how to set parameters 4 and 5 for correct noise estimation (see also Note 4).

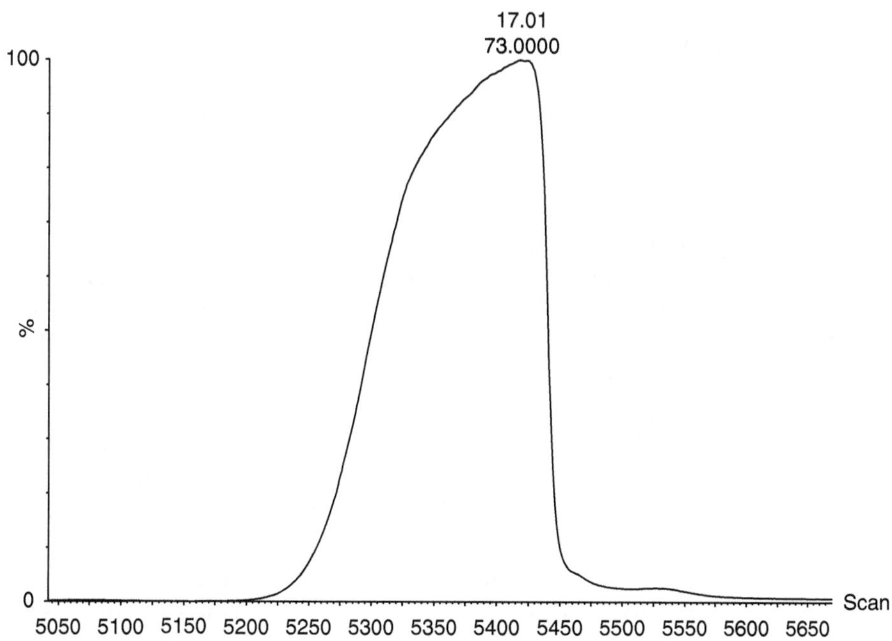

Fig. 6. Example of severe saturation. The resulting mass spectrum of this compound should be inspected to determine what amplitude threshold is acceptable.

(determined in the mass dimension) is used for parameter 6 (see Note 5). It is beneficial to evaluate more than one occurrence of saturation.

If no saturation occurs fill in a value exceeding any mass peak amplitude present.

3.6.3. Smoothing the Data

Parameter box "9. Average peak Width at Half Height (Scans)" should hold a value which is determined at half of the highest amplitude of a mass peak. The number of scans across at that height is the desired value. A number of mass peaks, which are not saturated, should be used for this purpose. This value is used to construct a binomial digital filter for smoothing of the dataset as well as the calculated noise (see also Note 6).

3.6.4. Peak Finding Using Calculated Local Noise

The peak finding algorithm in MetAlign has been described in ref. (11). The noise estimation in Subheading 3.6.1 is used locally to find out what is signal and what is the baseline and noise. If the difference in amplitude between any two consecutive data points on one side of a potential signal is greater than "7. Peak Slope Factor (× Noise)" times noise, the software tries to reconstruct the potential signal. By defining what parts of a mass trace is baseline and noise and what is signal, a series of linear corrections will eliminate the baseline. The value in parameter box "8A. Peak Threshold Factor (× Noise)" is applied as a local "times noise" threshold to eliminate noise. A second elimination of noise is achieved by an absolute threshold given in parameter box "8B. Peak Threshold (Abs. Value)." An example of where and how to determine this last threshold is given in Fig. 7.

3.6.5. Option: Keeping the Peak Shape

The check box "10. Keep Peak Shape (no alignment)" is only operational in the nominal mass mode.

If this box is unchecked the end result of PART A is a baseline-corrected noise-eliminated peak-picked dataset without peak shapes. Alignment can be done with this type of data.

If this box is checked the end result of PART A is a baseline-corrected noise-eliminated dataset containing the full peak shapes. Alignment can not be done with this data. This data can be used in deconvolution programs, such as AMDIS (16).

3.7. Executing the Baseline Correction and Storage

The baseline correction and preparation for alignment is executed by the button "11. Run Baseline Correction." This button sequentially does all the datasets in group 1 and group 2. A baseline correction and noise elimination in the time dimension set by the parameters in PART A and following the configuration set previously through button 1A is performed. In the case of Leco GCMS data in netCDF format only, an additional prior baseline correction in the mass dimension is done in the background (11).

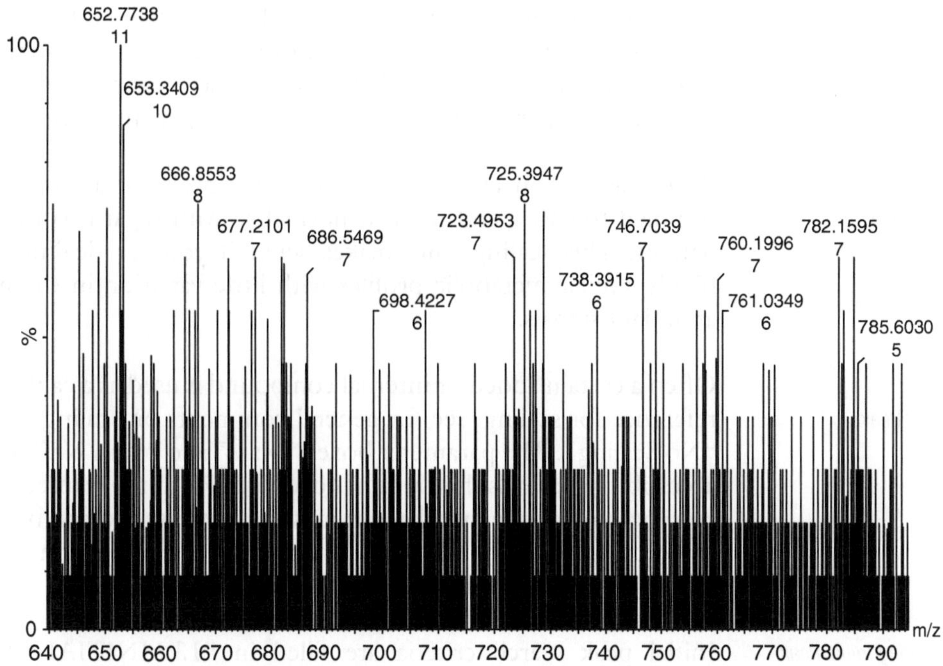

Fig. 7. Example of an empty part of a chromatogram in the higher mass range. A absolute threshold value for parameter 8B can be estimated here, for example 8B = 15.

For nominal mass mode, two subfolders are found in the "Final Results Folder" (see Subheading 3.3). Subfolder "Nominal" contains the original data in the output format defined (if Leco GCMS data, then a baseline correction was performed in the mass dimension). Subfolder "Baseline" contains the calculated "reduced" data in the output format defined (see Subheading 3.3).

For accurate mass mode one subfolder is found in the "Final Results Folder". Subfolder "Baseline" contains the calculated "reduced" data in the output format defined. The masses have been averaged over the peaks in the amplitude range defined through button 1B (see Subheading 3.4.1).

Execution also creates in the "Baseline" folder *.redms* files for nominal and *.redms_acc* files for accurate mass data, when parameter 10 is unchecked. These small files are used in the alignment (PART B) and identification software modules (see below).

3.8. Setting Up Scaling and Alignment

"PART B: SCALING AND ALIGNING DATA SETS" is done on output of Subheading 3.7.

3.8.1. Scaling the Datasets

There are three options in PART B in box "12. SCALING OPTIONS" for scaling the data. This is done prior to alignment and is not visible after baseline correction. The three options are as follows:

No Scaling

This option is the most frequent choice. It is in principle always best to perform the experiments in such a way that scaling is not necessary. This will avoid problems such as (a) scaling of noise and (b) dealing with saturation in which case the original height of a peak can not be known and therefore not scaled properly. A scaling can always be performed afterwards in an alignment output.

Auto-scaling on Total Signal

With this option all amplitudes of mass peaks of a dataset are summed together and used to normalize with regard to the first dataset. This scaling only makes sense if you are dealing with highly similar metabolic profiles with little variation in the more abundant signals.

Scale on Marker Peak

Often a certain added or internal compound is used as a calibration reference for scaling. see for general comments on scaling section "No Scaling". The parameter boxes "Mass" and "Scan Nr." will be accessible as soon as this option is checked. Fill in the wanted reference peak using the first dataset: mass at scan (use the first baseline corrected dataset for this). After alignment the correction will be performed on all data.

3.8.2. Setting Initial Peak Search Criteria for Alignment

Initial peak search criteria are filled in "13. INITIAL PEAK SEARCH CRITERIA." For the alignment an initial window (two times "Max. Shift") in the time domain must be defined. This window tells MetAlign where the alignment algorithm can look for the same mass peaks in different files. Two adjacent regions can be defined. However, most metabolomics experiments use more or less linear gradients. Therefore, in most of the cases one region (1st) is sufficient. In a user-defined region the window will expand linearly with scan number analogous to retention time shifts increasing with the time axis. To define this linear behavior two points are needed: "Begin of 1st Region" with coordinates ("Scan Nr." "Max. Shift") and "End of 1st Region" with different coordinates ("Scan Nr." "Max. Shift").

Normally it is advisable to fill in values for "Max. Shift" that are twice the maximum expected shift. The user must define this shift. If following Subheading 2.2, an overlay of the mix samples may give a nice indication for these parameters. Inspection of shifts occurring in the beginning as well as end of the chromatogram is needed to fill in appropriate values for "Begin of 1st Region" and "End of 1st Region."

3.8.3. Choosing Between a Rough and Iterative Alignment

In "TUNING ALIGNMENT OPTIONS AND CRITERIA" two choices are available with regard to alignment. Both options are depicted in Fig. 8 (see Subheadings "No Pre-align Processing (Rough)" and "Pre-align Processing (Iterative)" and also Note 7).

No Pre-align Processing (Rough)

Rough alignment (gray arrows in Fig. 8): this type of alignment can be used for any alignment and will always give a result. The alignment is restricted by "+-max shift" (see Subheading 3.8.2).

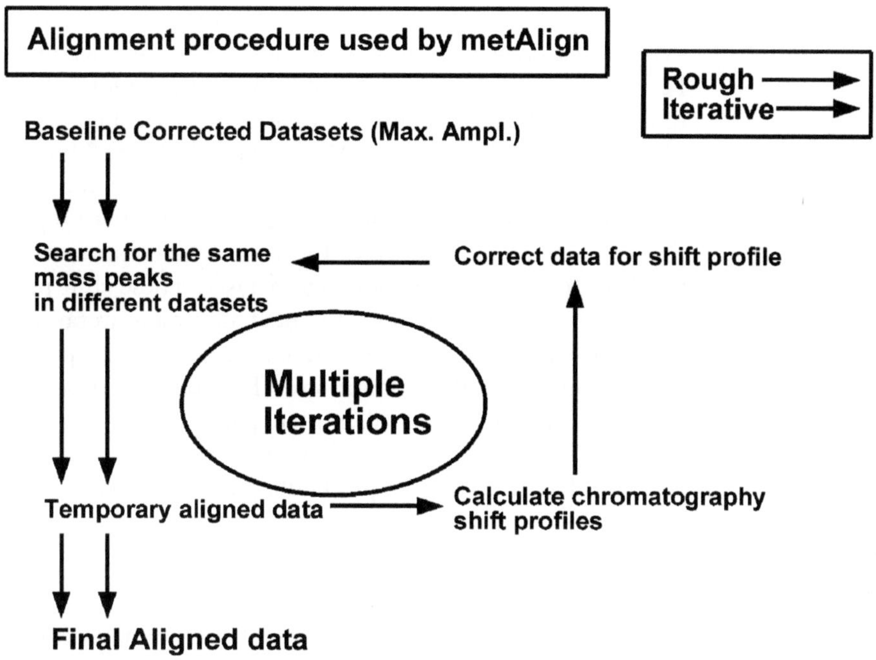

Fig. 8. Schematic overview of the alignment procedures used by MetAlign. "No Pre-align Processing (Rough)" is shown by *gray arrows*. "Pre-align Processing (Iterative)" is shown by *black* arrows.

This option is useful for:

(a) Relatively empty chromatograms of datasets where the size of "Max. Shift" is less important.

(b) Chromatograms with large empty regions (front, middle, back).

(c) Reproducible chromatography of complex data with only very small shifts.

(d) In general when the iterative alignment fails.

Pre-align Processing (Iterative)

Iterative alignment (black arrows in Fig. 8): this type of alignment is used most often in metabolomics, where the data can be characterized as complex with compounds evenly distributed over the chromatogram. This mode requires additional parameters which are opened up in the "Calculation Criteria for Chromatography Shift Profiles" as soon as the iterative option is chosen.

The parameter "15. Maximum Shift per 100 Scans" should be given to limit (positive and negative) the first derivative of the function $y = func(x)$, where x (scan) corresponds to a scan in the first defined dataset and y is the shift in scans in a dataset with regard to the first defined dataset. In effect large calculated local shifts (absolute value) are omitted from the shift profile estimations if they exceed "Maximum Shift per 100 Scans."

To be able to calculate reference points for $y = \text{func}(x)$, only mass peaks present in all datasets are taken into account ("Mass Peak Selection"). For the first iteration ("1st Iteration") the number of masses and their minimum amplitudes are defined resp. by "17. Min. Nr. of Masses" and "16. Min. Factor (× Noise)." If in a window of scan x+− "Max. Shift" the minimum criteria 16 and 17 are met, then from the alignment the average delta scans with regard to the first dataset can be calculated. Over the whole chromatogram this will lead to a shift profile for each dataset. These shift profiles will be used as starting points for the next alignment iteration.

Subsequent iterations will make "Max. Shift," parameter 16 and 17 smaller in the background. The iterative process will halt when "Max Shift" is smaller than parameter "9. Average peak Width at Half Height (Scans)" and parameter 16 and 17 are equal to their "Last Iteration" equivalents (see also Table 1).

3.8.4. Selecting Minimum Occurrences of Aligned Peaks

This option ensures that a selection is performed on the aligned output:

18: "max=?" indicates the present number (?) of datasets in group 1.

19: "max=?" indicates the present number (?) of datasets in group 2.

Parameter box "18. Group 1:" the minimum number of datasets in group 1 having a particular mass peak. Parameter box "19. Group 2:" the minimum number of datasets in group 2 having a particular mass peak. If none of the two conditions for 18 and 19 were met, the mass peak is deleted from the alignment (see also Note 8).

3.9. Executing Scaling, Alignment and Storage

The button "20. Run Scaling and Alignment" does the scaling and alignment of all preprocessed datasets derived from group 1 and group 2 (i.e., all *.redms* or *.redms_acc* files). This is done using the settings in PART B. A file called *End_result.rap* and its derivatives are stored in a subfolder of the "Final Results Folder" called 1-2_abs (if "26. FILTER ON CONDITION" "Group 1 > Group 2" is checked) or called 2-1_abs (if "26. FILTER ON CONDITION" "Group 2 > Group 1" is checked).

3.10. Outputting Aligned Data

Clicking on the button "21. Detailed Ascii Ouput etc" starts the View_data.exe subprogram as shown in Fig. 9. The "Browse" button offers the possibility of loading a different *.rap* file (see Note 1) the default is the current alignment. There are three output options, each with the possibility of making a subselection of mass peaks:

1. A selection can be made using a window for the "Mass" ("LOW" and "HIGH") as well as "Retention" ("LOW" and "HIGH") (minutes).
2. A threshold can be given as a factor times local noise (parameter box "Peak Threshold factor (× noise)"); default is parameter 8A.

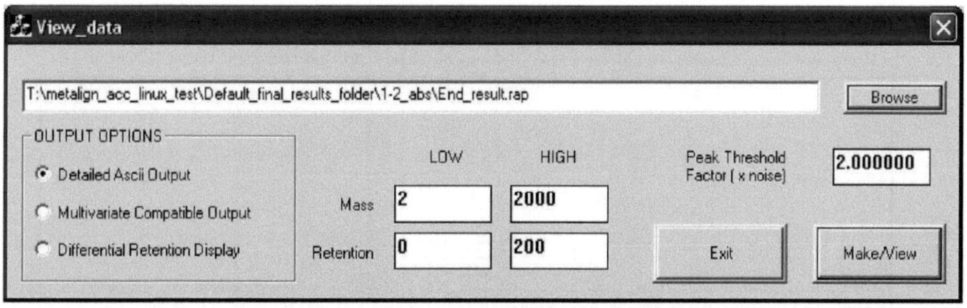

Fig. 9. The View_data interface for creating output from aligned data (after button 21).

The three options are activated, when clicking on the "Make/View" button:

1. "Detailed Ascii Output" gives a text file containing the selected information; this option can be used to look up the alignment of a particular mass peak of interest (see Note 9). This text file is opened by the Microsoft Windows Wordpad.exe (see Note 3).
2. "Multivariate Compatible Output" gives a comma separated value file (*.csv*). This file is Microsoft Excel compatible (English version). Each Excel cell is separated by a comma (see also Note 10). The *.csv* files are stored in a subfolder of the "Final Results Folder" called 1-2_abs (if "26. FILTER ON CONDITION" "Group 1 > Group 2" is checked) or called 2-1_abs (if "26. FILTER ON CONDITION" "Group 2 > Group 1" is checked).
3. "Differential Retention Display" gives a graphical representation of the differences in retention between the first file in group 1 and all others.

3.10.1. "Differential Retention Display" Instructions

Difference in retention with regard to the first dataset in group 1 is displayed for all files. Red shaded points are from datasets of group 1; blue shaded points are from datasets of group 2. Black points ("Pre-align Calibration Points" = shift correction profile points) and lines ("Pre-align Estimate" = shift correction profile) indicate resp. calculated retention differences and interpolations and extrapolations between the black points (see Fig. 10). Checkboxes on the left can be used to include or exclude data from the view.

"Data mode" gives the option to view "All Data" simultaneously or "File by file." In the latter case, "Select a Group," "File Number" and buttons "Up" and "Down" will work and can be used to view the data per file. "Display Mode" toggles between difference mode in scans ("Scan") and retentions ("Retention"). "View Graph data" needs you to select data points first and will return by opening a Microsoft Windows Wordpad.exe (see Note 3) text file containing your selection. This selection is done by a single click on the white window, then a double-click and hold-down-and-drag on

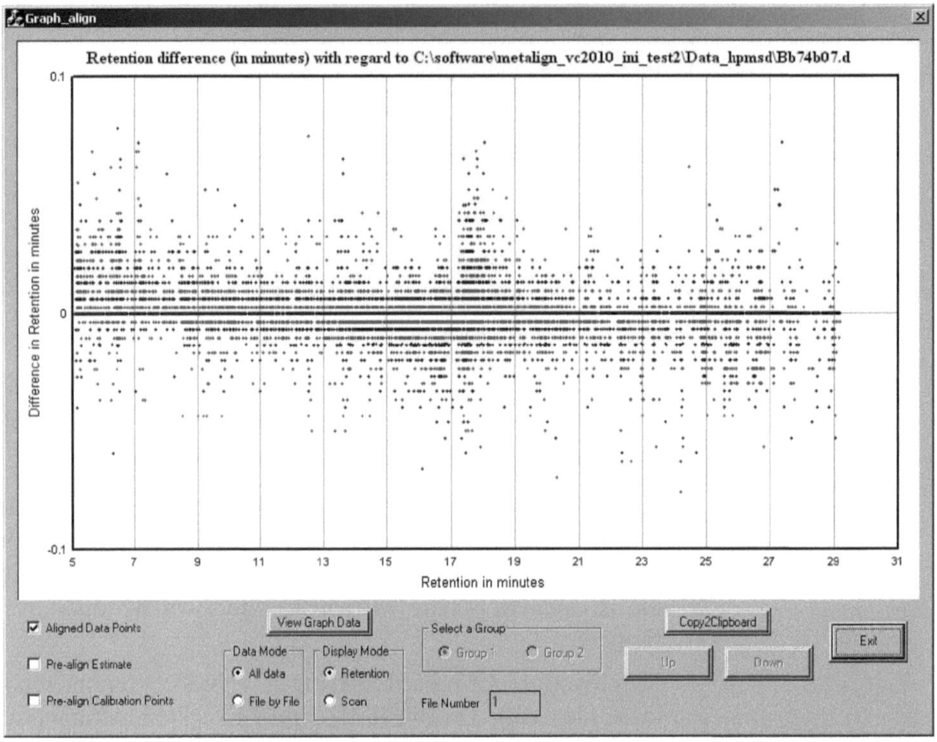

Fig. 10. The Graph_align interface opened by the "Differential Retention Display" option in "View_data" (button 21).

the second click to make a rectangular selection. This then is expanded on release. Right-click returns the previous total view (see Notes 11–13).

3.11. Setting Up Peak Selection and Export When Having Two Groups Defined

If two groups were defined in Subheading 3.5, then PART C "PEAK SELECTION AND EXPORT TO MS SOFTWARE FORMAT FOR VISUALISATION" will be available (see Fig. 1). Using "PEAK SELECTION CRITERIA" and "26. FILTER ON CONDITION" differences between both groups can be selected. The output is either group 1 minus group 2 (in 1-2_abs and 1-2_rel) or vice versa (in 2-1_abs and 2-1_rel).

3.11.1. Peak Selection Criteria

Four parameter boxes can be filled in this box:

1. Parameter box "22. Significance Percentage": A minimum significance percentage can be set here. For example: a criterion $p < 0.01$ would correlate to 99%. 99 should then be filled in.
2. Parameter box "23. Minimum Ratio between Means": A minimum ratio between the means of the two groups is set here.
3. Parameter box "24. Minimum S/N Ratio": A difference in means should given, which is X times noise; X should be entered.

4. Check box controlling parameter box "25. Either in Gr. 1 or Gr. 2: >=": Checking this opens the parameter 25 box and activates this option in the selection. Parameter box 25 in turn is dependent on "26. FILTER ON CONDITION." Parameter check box 25 is used to select on compounds present in one group and absent in the other. Present is now defined by > ("parameter 24" times noise) in all datasets in one group; absent is now defined by < ("parameter 24" times noise) in all datasets in the other group. Furthermore, for selection, all mass peaks present in one group should be > ("parameter 23" times the mean of the other group). Finally, parameter 22 is defined as in A. Consider a window of root (parameter 9) scans moving through your chromatogram. If after selection done on the basis of parameter 22, 23, and 24 there are at least "parameter 25" mass peaks within the window, then they are retained; if not, then all peaks in the window are deleted from the selection (see Note 14).

3.11.2. Group Selection

"26 FILTER ON CONDITION" provides the possibility of selecting on mass peaks for which "Group 2 > Group 1" (higher in group 2) or "Group 1 > Group 2" (higher in group 1).

3.12. Executing Peak Selection and Storage

The button "27. Run Peak Selection" executes PART C and generates output files.

A file called *Stat.rap* and its derivatives are stored in a subfolder of the "Final Results Folder" called 1-2_abs if ("26. FILTER ON CONDITION" "Group 1 > Group 2") or 2-1_abs if ("26. FILTER ON CONDITION" "Group 2 > Group 1").

Difference datasets are generated for overlay purposes according to the format selected in Subheading 3.3. Their names are equal to the original files. Retentions and scan numbers are also identical to the original data. Amplitudes are the absolute differences in means between the two groups (stored in either 1-2_abs or 2-1_abs).

Ratio datasets are also generated for overlay purposes according to the format selected in Subheading 3.3. Their names are equal to the original files. Retentions and scan numbers are also identical to the original data. Amplitudes are the 1,000× ratios between means for the two groups (stored in either 1-2_rel or 2-1_rel). Only one mass per scan is displayed; this is always the mass giving the largest ratio.

3.13. Outputting Differences in Aligned Data

Clicking on the button "28. Detailed Ascii Ouput etc" starts the View_data.exe subprogram as shown in Fig. 11. This section is highly similar to Subheading 3.10. The difference lies in "Minimum average amplitude (absolute)" and the option "Masslynx Include List Output." The first parameter box is an absolute threshold which may be used additionally to Subheading 3.11.1. The "Masslynx Include List Option" can only be used for running accurate mass

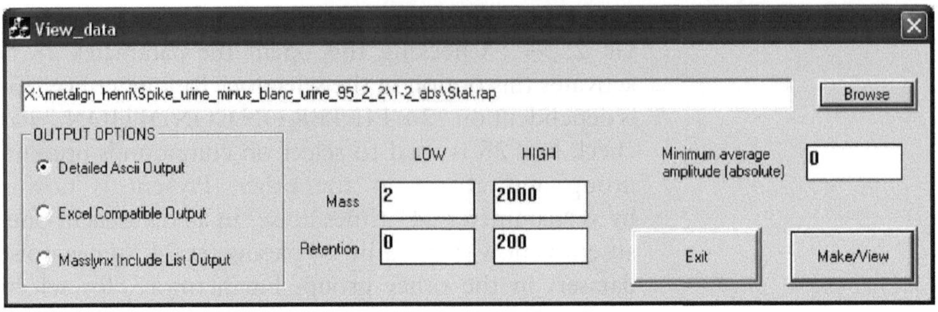

Fig. 11. The View_data interface for creating output after selection of differences in aligned data (PART C) (after button 28).

Masslynx data in nominal mode and creates a so-called "Include List," which can be used for MS-MS triggering within Masslynx.

3.14. Total Processing

Clicking on button "29. Total processing" will sequentially execute button 11 in PART A, button 20 in PART B and button 27(if applicable) in PART C.

3.15. Saving and Exiting

Clicking on button "30. Save and Exit" exits MetAlign and saves the settings of the ms.exe interface.

3.16. Additional Software Tools for MetAlign Output

The modules rap2subrap.exe and GM2MS.exe are available in the MetAlign download and can process MetAlign output.

3.16.1. Reducing the Number of Datasets and Recombining Groups of Datasets After Alignment

Rap2subrap.exe is started by double-clicking this application in your MetAlign folder. The module rap2subrap.exe (see Fig. 12) was programmed to take an "End_result.rap" and the data set listings (defined in Subheading 3.5) from a "Final Results Folder" and reorder and reduce this into a new "Final Results Folder." So, if the listing of files for processing in MetAlign contained multiple groups of data files (in one or two data lists) and you would want to take two subgroups out of the completed alignment for processing in PART C of MetAlign, this program is the solution. see Note 15 for an example of use.

The edit box "Folder containing alignment file (.rap):" can be filled in by selecting the appropriate "Final Results Folder" (with End_result.rap etc.) using the top "Browse" button (see Note 2).

The edit box "Output file path:" can be filled in by creating a new "Final Results Folder" and selecting it using the bottom "Browse" button (see Note 2).

The box "List_group1 and List_group2 in One List" contains the parameter boxes "Define group1:" and "Define group 2:" in which the definition of the two desired groups should be encoded. The encoding numbers, here indicating the files, suppose that List_group1 and List_group2 are combined in one list with one consecutive numbering; if two files exist, then these are automatically

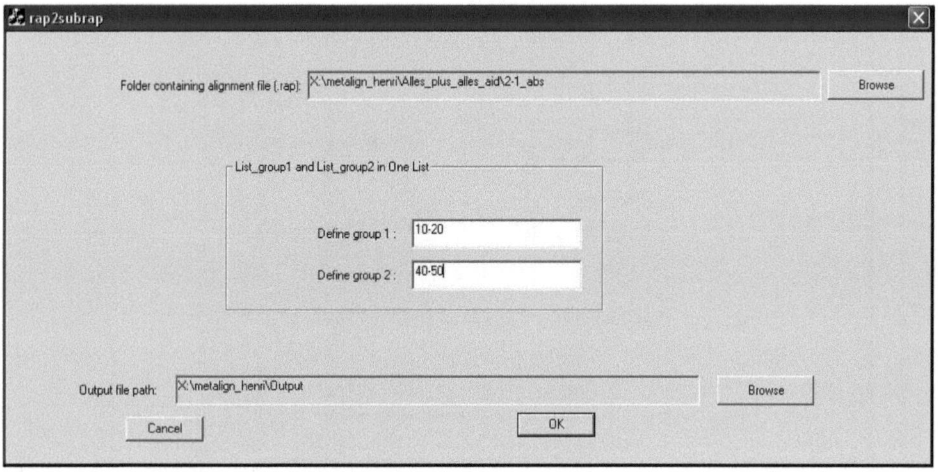

Fig. 12. The interface of the MetAlign related tool, rap2subrap.exe.

combined together in the background. Encoding a group can be done by any combination of "," and "-": example = 13-23,45-48,50-52,53,54,2. The encoding order is not important as long as it corresponds to the files in the dataset list (see also Note 16).

The "OK" button runs the program. The new "Final Results Folder" resulting from rap2subrap.exe can be imported (see Note 2) and processed with PART C of MetAlign. This results in an efficient way of avoiding new alignments on the same data.

3.16.2. Converting a Part of a Tab-Delimited Spreadsheet to a MS Data File

GM2MS.exe is started by double-clicking this application in your MetAlign folder. The module GM2MS.exe (see Fig. 13) was programmed to take a tab-delimited Excel format text file (see Note 17). By clicking on the "OK" button the program is run. A number of user-defined files (columns) is then averaged and converted to a MS data file format.

INPUT

The "INPUT" box holds the information for the conversion of the input to the output.

The edit box "Tab delimited file (txt):" can be filled in using the top "Browse" button (see Note 2).

The "Format issues tab delimited file" box has a number of edit boxes to be filled in:

(a) The first necessary descriptors are "Column number for scan:" "Column number for mass:" "Column number for first ampl:" respectively describing where to find scan, mass, and first file information.

(b) The edit box "Reverse of ampl……log transformation" should contain the type of log that was used to transform the amplitudes previously. This will then be reversed on processing.

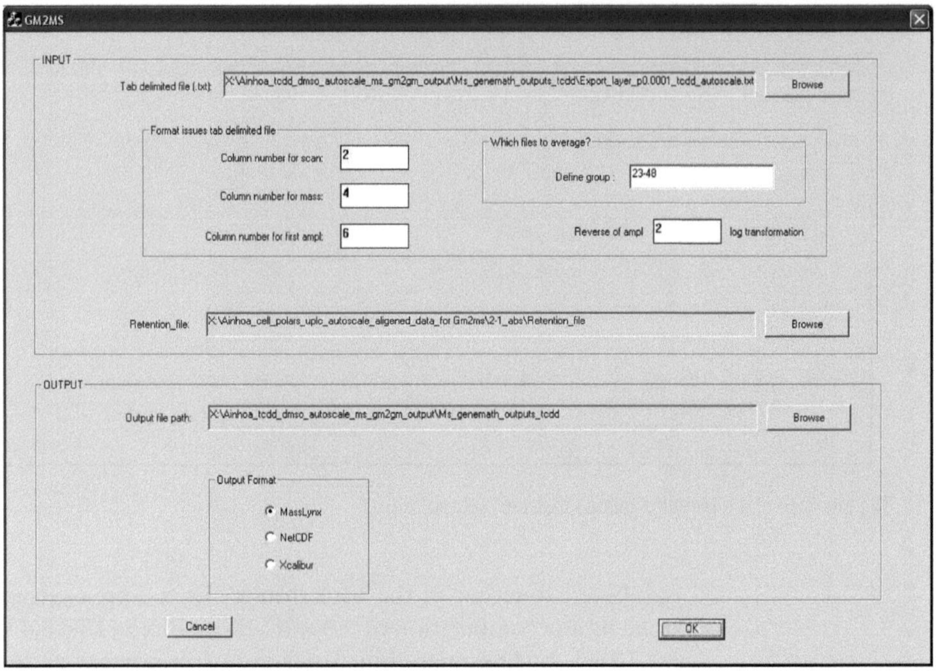

Fig. 13. The interface of the MetAlign related tool, GM2MS.exe.

(c) The "Which files to average?" box contains the "Define group:" edit box in which the definition of the files should be encoded. Encoding a group can be done by any combination of ',' and '-': example = 13-23,45-48,50-52,53,54,2. The order is not important (see also Note 16). The "Column number for first ampl: "should be considered as 1 in the encoding."

The edit box "Retention_file" can be filled in using the middle "Browse" button (see Note 2). This file is used to extract the retention information. The first file is used as the template.

OUTPUT

The "OUTPUT" box is used to configure the output files of this conversion to MS format. The edit box "Output file path:" can be filled in by creating a new folder and selecting it using the bottom "Browse" button (see Note 2). An MS file will be created in this folder using the "Output Format" option. The prefix of the name will be identical to the input file; the changed suffix will indicate the format (see Note 18).

4. Notes

1. Using a "Browse" button will open up a new window for selecting files or folders (see Fig. 14). Depending on the function one or more files or folders may be selected. Start by defining

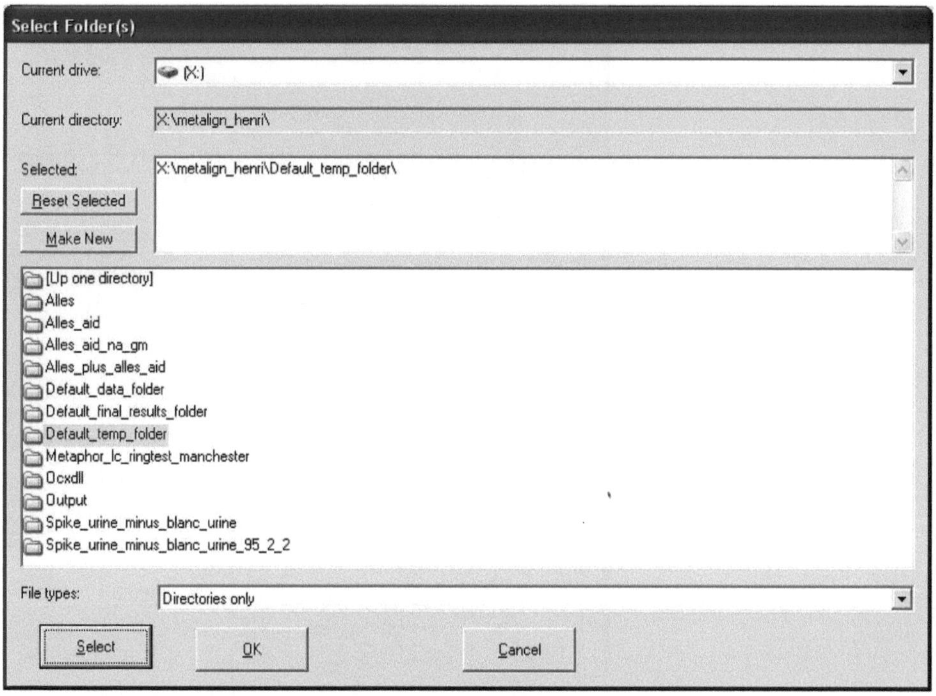

Fig. 14. Example of selecting a Folder through a "Browse" button (see Note 1).

the drive to look in by clicking on the arrowhead pointing down to the right of "Current drive." In the drag-down list you can define the drive by clicking on your choice. The selected drive should appear in the "Current drive" box. "Current directory" now defines the folder you are in and the big list box in the middle shows all available folders and files ("File types" shows the type of mask applied to the "Current directory"). By double-clicking on a folder you can go into a subfolder; if you want to go up a folder double-click on the "[Up one directory]" (top left in the box). You can add a folder by using the button "Make new." Change to the folder you need. You can now highlight files or folders as normally done in Windows Explorer. When highlighted, click on the "Select" button. The selected files/folders should appear in the "Selected:" box. Clicking on the "OK" button finalizes the choice and exits this module.

2. Using the top "Browse" button in "Start from a Previous Metalign Session" box in the configuration module followed by selecting a previous "Final Results Folder" (previous session) gives the window appearing in Fig. 15. The top box gives the previously exported subfolders present. The "Options" box gives different levels of interaction. Option "Parameters only" loads only the parameters. Option "Parameters + reduced ms data files" loads all export from PART A in the MetAlign interface (Baseline subfolder). Option "Parameters + previously aligned files" loads all export from PART B in the MetAlign

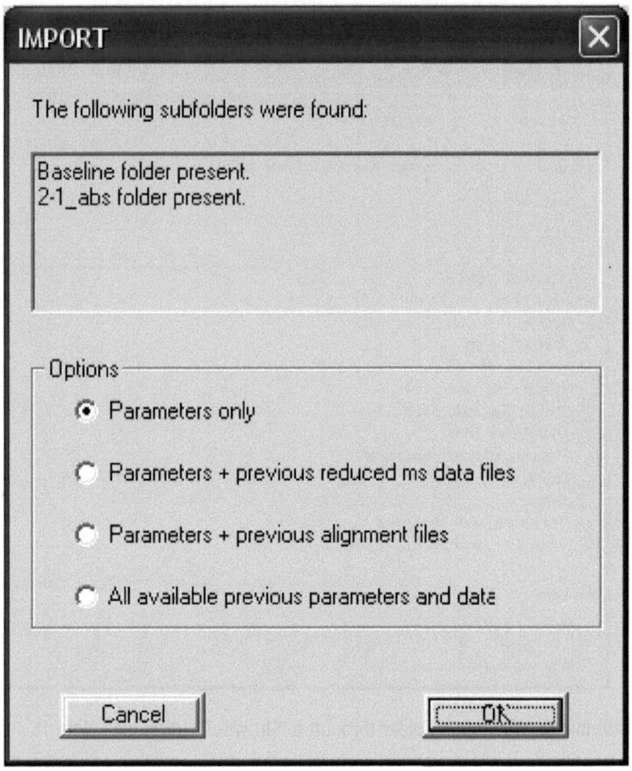

Fig. 15. The options interface when importing a previous MetAlign session (button 1A in ms.exe and then top "Browse" button).

interface (2-1_abs subfolder). Option "All available previous parameters and data" loads all export from PART A and PART B in the MetAlign interface (Baseline and 2-1_abs subfolders). Loading can take several minutes.

3. The program Wordpad.exe should be found automatically if it is present in the default folder (for example: C:\Program Files\Windows NT\Accessories\wordpad.exe in English version of Windows XP). If this is not the case and an error occurs, you can manually fill in the correct path to Wordpad by editing programs in your MetAlign folder. The third line should contain an equivalent of "C:\Program Files\Windows NT\Accessories\wordpad.exe." It is possible to use a different text editor in the same way.

4. Noise is estimated as a function of mass and time. MetAlign uses parameter 5 to estimate the chemical noise contribution. It assumes as a rough approximation that noise is "chemical noise" times "background chemical concentration" plus detector noise. The noise estimate is important in determining what is peak and what is baseline. 1% of the total number of scans

(but minimally 30 scans) preceding parameter 5 are used to calculate a chemical noise contribution. Preferably this 1% at the end should be empty with respect to mass peaks; if, however, coincidentally a peak is present, the program tries to take this into account.

5. A common mistake is that parameter 6 is determined from the TIC instead of from high mass peak(s) from a mass spectrum.
6. Parameter 9 is also used in the background in the alignment.
7. Iterative alignment fails if controls do not have enough signal or if excessively large empty regions occur at the beginning or end of the chromatogram. It can work nicely for temperature-dependent shifts in very complex data. It will fail for larger pH-shifts in complex data. pH-shifts in general disrupt retention relations between compounds and therefore disrupt iterative alignment. In the rough alignment pH-shifts are less important, but will necessitate larger "Max. Shift"-values and can lead to swapped peaks if the data are too complex.
8. Only one of the two conditions has to be met.
9. Do not attempt to print or display all masses in ASCII files obtained from very large alignments.
10. It might be convenient to export "Multivariate Compatible Output" using a tab instead of a comma as a separator. This can be done by manually editing "programs.ini" (MetAlign folder) and substituting the comma by a tab.
11. Only points also present in the first dataset will be displayed because otherwise delta retention can not be calculated.
12. "Copy2clipboard" copies the white window to the clipboard. Paste will then copy the figure into a user-defined file.
13. For a good alignment in iterative mode prealign calibration points should be distributed over the whole chromatogram.
14. This option (parameter 25) builds on the assumption that compounds may have more than one mass peak and that in difference mode you will want to see more than one peak per compound. For GC-MS this is obvious and a value of 5 is recommended. For LC-MS a value of 1 or 2 is suggested if this parameter is used.
15. Example of the use of rap2subrap.exe: 100 files in group1 have been preprocessed and aligned and stored in a "Final Results Folder." Processing the output with rap2subrap could create a new "Final Results Folder" with two groups (file 10–20) and file (50–60), which can then be run through PART C after importing through config.exe (see Note 2).

16. Easy handling/counting can be achieved by copying and pasting the contents of list_group1.txt and list_group2.txt consecutively into an excel file. The first file in list_group1.txt is 1.
17. GM2MS.exe takes tab-delimited files instead of comma delimited files (*.csv*). Tab-delimited is the export format for many multivariate statistics programs, such as for instance here Genemaths XT (17).
18. Consecutive conversions using the same original *.txt* file will result in overwriting of each conversion. Rename the original file if it is needed more than once.

Acknowledgements

This work was supported by the Dutch Ministry of Agriculture, Nature and Food Quality, Strategic Research Funds RIKILT-WUR (project 77232903), Statutory Research Tasks (theme 3): veterinary drugs (project 87203001), The Netherlands Toxicogenomics Centre (NTC), contract AIR3-CT94-2311 (European Commission DG XII), and the EU-Framework VI programme: EU-METAPHOR (FP6: FOOD-CT-2006-036220), EU-NOFORISK (FP6: FOOD-CT2001-506387), EU-GMOCARE (QLK1-1999-00765). Datasets used in the development of MetAlign are from these projects and research consortia. Ric de Vos and Yury Tikunov of Plant Research International (Centre of Biosystems Genomics) are thanked for critical evaluation using their own data in the validation process of MetAlign.

References

1. http://www.metalign.wur.nl/UK/Download+and+publications/.
2. Tolstikov, V.V., Lommen, A., Nakanishi, K., Tanaka, N., Fiehn, O. (2003) Monolithic Silica-Based Capillary Reversed-Phase Liquid Chromatography/Electrospray Mass Spectrometry for Plant Metabolomics. *Anal. Chem.*, **75**, 6737–6740.
3. Vorst, O., de Vos, C.H.R., Lommen, A., Staps, R.V., Visser, R.G.F., Bino, R.J., Hall, R.D. (2005) A non-directed approach to the differential analysis of multiple LC MS-derived metabolic profiles. *Metabolomics* **1**, 169–180.
4. Tikunov, Y., Lommen, A., de Vos, C.H.R., Verhoeven, H.A., Bino, R.J., Hall, R.D., Lindhout, et al (2005) A Novel Approach for Non-targeted Data Analysis for Metabolomics. Large-Scale Profiling of Tomato Fruit Volatiles. *Plant Physiol. Break Through Technologies Section* **139**, 1125–1137.
5. America, A.H.P., Cordewener, J.H.G., Van Geffen, H.A., Lommen, A., Vissers, J.P.C., Bino, R.J., Hall, R.D. (2006) Alignment and statistical difference analysis of complex peptide data sets generated by multidimensional LC-MS. *Proteomics*, **6**, 641–653.
6. Keurentjes, J.J.B., Jingyuan, F., de Vos, C.H.R., Lommen, A., Hall, R. D., Bino, R. J., van der Plas et al (2006) The genetics of plant metabolism. *Nature Genetics (Technical Report)* **38**, 842–849.
7. Lommen, A., van der Weg, G., van Engelen, M. C., Bor, G., Hoogenboom, L.A.P., Nielen, M.W.F. (2007) An untargeted metabolomics approach to contaminant analysis: Pinpointing potential unknown compounds. *Analytica Chimica Acta*, **584**, 43–49.
8. de Vos, C.H.R., Moco, S., Lommen, A., Keurentjes, J.J.B., Bino, R.J., Hall, R. D. (2007) Untargeted large-scale plant metabolomics

using liquid chromatography coupled to mass spectrometry. *Nature Protocols*, **2**, 778–791.

9. Ducruix, C., Vailhen, D., Werner, E., Fievet, J.B., Bourguignon, J., Tabet, J.-C., Ezan, E., et al (2008), Metabolomic investigation of the response of the model plant Arabidopsis thaliana to cadmium exposure: Evaluation of data pretreatment methods for further statistical analyses. *Chemometrics and Intelligent Laboratory Systems*, **91**, 67–77.
10. Matsuda, F., Yonekura-Sakakibara, K., Niida, R., Kuromori, T., Shinozaki, K., Saito, K. (2009) MS/MS spectral tag-based annotation of non-targeted profile of plant secondary metabolites. *The Plant Journal*, **57**, 555–577.
11. Lommen, A. (2009) MetAlign: an interface-driven, versatile metabolomics tool for hyphenated full-scan MS data pre-processing. *Anal. Chem.*, **81**, 3079–3086.
12. See Masslynx manual: http://www.waters.com/.
13. http://www.unidata.ucar.edu/software/netcdf/.
14. See HP 5970 MSD manual: http://www.gmi-inc.com/Agilent-HP-5970-Mass-Spectrometer.html.
15. See Xcalibur manual: http://www.thermo.com.
16. Stein, S.E. (1999) An Integrated Method for Spectrum Extraction and Compound Identification from GC/MS Data. *J. Am. Soc. Mass Spectrom*, **10**, 770–781.
17. http://www.applied-maths.com/genemaths/genemaths.html.

Chapter 16

TagFinder: Preprocessing Software for the Fingerprinting and the Profiling of Gas Chromatography–Mass Spectrometry Based Metabolome Analyses

Alexander Luedemann, Luise von Malotky, Alexander Erban, and Joachim Kopka

Abstract

GC-MS based metabolome studies aim for the complete identification and relative or absolute quantification of metabolites in complex extracts from a large diversity of biological materials. The resulting high-throughput chromatography data files are typically processed following two complementary workflows, namely, fingerprinting and profiling. For fingerprinting studies all observed mass features, here called mass spectral tags (MSTs), are quantified in a nontargeted and (within the limits of the GC-MS technology) comprehensive approach. Fingerprinting allows for the discovery of MSTs, which, in the sense of a biomarker, indicate significant changes of metabolite pool sizes. The significance and relevance of such MSTs are typically tested in comparison to standardized reference samples. Only after this confirmation step are the relevant MSTs identified and the underlying metabolic biomarkers elucidated. Both the metabolite fingerprinting and profiling approaches are essential to modern biotechnological investigations. Studies which are aimed at establishing the substantial equivalence at metabolic level or aim to breed for optimum quality of human food or animal feed especially benefit from the potential to discover novel unforeseen metabolic factors in fingerprinting approaches and from the option to demonstrate unchanged pool sizes of known metabolites in the metabolic profiling mode. As GC-MS technology represents one essential element which contributes to investigations of substantial equivalence, we have developed a dedicated software tool, the TagFinder chromatography data preprocessing suite, which has all essential functions to support both fundamental workflows of modern metabolomic studies. In this chapter, we describe the TagFinder software and its application to the assessment of metabolic phenotypes in fingerprinting and profiling analyses.

Key words: Mass spectral tags, Nontargeted fingerprint analysis, Targeted profiling analysis, Peak extraction, Spectral reconstruction, GC-MS profiling, Chromatography data processing

Nigel W. Hardy and Robert D. Hall (eds.), *Plant Metabolomics: Methods and Protocols*, Methods in Molecular Biology, vol. 860, DOI 10.1007/978-1-61779-594-7_16, © Springer Science+Business Media, LLC 2012

1. Introduction

Typical GC-MS based profiling experiments which are employed for the assessment of differential patterns of multiple metabolite pools may represent large-scale screening experiments and comprise hundreds of chromatogram files. Each file may contain up to 100 or even more metabolic components separated in elution time and amenable to characterization by retention indices (RIs). As GC-MS based metabolite profiling is dependent on chemical derivatization to make nonvolatile compounds applicable to gas chromatography, each metabolite is represented by one ore more chemical derivatives or so-called analytes. Furthermore, each of these compounds is, after chromatographic separation, subjected to electron impact induced mass fragmentation. This process generates mass spectra, highly reproducible patterns of hundreds of mass fragments and respective mass isotopomers resulting mostly from the naturally occurring ^{13}C-isotope and other abundant elemental isotopes.

The preprocessing of the resulting information rich datasets has to serve two conflicting purposes. First, the observed mass fragmentation patterns need to be maintained as completely as possible to enable the best possible identification by mass spectral matching. Second, essentially only a single specific mass fragment in the optimum quantification range between lower and upper detection limits is required for relative quantification. Indeed, those high or low intensity mass fragments which are subject to increased technological noise at the analytical detection limits need to be removed prior to quantitative investigations. Furthermore, once a single or a few optimum mass fragments have been isolated, respective mass isotopomer distributions can be retrieved from the complete dataset and the path is open for stable isotope tracing studies and flux assessments.

When the TagFinder software project was initiated no software tool was available for the comprehensive data retrieval and selective quantification of either metabolite pool sizes or stable isotope flux. We now make publicly available a toolbox which allows automated and manually supervised comprehensive mass feature identification and extraction, including mass isotopomer distributions, from hundreds of GC-MS chromatography files. TagFinder provides standardized conversion into comprehensive numerical data matrices for subsequent statistical analyses (1, 2).

In the following section, we give a detailed description of the TagFinder toolbox and respective applications beyond the previously published application note (1). We first describe the *general architecture* of the software tool, then demonstrate and discuss the two supported metabolomic data processing workflows defined above: *Fingerprinting*, from chromatography files to standardized numerical data matrices for nontargeted metabolic marker discovery (see Subheading 3.1) and *Profiling*, from numerical data matrices to profiles of identified metabolites for metabolic phenotype assessment (see Subheading 3.2). Finally, we guide through the operating details of the TagFinder software (see Note 1) and comment on the choice of parameter settings (see Subheading 3.3).

2. Materials

2.1. TagFinder Software and File Formats

TagFinder (1) is a single user application written in the JAVA™ programming language. TagFinder utilizes the CDF chromatography data file interchange format (*.cdf* files), e.g., (3). TagFinder was initially developed for the chromatography data preprocessing of typical GC-time of flight (TOF)-MS based metabolite profiles (e.g., (4, 5)). However, essentially all GC-MS files can be submitted to TagFinder analysis provided the vendor- or mass detection-specific file formats are converted to the general CDF data-interchange format. So far, the *.cdf* file generating Andi MS export of the ChromaTof software (LECO Inc., St Joseph, USA) and the *.cdf* file generating AIA export of the ChemStation software (Agilent, Santa Clara, USA) have been tested. The TagFinder results are provided in an XML format or as more user-friendly comprehensive, tabulator (*tab*) delimited data matrices which can be submitted to visual statistical data mining software such as the TM4 multiexperiment viewer (6, 7). In the case of mass spectral exports or of processed MST information, the *msp* data format is used to allow uploading into the widely applied NIST mass spectral comparison software (8–11).

2.2. TagFinder Architecture and Size Limitations

The architecture of the TagFinder software (see Fig. 1) offers a graphical user interface (GUI) which enables control of the software functions by intuitive, user-friendly buttons and pull-down menus. Basic functions comprise all the tools and algorithms necessary for the general workflow from data import (*.cdf* files) to the output of numerical data matrices (*.tab* or *.msp* files) for fingerprinting analysis. Further functions of the Tagfinder program are added via the plug-in interface (see Note 2).

For data storage TagFinder creates and uses a workspace folder on the computer hard disk which may store the complete processing of experiments under evaluation. The folder can be named by the user. We recommend a unique identification of the processing job. Typically sets of 50–250 GC-TOF-MS chromatogram files with *.cdf* file sizes of 158.100 kB per chromatogram are recommended for efficient TagFinder analysis. TagFinder routinely creates and checks a peak data base file (*.tf* file) for each analysis job and provides a file identified by the extension *.props*, which lists and allows reloading of all workspace parameter settings of the current job.

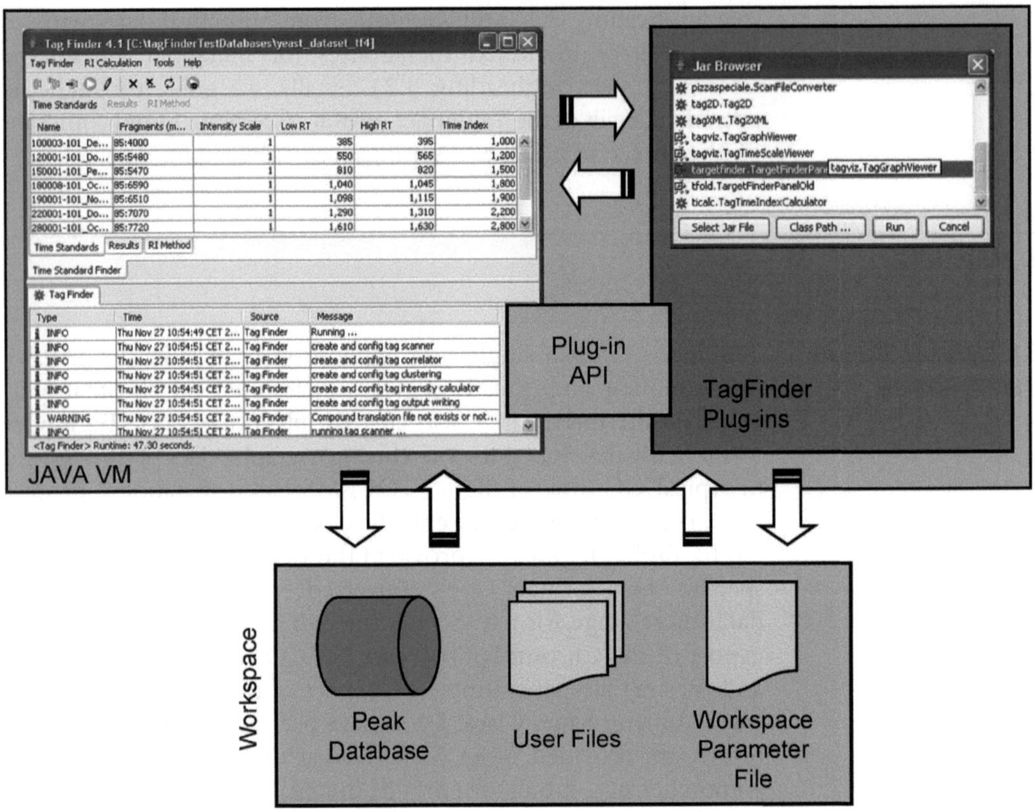

Fig. 1. Overview of the general TagFinder software architecture. The main graphical user interface (GUI) operates via buttons and pull-down menus. Plug-ins for specialized processing and visualizations are embedded in JAVA archive files and accessible through the so-called "jar–browser." TagFinder requires the generation of a workspace folder for each processing job. This folder can be named according to user requirements and should contain all relevant input files, e.g., the *.cdf* files and respective tabular peak lists. The same folder will contain the automatically generated peak database file (*.tf*) and subsequent optional files which may be generated during TagFinder processing, for example the workspace parameter settings file (.props).

We suggest that all additional input files and accessory data files which are used for or related to a TagFinder job and all intermediate files generated in the course of a TagFinder job should be conveniently stored and finally archived in the initial workspace folder. After initial workspace establishment, TagFinder requires an initial fingerprinting workflow which converts data from chromatography files into standardized numerical data matrices with sample annotations but without matched compound identities. This fingerprinting workflow is mandatory and must precede any compound identification in the subsequent metabolite profiling workflow.

2.3. System Requirements and Installation

2.3.1. System Prerequisites

1. Installed JAVA runtime environment 1.5 or more recent version.
2. Personal computer (PC) with an operating system suitable for JAVA (e.g., Windows, Mac OS, Linux).
3. 512 Mb RAM, recommended 1024 Mb RAM.

2.3.2. Installation Steps

1. Install a JAVA runtime environment.
2. Create an installation directory.
3. Unpack tag-finder.zip archive into this directory.

2.3.3. Running TagFinder

TagFinder runs under a JAVA virtual machine (VM). The memory usage needs to be specified and will be dependant on the available computer memory. To initialize the program use a command tool, and with the installation directory as the working directory, enter the following command:

>> java -cp .\TagFinder4.1.jar -Xms64M –Xmx512M tagfinder.TagFinderFrame

(PCs with a maximum of 512 Mbyte RAM);

>> java -cp .\TagFinder4.1.jar -Xms128M -Xmx1024M tagfinder.TagFinderFrame

(PCs with a maximum of 1024 and more Mbyte RAM).

The –Xms parameter defines the minimum memory allocation size and the –Xmx parameter defines the maximum memory allocation size. Note that the JAVA runtime environment is restricted to a maximum of 1024 Mbytes possible for the Xmx parameter. As an alternative a command file can be created to execute one of the command statements explained above. When starting TagFinder on a Windows PC use one of the batch files *runTF4.1-512 MB.bat* or run*TF4.1-1024 MB.bat*. Choose the memory parameters according to memory available on your PC.

3. Methods

3.1. The Fingerprinting Workflow

3.1.1. Generation of a Workspace

After opening the TagFinder software the user must either load an existing workspace or create a new workspace to begin a processing job. This process establishes a TagFinder folder dedicated to each job with an arbitrary, user-definable name. In addition, basic processing parameters of the workspace are defined, namely, the decimal precision of the RI system and the mass fragment range. We suggest using an *n*-alkane-based Kováts (12) or temperature programmed van den Dool and Kratz (13) RI system with 0.00 (1/100th) decimal precision and a 35–1,000 nominal mass range for yet unknown sample types or reference compounds or 70–600 amu for routine experiments. TagFinder creates a workspace file (*.workspace*, readable by a text editor program) and a database file (*spectra.tf*) within the workspace folder which stores and contains all settings and parameters selected through the TagFinder user interface. Current sessions can be stored prior to leaving the software. Upon reopening a workspace all previous settings and data are automatically reloaded and available. The TagFinder job folder can be conveniently used to build a user-definable folder system and to store all additional data files which may be relevant for the processing job and respective metabolomic experiment.

3.1.2. Data Import

TagFinder expects peak lists in *tab* delimited text format. Each peak list is required to correspond to a single chromatogram data file. The name of the peak list file should be unique and identical to the name of the vendor chromatogram raw file name and respective resulting *.cdf* file as this name is used for subsequent unambiguous sample identification. Each peak list file comprises rows which represent MSTs ranging from single observed mass fragments to lists of multiple coeluting mass fragments or even full, deconvoluted, mass spectra at given retention times (RTs) of the chromatogram files. A typical data format and required column header names are demonstrated by Luedemann et al. (1). The minimal requirement for a row entry is the fragment mass separated by a colon (:) from the measured intensity in the "Spectrum" column and an RT in the "Retention_Time" column. All other optional information gives TagFinder access to previous processing results from external software tools. For example, if a RI is calculated by an external software tool, TagFinder may use these data for subsequent processing using the optional "Time_Index" column of the peak list file. Externally deconvoluted mass spectra, for example deconvolutions of the ChromaTof software (LECO Inc., St Joseph, USA) or AMDIS (14) can also be processed observing the respective data formatting, as dedicated deconvolution algorithms may represent highly valuable data resources for qualitative investigations of metabolite inventories (15). For this purpose, the TagFinder peak list format provides a "Lib_Time_Index," "Lib_Match," and a "Lib_ID" column, the latter to accommodate a compound identifier such as given by the Golm Metabolome Database (16) or any other user-definable metabolite identification or name.

During data import the user can specify the minimum fragment intensity and RT range to restrict the final data size for the workspace file, which due to limitations of the operating system and JAVA run time environment may not exceed 2.04 GB. These data reduction options allow avoidance of low intensity data which are known to be subject to a high influence of technical noise or regions prone to chromatographic artifacts.

For those users who do not use external peak peaking or apply deconvolution software other than ChromaTof, the TagFinder software offers two built-in tools to create peak lists from chromatogram data files. First TagFinder uploads deconvoluted mass spectral lists which can be exported as a final processing result from the ChromaTof software, mass spectra in absolute peak intensities are accepted for the analysis of relative pool sizes, maximum normalized mass spectra can be imported into TagFinder for qualitative assessment of mass spectral deconvolutions. Second, TagFinder performs a comprehensive peak apex search and retrieval from baseline corrected chromatography files in the CDF interchange format. The first tool is a simple file converter which transfers the ChromaTof text format into TagFinder peak list format and includes matching information, as far as available from the

ChromaTof source files. The superior deconvolution properties of the ChromaTof processing for qualitative assessments (15) and the caveats of using deconvoluted data for quantitative investigations (4) have been described earlier (see Note 3).

3.1.3. Definition of Sample Attributes and Replicate Sample Groups

Sample attributes may be linked to the respective TagFinder imported peak lists and can be automatically added to the chromatogram header information of the final numerical data matrix. In addition, sample replicate groupings are used for some TagFinder options or may be highly useful for subsequent supervised data processing and mining methods applied to the numerical data matrix which is exported from TagFinder. Therefore, TagFinder allows manual editing and combination of replicate sample groups or the import of appropriate information from *tab* delimited sample annotation files. The required tabular sample annotation file must contain one column labeled "RAWNAME," which contains the exact names of all imported peak list files, which—as was stated above—are used as unique sample identifiers. Furthermore the sample annotation file may contain multiple columns each representing a different user-definable attribute. One of these attribute columns can be selected to represent the replicate sample group information and should contain the repeated sample group names, respectively.

3.1.4. RI Calculation

TagFinder aligns peak list files according to RI calculations based on RTs of authenticated internal reference substances, such as *n*-alkanes (5) or fatty acid methyl esters (4). RI calculation is a classical chemical standardization of variable retention behavior and substantially improves the alignment of observed mass fragments and MSTs between all constituent chromatogram peaks lists of a TagFinder processing job. Also, RI calculation allows the comparison of observed retention behavior in each new experiment to previously recorded reference and library information obtained from pure reference substances (17). RI alignment is typically sufficient for subsequent TagFinder processing and well aligned numerical matrix generation.

For RI standard finding in each chromatogram peak list and the subsequent chromatogram-wise RI calculation, TagFinder provides a tool which searches for the RTs of added internal RT standard substances. This time standard finder uses predefined and compound-specific mass fragments and respective normalized fragment intensities. The use of single mass fragments is recommended for efficient time standard finding. Partial or full mass spectra can be employed for the respective queries, but these bear the risk that peaks may not be recognized because deconvolution or apex retrieval may generate incomplete or split mass spectral entries in the chromatogram peak lists. The queries can be restricted to user-defined and adjustable windows of expected RTs. Queries are manually initiated and automatically retrieved RT results stored within a RI calculation file. This file contains all predefined RI values of the internal reference substances and the corresponding RTs from each of the chromatogram peak lists. This RI calibration file is best stored within the workspace folder of each separate TagFinder job. Ambiguities resulting in more than one hit for single RT standards are called to the attention of the user and can be solved automatically by accepting the most abundant hit or by refining the expected RT window. However, a user-supervised decision process is recommended as unexpected compounds with mass fragments similar to the applied reference substance class or contaminations by other than *n*-alkanes or *n*-fatty acid methyl esters may frequently occur. Once the RI calibration file is complete and accepted by the user, the RI calculation can be evoked through the time index calculation option. The calculation is an automated interpolation between the chosen RT anchors. The RI calculation process can also be used to employ constant laboratory contaminations or obligatory metabolites for retention standardization. Expected mass fragments or spectra and RT windows need to be adjusted accordingly. This approach is, however, rarely recommended as the amount of such compounds cannot be controlled and the variable loading of retention standards may contribute to the residual variability of RI standardization (17). For challenging alignments we recommend the testing of external mathematical alignment tools, as for example provided by the MetAlign software. RT optimized chromatogram peak lists or *.cdf* files can subsequently be uploaded and processed by TagFinder. Strong variation of compound loading, specifically discontinuous variations between lower detection limit to almost overloaded peaks, and attempts to compare highly diverse biological samples with differing qualitative and quantitative metabolite composition may, however, compromise mathematical peak alignment algorithms.

3.1.5. Mass Tag Scanning

Mass tag scanning or in other words the mass feature extraction by TagFinder is best performed after chromatographic alignment. Because conventional GC-MS data files are typically investigated at nominal mass precision provided and calibrated by GC-MS systems, the mass axis is not aligned by TagFinder. Higher mass resolution may be accommodated in TagFinder by rounding and multiplication by 10, 100 etc. to obtain integer values of the required precision. Given the restriction of workspace files to 2.04 GB this amplification of mass resolution will require more storage space and, as a consequence, the number of chromatogram files which can be processed will be reduced accordingly.

The mass tag scanning process of TagFinder screens independently for each nominal mass trace. Deconvoluted mass spectra and sets of uploaded coeluting mass fragments are decomposed in this process. For each single mass trace all fragment signals of a single nominal mass which are observed in the constituting chromatogram files of a TagFinder job are sorted independently by ascending RI. Within this sorted array of mass fragments RI gaps are monitored, which separate groups of closely coeluting fragments. Fragments of identical mass and similar elution behavior from separate chromatogram files are binned across all chromatograms. In this process each mass fragment bin receives an experiment specific RI window which characterizes the experiment specific variability of the RI behavior. Thus the mass fragment bins represent generalized single mass spectral tags and receive the properties mass unit (amu), minimum RI, maximum RI, and median RI.

Cases may occur where fragments of identical mass are found several times closely coeluting in the same chromatogram. For example, imperfect smoothing may result in multiple fluctuating peak apices. Also peak overloading may produce a plateau-peak with multiple local maxima. Using uploaded deconvolution data respective algorithms may offer multiple equivalent deconvolutions in identical retention windows. In such cases TagFinder performs an intensity aggregation, with optional summation, averaging or maximum intensity picking. The choice of aggregation mode should be dependent on the nature of the deconvolution algorithm. Smoothing and overloading artifacts may be accommodated by maximum intensity picking.

The main parameter for the RI gap finding process is the time scan width or allowed distance between retention bins defined by RI units. This parameter should be chosen high enough to cover the RI variation caused by variable compound loading and resulting peak intensity among the constituent chromatograms of a TagFinder job low enough to avoid unnecessary binning of chromatographically neighboring fragments into a broad unspecific mass tags. Each gas chromatographic variant, which may range from a fast GC temperature ramp with narrow peaks to a shallow temperature ramp with broad peak shape, requires optimization of the gap parameter. We recommend the selection of a chromatographic region with narrowly spaced peaks for optimization runs. Typically a slightly too broad gap is preferred because broad gaps will avoid miss alignment and the subsequent clustering process may still separate different but coeluting compounds.

Gap parameter settings will depend on the chromatographic and RI variants employed for GC-MS based metabolite profiling. For example, the method specifically described by Erban and coauthors (5) favors a range 0.5–1.5 *n*-alkane RI units as typical gap parameter settings. Low gap values are possible if high numbers of chromatograms (>50–100) are coprocessed in a TagFinder job or if the chromatogram files have highly similar qualitative and quantitative composition. Low numbers of coprocessed files < 25–50 or chromatograms exhibiting high quantitative variation require the choice of higher gap values. As each GC-MS experiment may differ with respect to the concentrations of hundreds of compounds, previously optimized parameter settings used for defined biological sample types and experiments should be archived but also critically revised and adapted upon reapplication in a similar but new TagFinder job.

3.1.6. Time Grouping and Clustering of Mass Tag Bins

As has been described previously TagFinder first decomposes fragments which may have been found to coelute in single chromatograms for the purpose of mass fragment-wise alignment and binning according to RI windows. After this procedure, TagFinder reconstitutes the resulting *mass fragment bins* into groups of different but coeluting mass tag bins exhibiting overlapping RI windows. These provisional groupings of coeluting mass tags are in the following called *time groups* and representative mass spectra are reconstructed. The basic criterion of this spectral reconstruction is the grouping of all mass tags which have identical or similar median RIs and overlapping RI windows. Small RI variations of single mass fragments occur and may become apparent by deviations of minimum and maximum RI where low intensity fragments typically have a smaller RI window width. Therefore, the grouping algorithm first sorts mass tag bins according to ascending median RI then by ascending minimum RI. Consecutive time groups are split, if the median RI of the preceding mass tag bin is smaller than the minimum RI of the following mass tag bin. The resulting time group partitioning sorts all mass tag bins which represent the same compound into the same time group. In a first and simplified approach implemented in TagFinder mass spectra of time groups can be reconstituted using robust averaged intensities of multiple or essentially all chromatograms of a TagFinder job. It is easily conceivable that mass spectral reconstruction based on multiple deconvolutions of individual chromatogram files or based on multiple peak height retrievals from many chromatograms may be superior compared to reconstructions obtained only from single chromatogram files. Also the identification of a time group (see Subheading 3.3) can be performed once for the complete data matrix instead of repeatedly for each single chromatogram data file.

Besides these obvious advantages of time grouping, the procedure also has a potentially severe disadvantage. Namely, a single time group may contain more than one coeluting compound (see Fig. 2). This phenomenon will become more severe with the increasing probability of coeluting compounds brought about either by TagFinder coprocessing of high numbers of chromatography files or by attempts of joined analyses of metabolically diverse sample types. As a consequence time groups of large or diverse experiments will contain an increasing number of nonspecific mass fragment bins which are aggregated from more than one coeluting

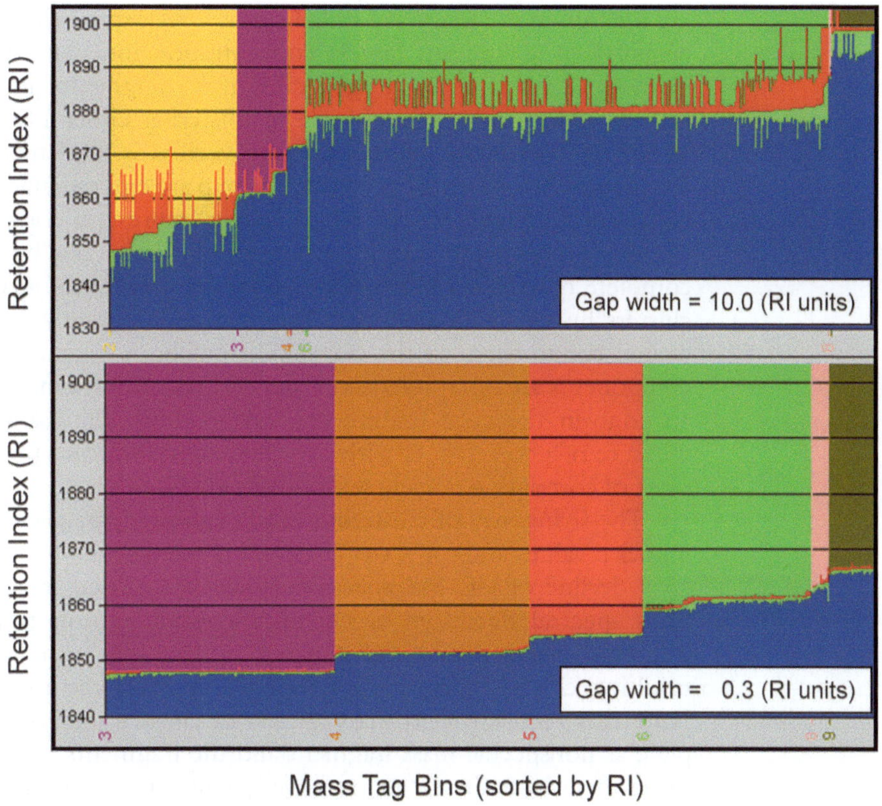

Fig. 2. Visualization mass tag bins arrayed by RI and sorted into time groups. The array of mass tag bins was according to ascending median and ascending minimum RI. Median RI is indicated at the *border of red* (maximum RI) and *green whiskers* (minimum RI). The variable width and gap parameter dependency of mass tag bins is exemplified. The split into consecutive time groups is indicated by color fields on the top-half of each panel. Two gap parameter settings are exemplified, gap width = 10.0 RI units top panel, gap = 0.3 RI units lower panel. The lower panel exemplifies the optimum choice of the gap parameter for the underlying dataset. Most time groups have homogeneous median RI and RI width. Some coelution effects cannot be avoided (*dark green* and *pink time group*). Choice of an extremely high gap value compromises TagFinder analyses as mass tags of neighboring time groups are aggregated. Broad and nonhomogeneous RI width may result. Visualization was performed by the tagviz.TagTimeScaleViewer plug-in.

compound. In extreme cases, only a few compound-specific mass fragment bins may remain for accurate identification and selective quantification of coeluting compounds. The occurrence of nonspecific and even identical mass fragments from different metabolites is caused by the high structural similarity of many metabolites. GC-MS fragmentation reflects essentially the set of possible substructures of a metabolite where selective mass fragments are generated only from the unique substructures of each metabolite.

In order to enable the retrieval of selective mass tag bins from composite time groups we implemented a clustering algorithm which screens the mass tag bins of each time group for those sets of fragments which exhibit optimal correlated quantitative behavior across all chromatograms of a TagFinder job. These sets of mass tag bins are in the following called *clusters*. We exploit the GC-MS property of a constant largely concentration independent fragmentation process, which generates mass fragments in highly reproducible relative quantities. For clustering either Pearson or Spearman correlation of the intensity vectors is implemented in TagFinder. All constituents of a time group are entered into a complete correlation network where edges represent correlations coefficients whereas vertices represent the mass tag bins. The complete correlation network of a time group is taken apart into clusters by a core finding algorithm (18). Only highly interconnected components of the correlation graph are retained. These components are interpreted as clusters. Thresholds for maintaining an edge in the core finding process are both the significance measure of a correlation, with p ideally set <0.001, and the coefficient of correlation, ideally set to >0.8 or higher.

The stringency of clustering can be adjusted (by allowing for higher p values and/or lower correlation coefficients) to the technical decline of GC-MS system performance caused by aging of mass spectral detectors or chemical contaminations requiring frequent cleaning and maintenance cycles. Like time groups, resulting clusters can be used for mass spectral reconstruction. Note that reconstructed mass spectra of clusters may be incomplete as nonspecific mass fragment and the fragments at upper or lower detection limits are removed, while reconstructed mass spectra of time groups may be composite. Also note that time groups and clusters are characterized by size (i.e., the number of constituent mass tag bins) and count of observations (i.e., the number of chromatogram files with a mass tag signals above the selected intensity threshold for data uploading and processing). Clusters are, in addition, characterized by a score value of the core-finding process. Mass tag bins which do not fall into a cluster may either be discarded as these may contain highly noisy data or maintained for fingerprinting analyses, as rare cases exist where a metabolite may be represented only by a single mass fragment.

3.1.7. The Mass Spectral and Numerical Data Matrix Output

The mass spectral reconstructions of time groups and clusters can be exported together with respective RI information in .msp format for uploading into the NIST mass spectral matching and interpretation software, for visual inspection and manual mass spectral comparisons. The final numerical data matrix of the TagFinder processing is written into a *tab* delimited text file (*.tab* file) which can be uploaded to the Microsoft (MS)-EXCEL table calculation program or to any other more refined software tool for statistical assessments. A .xml version of the *.tab* file can be generated using the *tagXML.Tag2XML* converter of the plug-in collection. The tabular matrix contains all nonnormalized intensity data of each mass fragment observed in the chromatogram compendium of the current TagFinder job. Mass tag bins are arranged in rows complete with mass to charge ratio, median, minimum, maximum RIs and RI width, time group and cluster assignments. In cases of preprocessed and deconvoluted data the most frequent matches, the best mass spectral match and the match with lowest RI deviation are added. In many cases mass spectral hits may be recorded as non-human-readable mass spectral identifier codes which are transferred form customized mass spectral libraries. In order to translate such identifier codes into clear chemical names a compound translation file may be specified and selected from the TagFinger output settings prior to the generation of the final *.tab* file. The specified translation file must also be in *tab* delimited text format. The first column of this file must contain the exact identifier code of the respective mass spectral or compound library entries and the second column should list the respective preferred clear name. In the first row of the translation file a user-definable header row is expected.

At this step the TagFinder chromatography data preprocessing for nontargeted fingerprinting purposes ends. Typically visual inspection of color-coded (e.g., red-green) heat maps of the *.tab* file and visual inspection of time group or cluster based mass spectra may aid the judgment of the chosen TagFinder settings. Also nonsupervised statistical methods applied to the *.tab* file, such as principal component analyses (PCA) or supervised methods utilizing the sample group information, such as analyses of variance (ANOVA), will confirm the validity of the data processing. The combined assessment using these methods will allow good judgment of the overall data quality and may indicate an iterative optimization step of the TagFinder processing. In any case, a manually optimized *.tab* file should be exported, finalized and stored in the respective workflow folder for subsequent applications. At this point the user will need to accumulate experience with the respective analytical variant of the metabolomic laboratory and the sample types under investigation.

3.2. The Profiling Workflow

3.2.1. General Considerations of the TargetFinder Panel

The TagFinder software suite supports a second workflow—the metabolite profiling. This workflow starts with a numerical data matrix, previously and in the following called *.tab* file, which has been generated and optimized in the fingerprinting data preprocessing step. The profiling work flow enables the iterative identification of metabolites which are present in the *.tab* file and extracts those analytes, i.e., chemical derivatives of metabolites, and respective mass tag bins which are—within the context of the respective TagFinder job—best suited for the quantification of this metabolite. The profiling workflow can either aim for a comprehensive identification of all known MSTs and analytes or extract only relevant information of predefined, targeted metabolites. Comprehensive identification is time consuming and may not always be necessary as metabolite identification can either be driven by statistical data mining steps, which pinpoint time groups, clusters or single mass tag bins as potential biomarkers, or identification can be motivated by pathway knowledge and iterative screening of linked metabolites for significant changes or equally relevant unchanged pool size.

Assignment of compound identity has not yet been fully automated for GC-MS based or any other metabolomic technology. As both false positive and false negative assignments are possible the metabolomic scientist is well advised to perform manually supervised compound to peak assignments. Also external standardization with chemically defined reference mixtures of expected metabolites will contribute substantially to the confirmation of difficult to separate and hard to distinguish metabolite isomers, such as the typical epimeric structures of sugars, e.g., glucose, galactose, or mannose, etc. (17).

The Tag Finder software supports semiautomated mass spectral and retention index based matching. The matching procedure is built on MS/RI reference libraries which are exported from the Golm Metabolome Database compendium (16). Such files are available upon request for noncommercial applications from the authors of this chapter. The implemented matching procedure is applied to reconstructed mass spectral data embedded and extractable from the *.tab* file of the previous data preprocessing workflow. Both time group- and cluster-spectra can be matched. Results are presented to the user for manual confirmation or rejection.

Peak annotation is performed within the TargetFinder panel in sessions which can be stored and reloaded for iterative metabolite to peak annotation. Annotation results are used to retrieve and subset data from the initial comprehensive *.tab* file of TagFinder jobs. Only those mass tag bins which have a metabolite or yet unknown analyte MST identification assigned can be extracted in this step for subsequent compound specific visualization or statistical analyses. Furthermore, metabolite to peak annotations can be merged with the preexisting *.tab* file and the extended *.tab* file converted into an xml-format.

3.2.2. The TargetFinder Panel

The TargetFinder panel (see Fig. 3) is evoked by opening the *tagtools.jar* file from the jar-browser which can be activated through the external-tools-button. Within the *tagtools.jar* file, the "targetfinder.TargetFinderPanel" option starts the session. The TargetFinder has two folders, first the "Targets" folder, which operates the MS/RI library containing the targeted MSTs (targets) from the Golm Metabolome Database, second the "Match Results" folder which displays the potential matches, provides a visualization tool and allows hit selection.

3.2.3. The Targets Folder

The "Targets"-folder allows loading and saving of target lists, i.e., *tab* delimited *.txt* files which contain the targeted MS/RI information. Sessions can be saved as .tfs-files and reloaded complete with both the previously used target lists and the full matching results.

a

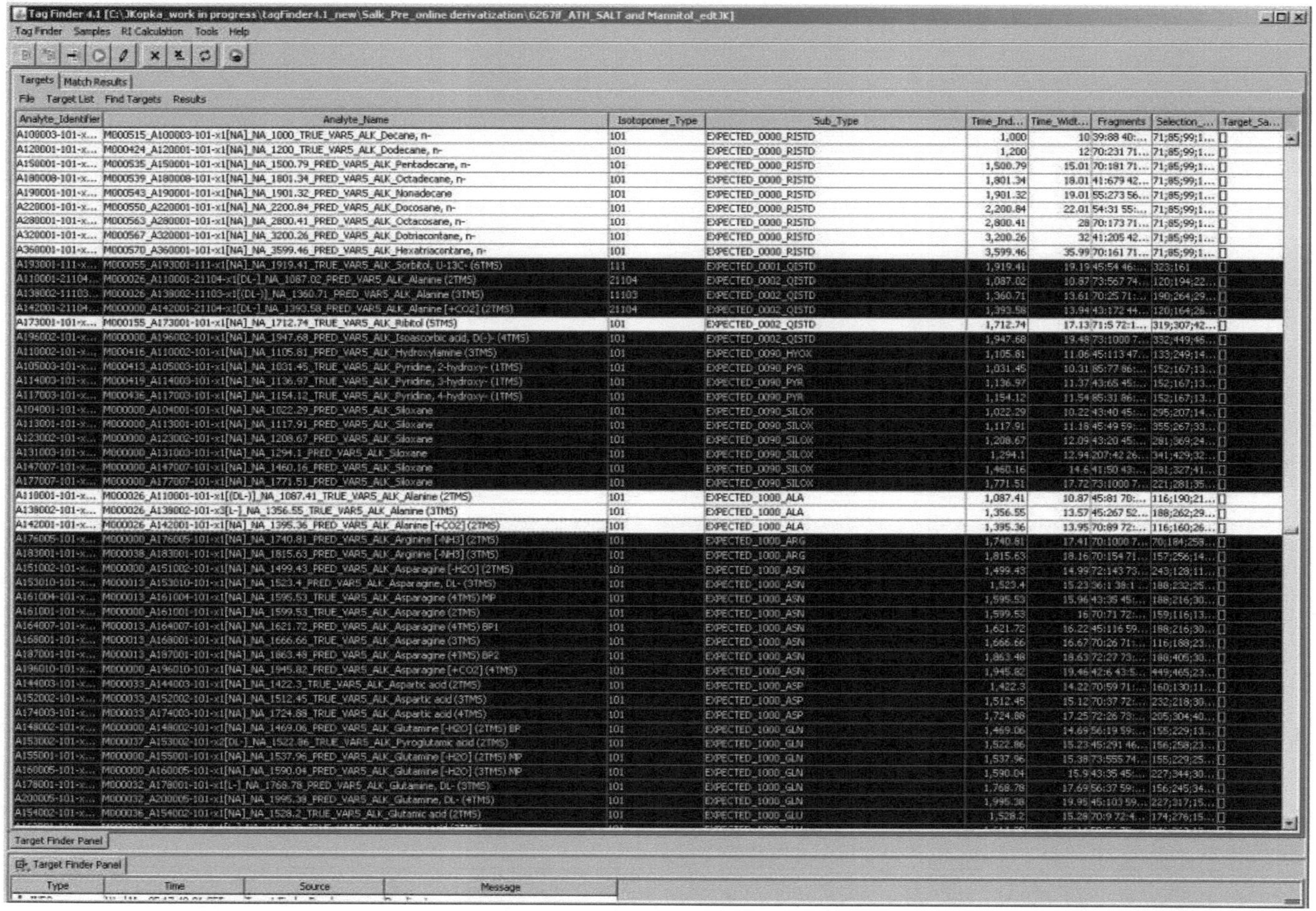

Fig. 3. (**a**) The TargetFinder Panel. The TargetFinder panel is split into two folders. The "Targets" folder organizes the mass spectra and retention index reference library and the matching processes. (**b**) The TargetFinder Panel. The TargetFinder panel is split into two folders. The "Match Results" folder visualizes the matching results and enables manually supervised metabolite/analyte to peak annotation.

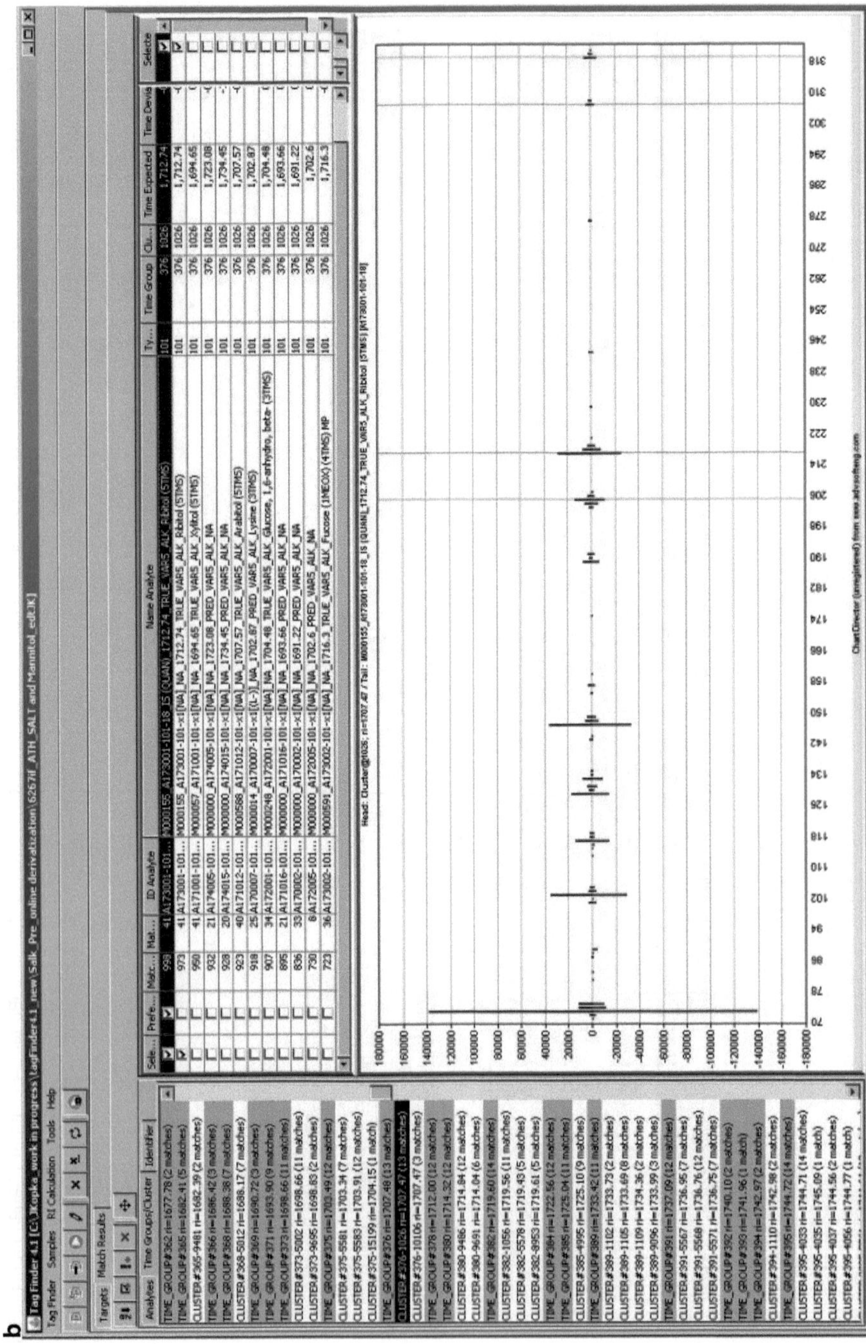

Fig. 3. (continued)

Furthermore, all previously mentioned export functions of metabolite to peak annotations are accessible from this folder through the "Results" pull-down menu. Target lists have a simple tabular format with the following required column headers, Analyte_Identifier, Analyte_Name, Isotopomer_Type, Sub_Type, Time_Index_Expected, Time_Width_Expected, Fragments, Selection_Masses, Target_Samples. In detail, the "Analyte_Identifier" and "Analyte_Name" column list the identifier and clear name of the targeted analyte and MST. The "Isotopomer_Type" column allows differentiation of stable isotope labeled MSTs from MSTs with ambient isotopomer composition. This column was designed to support the targeted analysis and detection of stable isotope labeled internal standards. The "Sub_Type" column is open for user-defined classifications, for example MST classes according to type of chemical derivatization, e.g., methylester, trimethylsilylester, *t*-butyldimethylsilylester, or type of mass spectrometric technology, e.g., quadrupole, time-of-flight, or ion-trap, used. "Time_Index_Expected" and "Time_Width_Expected" list the expected RI and the RI window within which the mass spectral matching is performed. Typically ±1% of the expected RI can be recommended (17). Because the variation of retention behavior appears to be a compound specific rather than a general property, the RI window and the expected RI can be user-defined for each analyte. A missing entry in RI-columns is indicated by "0" and causes extension of the mass spectral matching procedure to all time groups or clusters of the selected *.tab* file. The "Fragments" column harbors the mass spectrum which is formatted in colon separated *m/z*:intensity pairs. The target spectra can be either maximum normalized or nonscaled. In the column "Selection_Masses," a subset of fragment masses can be defined, separated by semicolons, which may have been found to be unique in earlier experiments or which are required to compare current results to earlier studies which may have been based on single or few mass fragments. These fragment masses are indicated upon visualization of matching results and may be selected for retrieval and subsequent quantitative analyses. Finally, the column "Target_Samples" can be employed to restrict the matching procedure to a single chromatogram peak list, for example a chemically defined reference sample, or to a subset of chromatogram peak lists. This feature is useful if the identification of compounds is exclusively driven by matching to chemically defined reference samples. Furthermore, flux and tracing experiments generate mass spectra with inherently fluctuating mass isotopomer distributions. These fluctuations interfere with current state-of-the-art matching algorithms. In such cases, parallel analysis and processing of ambient biological reference samples is recommended. Thus, the aligned biological reference samples can be used for the matching procedure of the TargetFinder whereas the mass isotopomer distributions are retrieved from all coprocessed samples.

Once the target list is uploaded and the targets of interest selected by the "Target List" pull-down menu, the *.tab* file under investigation may be uploaded using the respective function of the same pull-down menu. Also activated through the "Target List" pull-down the "Setup Target Finder" option allows specification of the matching details. The matching procedure can be optional either for all or for specified samples only (chromatogram peak lists). All masses, i.e., the full spectrum, can be used for matching, or only those m/z values defined in the "Selection_Masses" column. Matching can be applied to time groups or clusters and may be restricted to a minimum number of required fragments or limited by a matching value threshold. The match value is based on the dot product and converted to integers via the formula (1 − dot product) × 1,000, where the value 0 indicates dissimilarity and the value 1,000 complete identity. The number of matching fragments may be set to a threshold of 3–5, whereas the match value threshold should be approximately 100. Both proposed thresholds optimize for least false negative hits because ambiguous matching results can be sorted and visualized in the "Match Results" folder described in the following paragraph. The matching procedure is initialized by activation of the "Find Targets" button accessible by the "Find Targets" pull-down menu. Results of one or combinations of multiple sequential matching procedures are displayed in the "Match Results" folder.

3.2.4. The Match Results Folder

The "Match Results"-folder displays the hit lists of matching results for visual inspection and manually supervised metabolite/analyte to peak annotation. Three subwindows show the list of matched items (to the left), the sorted or ranked hit list corresponding to each item (top right) and the matched mass spectra in head to tail view (bottom right) with the reference spectrum displayed below in red. Five functions are provided via buttons in the top left corner. Buttons from left to right enable (1) a customizable sorting procedure, (2) general selection procedures, among others an automated selection of the first hit of time group matches or cluster matches according to the sorting defined under (1), (3) an automated search for matching conflicts, (4) a general clear option of the "Match Results"-folder, and (5) a scaling option for the visualization of matched mass spectra.

The subwindow showing the matched items to the left allows manually supervised annotation according to analyte identifier (Analytes), according to time group and respective clusters found within the time groups (Time Groups/Cluster) and according to other criteria provided by the uploaded target list, such as analyte name, mass isotopomer classification or the customizable "Sub_Type" column (Identifier). The experienced user can follow in essence one of two schemes of iterative annotation. The first workflow enables the search for and subsequent annotation of time groups and clusters which may have been indicated by statistical analyses to represent potential biomarkers or to be significant for the current TagFinder job and investigation. The second workflow enables metabolite or pathway driven annotation, for example according to metabolite/analyte name or according to subtypes which may contain collections of metabolites/analytes which belong to a common pathway.

The ranked hit list subwindow in the top right corner presents the hits of each item, for example the best metabolite/analyte hit of each time group and cluster or the best time group and cluster for each metabolite/analyte identifier. The sorting of these hit lists can be customized as described above. Selection of validated hits can be automated (see above). Wrong or missing automated annotations should be checked and corrected manually by clicking or unclicking the "Selected" and "Preferred" boxes. These two levels of selection allow the choice of more than one hit representing a metabolite/analyte. By choosing "Preferred" the preference of the respective hit is indicated. This option enables, for example the selection of both the time group and the cluster representing a targeted analyte and allows the choice of the annotated cluster for quantification. In cases of multiple analytes which may all represent the same metabolite, the preferred analyte and cluster may also be indicated.

The criteria which are used to match compounds by the TargetFinder are the match value in descending order, the number of common mass fragments comparing hit and target spectra in ascending order, and the RI deviation best approximating zero. The RI deviation thresholds, approximately 1.0%, are already well characterized (16). The number of matching mass fragments may be set to 5. This threshold will vary as increasing numbers of chromatogram peak lists subjected to a TagFinder job will enhance the occurrence of coeluting compounds and thereby reduce the number of available residual unique mass fragments. Also no rule of thumb is currently available for the thresholding of the matching value because (1) depending on the TagFinder job, time groups bear the risk of being composite or in other words chimeric and (2) clusters are in essence incomplete mass spectra because unspecific and noisy mass fragments which may compromise compound quantification are removed from valid clusters by correlation analysis and core finding.

3.3. Operation of the TagFinder Software

Two essential workflows are supported by TagFinder, first the nonsupervised and (within the limits of the GC-MS technology) comprehensive generation of a complete and chromatographically aligned numerical data matrix for large sets of chromatogram files, and second the manually supervised and partially automated metabolite to peak annotation. In all aspects, the full access to all primary data for the use of stable isotope labeled compounds is maintained and the basis for flux analysis (19) and quantification based on mass isotope ratios (20) maintained. In the following section, we give detailed instructions for the operation of the TagFinder software and for the optimization of the TagFinder parameter settings.

3.3.1. Creating a Workspace

1. From the *TagFinder* (see Note 4) menu click *Create Workspace* to open the *Create Workspace* dialog.
2. Click the *Select Path* button to define the workspace path.
3. Edit the *Time Index Scale* and *Fragment Mass Range* parameters according to your experiment (see Note 5).
4. Click the *Create* button.
5. A property file will be created under the defined workspace directory path. Modification of this file with external tools or movement of this file to a different folder location will compromise the TagetFinder job.

3.3.2. Creating a Sample Annotation File

1. Open an editor capable of producing *tab* delimited text files, for example Microsoft (MS)-EXCEL.
2. Define annotation column headers using the first row of the table. Take care to place the sample names without file extensions into the first column with the header "RAWNAME."
3. Write sample names into the first column, start a new row for each sample.
4. Write respective sample group names into second column (recommended) or any other defined column.
5. Add additional sample annotation data and alternative classifications to the remaining column(s).
6. Save the table as a *tab* delimited text file (*.txt*).

3.3.3. Import Procedures for ChromaTOF (LECO Inc.) Peak List Data Files

1. From the *Tools* menu select *ChromaTOF Text Converter*.
2. Select the files to be converted in the file selection dialog.
3. Select the directory into which the converted files will be placed.
4. Processing messages are reported within the message console of the TagFinder main window. For further detailed information, a processing log file is created at the target directory.
5. Continue with the steps listed under *Import Peak Lists* below.

3.3.4. Import Procedures for CDF Raw Data Files

1. Make sure that the *.cdf* data files have been properly smoothed and base line corrected using the vendor or alternative software. For using MetAlign for this purpose see Note 3 and Chapter 15.
2. From the *Tools* menu select *Peak Finder*.

3. Select the *Files* card.
4. Click the *Add* button to add all required *.cdf* files to be processed into the *Input Files* list box.
5. Specify the target directory into which the resulting peak list files will be placed.
6. Select the *Peak Finder* tabulator.
7. Specify the smoothing parameter in the *Smooth Width Apex Finder* field. This parameter defines the number of scans used by the box filter employed to reduce any residual apex noise prior to the peak finding process. The smoothing width will depend on the scanning rate of the *.cdf* files.
8. Specify the intensity threshold in the *Low Intensity Threshold* field. This option allows data reduction based on an assessment of the residual technological noise of the GC-MS system after smoothing and baseline correction.
9. The *Max Merging Time Width* option may remain unselected. The optional setting is 1/scan rate, i.e., 0.05 for chromatogram files recorded with 20 scans/s.
10. Select the *Large File Mode* option if the size of the processed *.cdf* files is >100 Mb to avoid "out-of-memory" error messages.
11. Click the OK button to start the peak finding process.
12. A progress bar dialog appears. *Stop* aborts the process. The processing may take considerable time depending on the number and size of the *.cdf* files.
13. All processing messages are reported in the message console of the TagFinder main window.
14. Continue with steps listed under *Import Peak Lists* below.

3.3.5. Import Procedures Using MetAlign Base Line Processing and Peak Searching

1. From the Tools menu click *MetAlign Base Line Processing* (see Note 3).
2. A configuration dialog appears. Select the *Path Settings* card. In field *Path to MetAlign Binaries*, specify the path to the folder into which the MetAlign executable files were installed. In field *Path to MS Data Files*, specify the path to the folder which contains the *.cdf* files for processing, under Path to *MetAlign Temp Folder* specify the MetAlign temporary output folder, under *Path to MetAlign Output Folder* specify the MetAlign processing output folder.
3. Select the *Baseline Processing* card. Specify parameters for base line processing and peak search. Consider the processing information provided via the MetAlign (21–23).
4. Click *OK* to start the process. A progress bar dialog will appear. *Stop* aborts the process.

5. The resulting peak list text files for import can be found in the *Baseline* subfolder of the *MetAlign Output Folder* you have specified in the MetAlign settings.
6. Continue with the steps listed under *Import Peak Lists* below.

3.3.6. Import of Peak Lists

1. From the *TagFinder* menu select *Import Peak Lists.*
2. Confirm to discard any preexisting data if you should decide to reimport peak list data.
3. The *Peak List Import* dialog appears. Specify the lowest intensity threshold in the *Low Intensity Threshold* field and the retention time range in the *Start Time/End Time* fields. This option allows the preprocessing of *.cdf* files with high sensitivity, i.e., a low noise threshold, and enables iterative optimization of the low intensity threshold upon import. Also size reduction of the TagFinder job is possible to accommodate extremely high numbers of peak lists at the loss of low intensity values (see Note 6).
4. Click the *Files* button and select the peak list files for import.
5. Run the import process.
6. A progress bar dialog will appear. *Stop* aborts the process.
7. Processing messages are displayed in the message console of the TagFinder main window. A completed peak list import procedure creates a peak database file in the workspace directory.

3.3.7. Definition of Sample Groups

1. From the *Samples* menu select *Set Sample Groups.*
2. The *Set Sample Groups* dialog will appear.
3. Select the *Sample Groups* button.
4. Select the sample annotation file from the dialog.
5. A selection dialog box allows the selection of the proper column of sample names from the sample annotation file. Select the column header.
6. Next a selection dialog box enables selection of the column which contains the sample group information. Select the column header.
7. Verify the assignment of the sample groups in the displayed sample table and click the *Apply* button. The table will be re-sorted lexically by group name and sample name.
8. Click the *Close* button to close and exit the dialog window.

At this point of the TagFinder job, all sample information and respective peak intensities which have been recorded are uploaded to TagFinder and are ready for further processing.

3.3.8. Retention Index Calculation

Creating a Retention Time Standard Reference File

1. From the *RI Calculation* menu select *Time Standard Finder* to open the time standard finder panel.
2. Select the *Time Standards* card.
3. From the *Time Standards* menu select *Add Time Standard Query* to add a new row to the existing or empty table.
4. Edit the name of the time standard spectrum in the *Name* column. The name must be unique within this time standard reference file.
5. Edit the list of query fragments in the *Fragments* column. The mass spectrum format of an entry is mass:intensity and each list entry is separated by a space character (only integer values are permitted). We recommend base-peak normalized intensities for this purpose.
6. Edit the intensity scale in the *Intensity Scale* column. The scaling factor will be applied to the queried fragment intensities for the time standard search and can be adjusted separately for each time standard.
7. Edit the expected retention time interval in the *LowRT/HighRT* columns.
8. Edit the retention time index into the column *Time Index*. This value is predefined according to the preferred retention index system and used by TagFinder to calculate the RI according to a linear interpolation model.
9. Continue to complete the list of retention time standards.
10. From the *Time Standards* menu select *Save Time Standard List* to save the retention time standard reference file as a *tab* delimited *.txt* file. These files can be reused for subsequent TagFinder jobs, provided the chromatography settings remained unchanged.

Loading and Editing of Retention Time Standard Reference Files

1. Select *Open Time Standard List* to load a retention time standard reference file.
2. Select add or remove *Time Standard Query* or edit existing rows of the *tab* delimited *.txt* file.
3. Save *Time Standard Query* (see above).

Creating a Retention Index Calculation Method

1. Create or load a retention time standard reference file (see the subheading above).
2. Select the *RI Method* card.
3. From the *RI Method* menu select *Init RI Method* to create a RI method with initialized expected time standard entries of each chromatogram sample. The time standard list must be finalized before initializing the RI method. Adding of retention time standards to an opened method is impossible.
4. Select the *Time Standards* card.

5. From the table entries select a single query for sequential analysis or select all time standard queries.
6. From the *Time Standards* menu select *Run Time Standard Finder* to execute the time standard finder query, or select the context menu by right mouse click into the table.
7. Refer to the TagFinder message console for the summary information of the procedure. Select the *Results* card for visual inspection of the retention time standard hits, unique and multiple, possibly ambiguous results are displayed also check for completeness of hits. If results are ambiguous or incomplete return to the respective retention time standard entry and modify the expected retention time window and/or the intensity scaling factor. Both threshold setting allow reduction of unwanted and unspecific hits or a more sensitive search for time standard compounds in not adequately standardized experiments.
8. The ambiguous results, i.e., multiple hits per single chromatogram/peal list are displayed in the bottom split of the result window. Manually select the proper time standard hit from the bottom split click *Move to Results* from the *Results* menu, or select the context menu by right mouse click within the table. Proceed to solve all ambiguous results of a single retention time standard.
9. From the *Results* menu click *Select All* and then select *Send to RI Method*, or select the context menu by right mouse click within the table.
10. Complete step 4–9 for all retention time standard queries.
11. Select the *RI Method* card.
12. From the *RI Method* menu select *Save RI Method* to save the RI method as a *tab* delimited *.txt* file.

Interpreting Time Standard Finder Results

Four cases of results may be distinguished: (1) *no results*, no peaks found, (2) *complete results*, for each sample exactly one peak found, (3) *samples with ambiguous results*, more than one potential hit found in at least one samples, and (4) *samples with incomplete results*, missing hits in at least one sample.

As experimental errors may occur the time standards are automatically searched but manually confirmed or edited be the TagFinder user. The expert user is advised to cautiously perform the retention time standard search procedure because (1) the process will reveal faulty chromatograms which may have been included in the TagFinder job by accident and (2) show potential retention time drifts in the course of larger chromatogram series which may afford splitting the TagFinder job into two or more. Make sure to assign exactly one peak per chromatogram peak list and retention time standard for the subsequent RI calculation.

Retention Index Calculation

1. From the *RI Calculation* menu select *Time Index Calculation*.
2. Select the RI method file within the file dialog and click *Open*.
3. A progress bar dialog will appear. *Stop* aborts the process.

3.3.9. Scanning of Mass Tag Bins

The mass tag binning process is highly parameterized (see Note 7). A default setting is included. Modifications to the parameters may be saved and reloaded as .props-files. As each GC-MS system may produce data with different properties, specifically different mass spectral scanning rates, mass ranges, peak width and peak separation, only general comments are made. Restrict the processing to a short selected chromatographic region with a set of closely eluting compounds and optimizes the gap parameter and the cluster settings for the respective type of GC-MS system.

1. From the *Tag Finder* menu select *Setup* to open the *TagFinder Settings* dialog (see Table 1 for a detailed description of the recommended settings).
2. Modify the parameters according to your experiment properties.
3. From the *Tag Finder* menu select Run to start the mass tag scanning process.
4. A progress bar dialog will appear. *Stop* aborts the process.
5. See the TagFinder message console for summary information.
6. A *.tab* file will be generated to contain the final numerical data matrix with assignments of mass tag bins, time groups and clusters as was described (see Table 2 for detailed descriptions). This file can be opened by MS-EXCEL or TM4 (2, 3) or evaluated by the *TagTimeScaleViewer* plug-in (see Note 8).

4. Notes

1. *Software Availability.* Exemplary TagFinder workspaces, test datasets, and target libraries are made available upon request for academic, noncommercial use (2).
2. *Notes on Running Plug-ins.* The TagFinder plug-in interface offers two classes of functions, so-called processes and panels. The "process" plug-in class is mainly designed for complex calculations requiring longer processing times and a minimum user control via the user interface. By contrast, the "panel" plug-in class is the basis of operations which require high GUI control and multiple user interactions. Also other enhanced graphical abilities, for example those required for data visualization, are provided as panel plug-in processes. Panels are presented and controlled within the TagFinder main window. Plug-ins need to be programmed in JAVA and have to be

Table 1
***TagFinder setup parameters for the tag finding process.* All parameters can be defined in the *TagFinder Settings Dialog*. Only the parameters for routine applications are explained. Default settings are provided within a .props-file which can be uploaded modified and saved for subsequent use**

Field	Description	
Tag scanning/time scanner		
Time scan width	The time scan width is the most important parameter for the tag search. This gap parameter defines the retention distance, expressed as RI units of two mass fragments with identical to be split into two separate, consecutive bins. This parameter requires optimization for the general chromatography variant employed by each laboratory and additional adjustment to retention drift and possible peak shape artifacts caused by column aging and accumulating contaminations.	
Gliding median group count	Set to 1. This option is not for routine use.	
Min fragment intensity	Excludes from the tag scanning process all mass fragments of intensity lower than the defined value.	
Force min tag width	Switch off; this option is not for routine use.	
Apply target scanning	Switch off; this option is not for routine use.	
Tag scanning/tag gen filter		
Tag mass	Suppresses the generation of tags for defined mass numbers. Definition of single mass values (i.e., 71) or single intervals (i.e., 75–80) or a list of intervals (i.e., 71; 75–80; 590–600) is possible.	
Fragment count	Generates only those tags which have at least the defined number of fragments.	
Sample count	Generates only those tags which have fragments in at least the defined number of chromatogram peak lists.	
Tag time width	Switch off; this option is not for routine use.	
Tag time index	Suppresses the generation of tags within defined RI intervals. Definition of single intervals (i.e., 1,495–1,550) or list of intervals (i.e., 1,195–1,220; 1,495–1,550) is possible.	
Tag scanning/intensity calculation		
Simple	Intensity aggregation	Define the mode of intensity aggregation: SUM_INTENSITY: returns the sum of the fragment intensities per sample, MAX_INTENSITY: returns the maximum fragment intensity per sample, typically used for peak apex data.
	Intensity range	Switch off; this option is not for routine use.
	Min intensity	Switch off; this option is not for routine use.
	Max intensity	Switch off; this option is not for routine use.
	Reverse out range intensities	Switch off; this option is not for routine use.
Tag scanning/intensity calculation		
Extended	Check sparse groups	Switch off; this option is not for routine use.
	Outlier check	Switch off; this option is not for routine use.
Tag correlation		
Correlation method	Defines the correlation method for the correlation network generation. Pearson: parametric, normal distribution of intensity data assumed. Kendall: nonparametric, no distribution of intensity data assumed.	

(continued)

Table 1 (continued)

Field	Description	
Maximum tag distance	The distances for the network edges are calculated by 1—correlation value. Defines the maximum tag distance threshold for which edges will be inserted into the network. "0" defines highest similarity; "1" defines maximum distance	
Significance level	Defines the significance level for the significance test of the correlation values for which edges will be inserted into the network.	
IQR check pair ratios	Applies an interquartile range estimation of fragment intensity ratios of each correlation pair (tags are correlated by sample intensities across all chromatogram peak lists).	
Maximum IQR pair ratio distance	Defines the maximum threshold for the interquartile range check for which edges will be inserted into the network.	
Minimum number of sample pairs	Defines the minimum count of intensity value pairs to use for correlation for which edges will be inserted into the network.	
Min sample group pair count	Set to 0. This option is not for routine use.	
Clustering		
Core adjacency option	Defines the core finding method, SAME_CORE: Interprets as a cluster all subnetworks of adjacent tag nodes at the same degree core level. MIN_CORE: Interprets as a cluster all subnetworks of adjacent tag nodes up to a defined minimum degree core level.	
Min core option	Selects the minimum core level by automated estimation or user defined input.	
Min core value	Stops graph traversal at tags with degree core < defined value. Usual set to 3–5. This value is only used if the min core option is the input value.	
Check score limit	Switch off. This option is not for routine use.	
Tag output		
Files	Tag output file	Specifies the file for the *.tab* file output.
	Sample annotation file	Specifies a sample annotation file (facultative).
	Compound translation file	Specifies a library match identifier to compound name translation file (facultative).
Tag output	Replace missing intensity	Inserts text which represents missing values into the sample intensity matrix.
	Scan for tags only	Creates a tag list without correlation and clustering.
	Ignore unassigned cluster	This option is not for routine use. Excludes from the output all mass tag bins which were not assigned to a cluster.
	Restrict by intensity rank	This option is not for routine use. Returns only mass tag bins up to a maximum intensity rank per cluster.
	Max intensity rank	This option is not for routine use. Defines the maximum intensity rank.
	Restrict by cluster size	This option is not for routine use. Returns only mass tag bins of clusters with defined minimum size.
	Min cluster size	This option is not for routine use. Defines the minimum cluster size.

Table 2
***Interpreting the TagFinder output.* The numerical data matrix generated by TagFinder, i.e., the *.tab* files, can be divided into five sections: (1) sample annotations, (2) mass tag binning and time grouping summaries, (3) mass spectral search and analysis results, (4) cluster assignments and (5) the fragment intensity matrix. The sample annotations are attached to the top of the data matrix above sample name header**

Field [unit]	Meaning
Mass tag binning/time grouping	
Tag time [time index unit]	Median time index of grouped fragments
Tag mass [mass number]	The mass trace number
Tag ID [sequential number]	A table-unique tag identifier
Tag sample count [count]	Number of samples from which fragments were grouped.
Tag fragment count [count]	Number of grouped fragments
Tag low time [time index unit]	Lowest time index of grouped fragments
Tag high time [time index unit]	Highest time index of grouped fragments
Tag width [time index unit]	Difference between lowest/highest time index
Time group [sequential number]	Time group number assigned by the time grouping algorithm
Mass spectral search results	
Hit at closest time index [text]	Compound name/identifier of MS library hit closest to the expected time index.
Time diff at closest time index [time index unit]	Absolute time difference between expected and mass tag bin time index of the closest RI library hit.
Time dev at closest time index [percentage]	Relative time deviation of expected time index of closest RI library hit compared to the mass tag bin time index.
Match at closest time index [ordinal number]	Match value of the closest RI library hit.
Hit at max match [text]	Compound name/identifier of the best matching MS library hit. In the target scanning mode the identifier of the target is placed here.
Time diff at max match [time index unit]	Absolute time difference between expected and mass tag bin time index of the best matching library hit.
Time dev at max match [percentage]	Relative time deviation of expected time index to the best matching library hit compared to tag time index.
Time at max match [time index unit]	Expected time of best matching RI library hit.
Match at max match [ordinal number]	Match value of the best matching MS library hit.
Hit most suggested [text]	Compound name/identifier of the MS library hit which was most frequently suggested.
Count most suggested [count]	Count of suggestions.
Cluster assignments	
Cluster [ordinal number]	Number of the assigned cluster, mass tag bins with the same number are in the same cluster.
Cluster score [number]	Degree core relative to the size of the cluster. The density of the extracted cluster network is indicated.
Intensity rank [ordinal number]	Inner cluster ranking by decreasing average sample intensity, maximum id rank = 1.
Cluster size [count]	Number of mass tag bins in a cluster
Cluster time [time index unit]	Median time index of all mass tag bins of the same cluster.

(continued)

Table 2 (continued)

Field [unit]	Meaning
Fragment intensity matrix	
Tag intensity [intensity counts]	Tag intensity calculated by fragment intensity ratios (for expert usage).
Sample value count [count]	Number of nonmissing values in the sample intensity matrix.
Max sample value count groups [percentage]	Percentage count of nonmissing intensity values in any sample group and maximum relative count.
Average intensity [intensity counts]	Average of all intensity values of the mass tag bin.
Further columns: sample intensity matrix [intensity counts]	The intensity value matrix. Column names represent the chromatogram (sample) names.

compiled into JAVA archive files (.jar-files). Plug-ins are loaded at runtime via user-demand triggering the plug-in import functionality. (a) From the *Tools* menu select *External Tools*; (b) the *Jar Browser* dialog will appear; click the *Select Jar File* button; (c) select the jar-file (JAVA-archive) of the plug-in tool and activate by clicking *Run*; (4) in the *Jar Browser* a list box will appear and a list of all available plug-ins will become accessible; (5) select the plug-in you want to run and click the *Run* button.

3. *Notes on data import.* We currently recommended data import of baseline corrected peak apex height, applying the second TagFinder option to respective *.cdf* files, which can be exported from most of the vendors' GC-MS acquisition software. The built-in apex finder may work on nonbaseline corrected data. However, baseline correction and noise reduction by smoothing is advised. We prefer to use vendor software for both tasks as each vendor has best assess to the nature and possible artifacts of the respective instruments' raw data and should, therefore, provide ideal algorithms and parameter settings tuned to the respective analytical equipment. A highly valuable alternative may be found in alternative metabolomic software for baseline processing and noise reduction, such as MetAlign which is freely available for non commercial use and can be easily downloaded and implemented (21–23 *see* Chapter 15). In the future, if suitable tools become available, peak area data may be imported instead of peak apex heights using the same data format.

4. *Names and terms* used in the TagFinder software are highlighted by bold format.

5. *Notes on the workspace initialization.* The definition of the Time Index Scale parameter is crucial and important, because

it defines the number of decimals used for your retention index system. The value 0 indicates that the desired RI system is ordinal and has integer RI values. The maximum number setting is 15. The *Time Index Scale* parameter and the *Fragment Mass Range* parameter strongly influence the size of the peak database file and should be set to the minimum requirements. Unnecessary mass range and scale settings will reduce the number of chromatogram peak lists which can be processed simultaneously within a single TagFinder job.

6. *Notes on the peak data import.* All peak data imported into TagFinder are written into a single binary file, called *spectra.tf.* The size of this file is limited by the JAVA file system to approx. 2.048 GB. Any attempt to import more data leads to the error message, "negative seek offset." In such a case, either reduce the number of chromatogram peak lists or modify the *Low Intensity Threshold* parameter—the frequency distribution of fragment intensities usually exhibit an exponential decline from low to high intensities—or modify using the *Start Time/End Time* import option the chromatographic range and possibly the fragment mass range.
7. *Notes on finding the proper settings for the scanning of mass tag bins.* Besides RI standard finding and RI calculation the choice of the *Time Scan Width* (gap) parameter is the most crucial and important task to perform before activating the tag finding process for the complete TagFinder job. The gap-parameter will strongly influence the grouping of the mass fragments into mass tag bins, and subsequently the time grouping, the alignment of the data matrix the aggregation of the intensity values and the clustering. Underestimation of the gap parameter results in partial grouping of fragments and artifact splits of mass tag bins. Samples with low amounts and high amounts of the same compound will exhibit a displaced row alignment within the numerical data matrix. Overestimation of the gap parameter will result in overly aggregated mass tag bins and time groups. The combination of fragment information from different but coeluting compounds is highly probable and signals of low intensity compounds will be lost. The characteristic row displacement of the gap parameter underestimation can be used to evaluate the proper settings of the time scan width. For that purpose, the TagFinder settings may be set to screen either one mass trace across the whole chromatographic range or all mass traces within a limited RI window. (a) From the *Tag Finder* menu select *Setup* to open the *TagFinder Settings* dialog. (b) For *Tag Scanning* select the *Time Scanner* tabulator. For the *Time Scan Width*, set a small value, i.e., 0.1 and then select the *Tag Filter* tabulator and disable all filters except for *Tag Mass.*

Type the mass number to process into the field. Type an *m/z* value, e.g., *m/z* 299 typical of the phosphoric acid 3TMS analyte, or use any other mass trace of compounds which may be expected in the chromatograms of the TagFinder job. (c) Disable correlation and clustering. For *Tag Output* select the *Tag Output* tabulator and click *Scan for Tags Only*. (d) Start the TagFinder process. (e) Open the tag output file for example with Microsoft MS-EXCEL or the TM4 software. Sort the table by increasing Time Group Number and decreasing AVG Intensity. (f) Refer to the *Tag Time* column and move to the expected RI of the testing analyte. All intensity values of the selected mass trace, in our example *m/z* 299, should be in one row. If you have displaced values in more than one row repeat the process by increasing the *Time Scan Width* gap parameter. (g) The choice of the gap parameter is strongly dependent on the performance of the GC-MS system, specifically GC-column aging and accumulating contaminations. An expert user may repeat the evaluation for different analytes, mass traces and check chemically defined reference samples for optimum alignment.

8. *Notes on the TagTimeScaleViewer*. For further rapid examination of mass tag bins use the *TagTimeScaleViewer* plug-in of the *tagtools* package (see Fig. 2). Tags can be sorted by ascending median RI (Tag Time) and plotted in a stacked bar plot. As is shown by Fig. 2 the correct alignment and overly extensive aggregation can be visualized by the TagTimeScale Viewer and judged by an experienced user. Steps in median RI and colored areas indicating the time group assignment match in cases of optimum alignment results. The expert user may visualize cluster results by a dedicated display option of the *TagTimeScaleViewer* plug-in (not shown).

Acknowledgements

This work received initial funding by the Max Planck Society and was subsequently supported by the EU as part of the Framework VI initiative within the plant metabolomics project META-PHOR (FOOD-CT-2006-036220). The authors acknowledge the long-standing support and encouragement by Prof. L. Willmitzer, Max Planck Institute of Molecular Plant Physiology (MPI-MP), Am Muehlenberg 1, D-14476 Potsdam-Golm, Germany. LvM and JK acknowledge the support by the EU GRASP project, ERA-Net Plant Genomics 0313996B, Research-Assisted Breeding for the Sustainable Production of Quality Grapes and Wines.

References

1. Luedemann, A., Strassburg, K., Erban, A., and Kopka, J. (2008) TagFinder for the quantitative analysis of gas chromatography – mass spectrometry (GC-MS) based metabolite profiling experiments *Bioinformatics* **24**, 732–737.
2. http://www-en.mpimp-golm.mpg.de/03-research/researchGroups/01-dept1/Root_Metabolism/smp/TagFinder/index.html
3. http://www.unidata.ucar.edu/software/netcdf/
4. Lisec, J., Schauer, N., Kopka, J., Willmitzer, L., and Fernie, A.R. (2006) Gas chromatography mass spectrometry-based metabolite profiling in plants. *Nat Protocols* **1**, 387–396.
5. Erban, A., Schauer, N., Fernie, A.R., and Kopka, J. (2007) Non-supervised construction and application of mass spectral and retention time index libraries from time-of-flight GC-MS metabolite profiles. In *Metabolomics: methods and protocols* (Weckwerth, W. Ed.). Humana Press, Totowa, pp 19–38.
6. Saeed, A.I., Sharov, V., White, J., Li, J., Liang, W., Bhagabati, N. et al. (2003) TM4: A free, open-source system for microarray data management and analysis. *Biotechniques* **34**, 374–378.
7. Saeed, A.I., Hagabati, N.K., Braisted, J.C., Liang, W., Sharov, V., Howe, E.A. et al. (2006) TM4 microarray software suite. *Methods Enzymol* **411**, 134–193.
8. http://chemdata.nist.gov/mass-spc/Srch_v1.7/index.html
9. Ausloos, P., Clifton, C.L., Lias, S.G., Mikaya, A.I., Stein, S.E., Tchekhovskoi, D.V. et al. (1999) The critical evaluation of a comprehensive mass spectral library. *J Am Soc Mass Spectrom* **10**, 287–299.
10. Halket, J.M., Przyborowska, A., Stein, S.E., Mallard, W.G., Down, S., and Chalmers, R.A. (1999) Deconvolution gas chromatography mass spectrometry of urinary organic acids – potential for pattern recognition and automated identification of metabolic disorders. *Rapid Commun Mass Spectrom* **13**, 279–284.
11. Halket, J.M., Waterman, D., Przyborowska, A.M., Patel, R.K.P., Fraser, P.D., and Bramley, P.M. (2005) Chemical derivatization and mass spectral libraries in metabolic profiling by GC/MS and LC/MS/MS. *J Exp Bot* **56**, 219–243.
12. Kovàts, E.S. (1958) Gas-chromatographische Charakterisierung organischer Verbindungen: Teil 1. Retentionsindices aliphatischer Halogenide, Alkohole, Aldehyde und Ketone. *Helv Chim Acta* **41**, 1915–1932.
13. Van den Dool, H., and Kratz, P.D. (1963) A generalization of the retention index system including linear temperature programmed gas–liquid partition chromatography. *J Chromatogr* **11**, 463–471.
14. Stein, S.E. (1999) An integrated method for spectrum extraction and compound identification from gas chromatography/mass spectrometry data. *J Am Soc Mass Spectrom* **10**, 770–781.
15. Lu, H., Dunn, W.B., Shen, H., Kell, D.B., and Liang, Y. (2008). Comparative evaluation of software for deconvolution of metabolomics data based on GC-TOF-MS. *Trends Anal Chem* **27**, 215–227.
16. Kopka, J., Schauer, N., Krueger, S., Birkemeyer, C., Usadel, B., Bergmueller, E. et al. (2005) GMD@CSB.DB: the Golm Metabolome Database. *Bioinformatics* **21**, 1635–1638.
17. Strehmel, N., Hummel, J., Erban, A., Strassburg, K., and Kopka, J. (2008) Estimation of retention index thresholds for compound matching using routine gas chromatography–mass spectrometry based metabolite profiling experiments. *J Chromatogr B* **871**, 182–190.
18. Batagelj, V., and Mrvar, A. (2004) Pajek – Analysis and visualization of large Networks. In *Graph Drawing Software* (Jünger, M., and Mutzel, P. Eds.). Springer Publishers, Berlin, Heidelberg, pp 77–103.
19. Huege, J., Sulpice, R., Gibon, Y., Lisec, J., Koehl, K., and Kopka, J. (2007) GC-EI-TOF-MS analysis of in vivo-carbon-partitioning into soluble metabolite pools of higher plants by monitoring isotope dilution after ($^{13}CO_2$)-labelling. *Phytochemistry* **68**, 2258–2272.
20. Birkemeyer, C., Luedemann, A., Wagner, C., Erban, A., and Kopka, J. (2005) Metabolome analysis: the potential of in vivo-labeling with stable isotopes for metabolite profiling. *Trends Biotechnol* **23**, 28–33.
21. http://www.pri.wur.nl/UK/products/MetAlign/; http://www.metalign.wur.nl/UK/
22. Lommen, A., van der Weg, G., van Engelen, M.C., Bor, G., Hoogenboom, L.A.P., and Nielen, M.W.F. (2007) An untargeted metabolomics approach to contaminant analysis – Pinpointing potential unknown compounds. *Analytica Chimica Acta* **584**, 43–49.
23. de Vos, C.H.R., Moco, S., Lommen, A., Keurentjes, J.J.B., Bino, R.J., and Hall, R.D. (2007) Untargeted large-scale plant metabolomics using liquid chromatography coupled to mass spectrometry. *Nat Protocols* **2**, 778–791.

Chapter 17

Chemical Identification Strategies Using Liquid Chromatography-Photodiode Array-Solid-Phase Extraction-Nuclear Magnetic Resonance/Mass Spectrometry

Sofia Moco and Jacques Vervoort

Abstract

The identification of metabolites in biochemical studies is a major bottleneck in the proliferating field of metabolomics. In particular in plant metabolomics, given the diversity and abundance of endogenous secondary metabolites in plants, the identification of these is not only challenging but also essential to understanding their biological role in the plant, and their value to quality and nutritional attributes as food crops. With the new generation of analytical technologies, in which liquid chromatography (LC)-mass spectrometry (MS) and nuclear magnetic resonance (NMR) play a pioneering role, profiling metabolites in complex extracts is feasible at high throughput. However, the identification of key metabolites remains a limitation given the analytical effort necessary for traditional structural elucidation strategies. The hyphenation of LC-solid phase extraction (SPE)-NMR is a powerful analytical platform for isolating and concentrating metabolites for unequivocal identification by NMR measurements. The combination with LC-MS is a relatively straightforward approach to obtaining all necessary information for structural elucidation. Using this set-up, we could, as an example, readily identify five related glycosylated phenolic acids present in broccoli (*Brassica oleracea*, group *Italica*, cv *Monaco*): 1,2-di-*O*-*E*-sinapoyl-β-gentiobiose, 1-*O*-*E*-sinapoyl-2-*O*-*E*-feruloyl-β-gentiobiose, 1,2-di-*O*-*E*-feruloyl-β-gentiobiose, 1,2,2'-tri-*O*-*E*-sinapoyl-β-gentiobiose, and 1,2'-di-*O*-*E*-sinapoyl-2-*O*-*E*-feruloyl-β-gentiobiose.

Key words: Metabolomics, Nuclear magnetic resonance, Mass spectrometry, Solid-phase extraction, Liquid chromatography, Hyphenation, Identification, Biomarker, Metabolite, *Brassica*

1. Introduction

It is estimated that over eleven megatonnes of cabbages and other brassicas were produced in 2008 in Europe (1), which makes this class of vegetables a significant segment of the vegetable market, representing an important group of crops in the dietary habits of Europeans. The alleged cancer-preventive properties of cabbage

Nigel W. Hardy and Robert D. Hall (eds.), *Plant Metabolomics: Methods and Protocols*, Methods in Molecular Biology, vol. 860, DOI 10.1007/978-1-61779-594-7_17, © Springer Science+Business Media, LLC 2012

and broccoli are associated with the occurrence of isothiocyanates in these plants which have been shown to have chemopreventive properties (2). Apart from isothiocyanates (which in the plant are actually stored in the form of glucosinolates), broccoli also contains a variety of glycosylated flavonoids and phenolic acids (3–6).

Being able to describe the variety of metabolites present in edible plants, in a fast and efficient manner, brings advantages in the characterisation of food and food products, as the presence of certain metabolites is related to quality traits such as taste, texture, smell, and also to nutrition. LC-MS and NMR are probably the most used analytical technologies in profiling metabolites in plants (7–10). Apart or combined, LC-MS and NMR cover a considerate part of the plant metabolome. Nevertheless, the full chemical description of a metabolite is scarcely achieved by profiling strategies where the aim is to cover as many metabolites as possible in a prompt manner. In fact, the structural elucidation of a metabolite is clearly a low-throughput procedure and involves the acquisition of analytical information from various sources and a great deal of puzzling effort.

A way to automate, at the analytical level, the identification of particular metabolites present in complex plant extracts, is to use a hyphenated set-up such as the LC-PDA-SPE-NMR/MS (11, 12). Such a system offers immense flexibility given by the hyphenation, i.e. configurations such as LC-PDA-MS or LC-NMR can be performed, as well as LC-PDA-SPE-NMR, or LC-PDA and NMR as separate techniques. Because the HPLC unit is the same for the various hardware conformations, the reproducibility of this type of separation is not an issue. Furthermore, this set-up allows metabolite profiling experiments and identification-focussed experiments to be carried out in analogous conditions.

In this study, we make use of the metabolite profiling method LC-MS to obtain a general overview of the semi-polar metabolome of broccoli and we demonstrate a strategy to isolate and identify, as an example, five chemically related endogenous metabolites in broccoli, i.e. the sinapoyl and/or feruoyl gentiobioses metabolites, by LC-PDA-SPE-NMR. The combined LC-PDA-SPE-NMR/MS analysis leads to the full elucidation of these five related phenolic acids.

2. Materials

2.1. Reagents

1. Acetonitrile for LC-MS gradient-grade (see Notes 1 and 2).
2. Methanol for HPLC isocratic-grade (see Notes 1 and 2).
3. 2-Propanol HPLC isocratic-grade (see Notes 1 and 2).

4. Ultrapure water with resistivity > 18.0 MΩ.cm at 25°C (being typically 18.2 MΩ.cm at 25°C) and TOC (total organic carbon) < 5 ppb. It should be used fresh.
5. Formic acid (FA) 99% ULC/MS (see Notes 2 and 3).
6. Sodium hydroxide (NaOH) pro analysi >99% (see Notes 2 and 3).
7. Deuterated methanol-d_4 with (HDO + D2O) < 0.03% (see Note 4).
8. Nitrogen gas to apply to the nebuliser of the mass spectrometer, to the SPE system and to the NMR, produced by a nitrogen generator.
9. Helium 6.0 for applying gas to the cryogenic unit of the NMR.
10. Compressed air produced by an air dryer unit for the NMR spectrometer.

2.2. Solutions

1. *Eluents used as mobile phase of the analytical column in the HPLC system.*

   Eluent A is a solution of 0.1 % (v/v) of FA in ultrapure water. Eluent B is a solution of 0.1% (v/v) of FA in acetonitrile. Eluent C is pure acetonitrile, used for column and system washing. All solutions and solvents are sonicated for 15 min before usage.
2. *5 mM sodium formate solution for MS external and internal calibration.*

   A solution of 0.2% (v/v) FA and 1% (v/v) 1 M NaOH in water/2-propanol 50/50 (v/v) is used as calibrant solution for the mass spectrometer.
3. *Eluent administered by the make-up pump for trapping chromatographic signals.*

   Solution of 0.1 % (v/v) of FA in ultrapure water (eluent D). This solution is sonicated for 15 min before usage.
4. *Eluents used as mobile phase to condition and equilibrate the cartridges in the SPE system.*

   For equilibration and washing, the protonated solutions are used: a solution of 0.1 % (v/v) of FA in ultrapure water (eluent D) and pure acetonitrile (eluent E). For compound elution, the deuterated solvent methanol-d_4 is used. All solutions and solvents are sonicated for 15 min before usage.
5. *Eluent used as mobile phase to elute the cartridges in the SPE system.*

   A deuterated solvent of choice should be used to elute compounds from the SPE cartridges. In this method, methanol-d_4 was used after 15 min sonication (eluent F).

2.3. Equipment

1. HPLC system composed by a vacuum degasser (G1379B), quaternary pump (G1311A; pump 1) and a standard autosampler (G1367B) 1200 series (Agilent) and a BSFU-O column oven (Bruker).
2. Photodiode array detector DAD (Bruker).
3. Online solid phase extraction (SPE) system Prospekt2 (Spark).
4. Pump K120 for administering the make-up eluent, eluent D (Knauer, pump 2).
5. Pre-column C18 (ODS Octadecyl), 2 mm diameter, 4 mm length (Phenomenex) and analytical columns Alltima C18 HP, 4.6 mm diameter, 150 mm length, spherical particles of 3 mm (Grace) and Alltima C18 HP, 2.1 mm diameter, 150 mm length, spherical particles of 3 mm (Grace) for the chromatographic separation.
6. SPE cartridges HySphere Resin SH, 2 mm inner diameter, 10 mm length, irregular shaped particles of 20–50 μm (Spark) for trapping compounds by the online SPE unit.
7. Membrane filters 0.2 μm OE66 (Schleicher & Schuell).
8. SPE cartridge Supelclean LC-18 Packing of 10 g (Supelco).
9. Time-of-flight mass spectrometer micrOTOF (Bruker) equipped with an ESI source (Bruker).
10. Gas-tight glass syringe with cemented needle 1005 LTN of 5.0 mL, with dimensions 22/2″/3 (Hamilton) and syringe pump 781100K (kd Scientific) for the administration of mass calibrant.
11. Nuclear magnetic resonance spectrometer Avance III with a 600 MHz/54 mm UltraShielded Plus magnet equipped with a CryoPlatform cryogenic cooling system, a BCU-05 cooling unit, a ATM automatic tuning and matching unit and a CryoFit cryoprobe flow conversion system insert of 30 μL (Bruker BioSpin).

2.4. Software Tools

1. Firmware versions of pump 1 (AgilentPump Control 3.2.45.1, A.06.04 (001)), pump 2 (K120 Pump Control, V03.20), autosampler (AgilentWellPlateAutosampler 3.2.45.1, A.06.03 (001)), column oven (LC Column + Oven Control 1.00, unknown), SPE (Prospekt2 Control 2.18.0.0/20080205, Ace MCCB: 1.02, Ace Modules: Ace RF 1.03, Ace TCM 1.03, Ace MPV 1.00, Ace CVML 1.02, Ace CVMR 1.02, Ace TASPE n/a, Hpd MCCB 1.22, Hpd Modules: Hpd Displ 1.21, Hpd Disp2 1.21), TOF-MS (MS Control Interface 3.2.44.0).
2. HyStar™, version 3.2—SR 2, build 44, created on 19 March 2007 (Bruker Daltonics).

3. micrOTOF™ Control, version 2.3, patch 1, build 40 (Bruker Daltonics).
4. DataAnalysis, version 4.0, build 234 (Bruker Daltonics).
5. TopSpin™ version 2.1, created on 24 October 2007 (Bruker BioSpin).
6. metAlign™ ((8); see Chapter 15).
7. PERCH version 2008.1 SA (PERCH Solutions Ltd.).
8. SciFinder Scholar™—Chemical Abstracts Service (CAS).
9. Excel 2003 (Microsoft Office).

3. Methods

3.1. Plant Material

The reference material proposed is broccoli and some examples are given from our own experiments. For this, we used broccoli (*Brassica oleracea* cv Monaco) plants grown in a Brittany field in France which were harvested in the Summer of 2006. Three biological replicates from 12 plants were collected and transported to the INRA-Bordeaux lab within 24 h. The plants were ground in liquid nitrogen and stored at −80°C before shipment in dry ice to the Wageningen University, Wageningen, The Netherlands, in the Spring of 2007, where they arrived still in perfect condition.

3.2. Metabolite Profiling by LC-PDA-TOF-MS

3.2.1. Sample Preparation and Extraction

1. Take the frozen broccoli powder and weigh 0.5 g of material.
2. Extract immediately with 1.5 mL methanol (final methanol concentration in the extract was approximately 75% (v/v)).
3. Sonicate all samples for 15 min, centrifuge for 5 min, filter through a 0.2-µm inorganic membrane filter (Minisart, Sartorius) and proceed to analysis (for sample stability see Note 5).

3.2.2. LC-PDA-TOF-MS Set-up Applied to Profiling Metabolites in Broccoli

1. Switch on the instrument units: degasser, pump 1, pump 2, autosampler, and PDA (the MS is always on stand-by when not used).
2. Verify that the column installed has 2.1 mm of internal diameter. Make sure that the PEEK tubing of the system is in the desired configuration (see Fig. 1), i.e. effluent from the column is connected to DAD flow cell and from this unit to the MS instrument. Disconnect the tubing into the MS source to the liquid waste. Verify that the liquid waste container in not full; otherwise, replace with an empty container.
3. Prepare fresh mobile phase eluents A, B and C.
4. At the computer, open HyStar. On the main *HyStar/Compass* window, click on *Hardware Setup* and choose the hardware

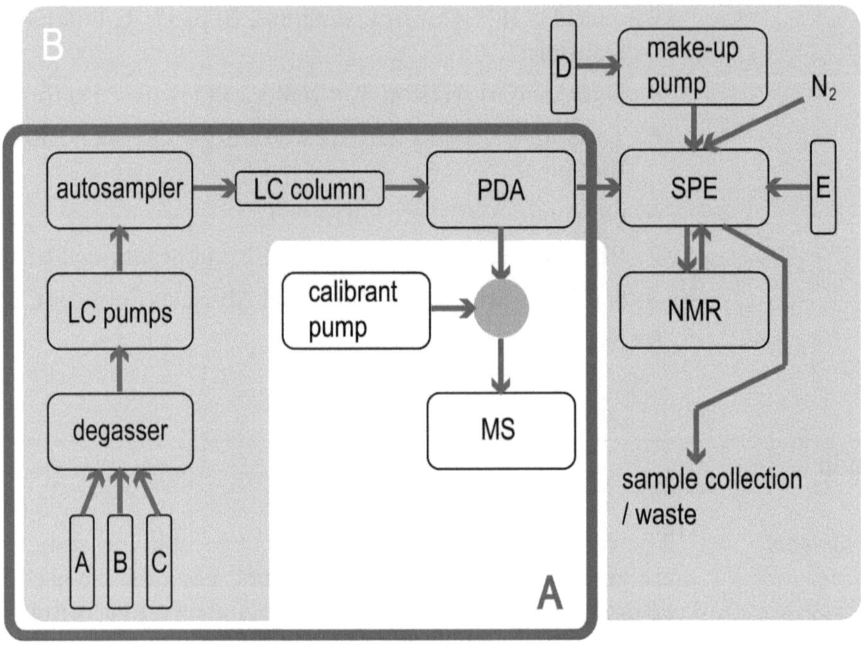

Fig. 1. Hardware configuration of a LC-PDA-MS-SPE-NMR used for metabolite profiling in the configuration A, LC-PDA-MS and in the configuration B, LC-PDA-SPE-NMR.

set-up which includes the pumps, the column oven, the autosampler, the PDA, and the mass spectrometer. Load this hardware set-up. Create a sample table by clicking *Sample Table* in the main *HyStar/Compass* window. The first sample in the list should be a water/methanol 50% (v/v) sample to be analysed with the same analysis methods to be used for the "real" samples (autosampler method, LC method and MS method). On the *Sample Table* main toolbar, click on *Acquisition*.

5. Open the valve of pump 1 before clicking on *Pump On* in the *Acquisition* window of HyStar. Purge all eluent tubing, even the ones not to be used, as this is important for the adequate functioning of the degasser (see Note 6).
6. Reduce the flow rate to a minimum value at 100% C and close the valve of pump 1. Increase, stepwise, the flow rate up to 0.2 mL/min at this eluent composition (100% C). This way the HPLC system is washed, including the analytical column, for 30 min. The pressure (in bar) indicated in the pump 1 module on the *Acquisition* window of HyStar should be comparable to previous values obtained under analogous conditions. If this is not the case, then extend the washing time (see Note 7).
7. Equilibrate the system by setting the eluent composition at the initial HPLC gradient conditions. Allow 20 min for system

stabilisation. Verify that the pressure obtained is comparable to previous values obtained under analogous conditions (see Note 8).

8. At the computer, go to the *Sample Table* and fill in the series of LC-PDA-MS measurements to be performed. In *General* (lower part of the window) fill in the sample name (*Sample Identifier*) and vial position (*Vial Position*; e.g. 1-A,4 is sample in tray number 1 on row A and column 4). Place the sample in the correspondent vial position in the sample tray inside the autosampler. Fill in the *number of injections* for this sample (in this case is 1 injection), *volume* of injection (in μL), set the equilibration time at the initial conditions of the gradient programme, *prerun* (in min), and browse to choose the subdirectory to store the results (*Result Data Path*, *subdirectory*).
9. To develop the method of analysis, go to *Methods* (lower part of the *Acquisition* window) and start by editing the LC method part, *LC*, click *Edit*. A window appears: *LC Method Part Editor*. To create and set up a method for the analysis of broccoli samples by LC-PDA-TOF-MS, the following parameters should be filled in:
   - In *LC Parameters*, fill in 60 min for *Total runtime* and *Data Acquisition* with a 0 min delay.
   - In the *LC pump* window, fill in the *Flow rate* of pump 1, 0.2 mL/min, with 400 bar of pressure limit, and the initial conditions of the *Solvents*: A, 5%, acetonitrile with 0.1% formic acid, B, 95%, ultrapure water with 0.1% formic acid, C, 0%, acetonitrile, and D, 0%. For pump 2, the *Flow rate* is 0 mL/min.
   - In the *Autosampler* window, check the *Parameters*: *loop capacity* of the autosampler is 100 μL; *seat capillary* 2.3 μL; *washing* in *flushport* (the needle is flushed with the current mobile phase by leaving the needle in the seat during 15 s *time*); *draw speed* 200 μL/min; *eject speed* 200 μL/min and *draw position offset* -8.0 mm (−10 mm is the minimum) refers to the lowest point of needle depth when injecting.
   - In the *Column/Oven* window, fill in the *Temperature* of the chromatographic separation: 25°C.
   - In the *UV/DAD Detector* window, fill in the slice width of the detector acquisition (200 ms or 5.00 points/s) and click on *Auto Zero*, so that the detector performs an autozero before starting acquisition for each sample. UV/Vis data over the whole spectrum (from 187 to 1,022.5 nm) can be acquired and saved under the *2D Parameters* but because this occupies too much disk space, it should be acquired only when absolutely necessary (see Note 9).

Table 1
Gradient programmes, 1 (used for LC-PDA-TOF-MS profiling) and 2 (used for LC-PDA-SPE-NMR), in time (min), in terms of % of eluents, A and B, for chromatographic separation

Gradient 1			Gradient 2		
t/min	A/%	B/%	t/min	A/%	B/%
0	5	95	0	15	85
45	35	65	35	25	75
46	95	5	36	95	5
49	95	5	39	95	5
50	5	95	40	15	85
60	5	95	50	15	85

- In *LC Timetable*, fill in the gradient programme, including a washing step and equilibration for the next injection (see Table 1, gradient 1).
- In *Signals1-8*, choose the following detector signals: signal 1, *LC 1 Detector (DAD)*, 280 nm; signal 2, *LC 1 Detector (DAD)*, 329 nm, signal 3, *LC 1 Detector (DAD)*, 360 nm and signal 4, *MS (micrOTOF series)*, base peak chromatogram of all polarities (BPC, All).
- Save the method. Browse to choose this method on the *LC part* of *Methods* (lower part of the *Acquisition* window).

10. In *Methods* (lower part of the *Acquisition* window) set-up, the method part of the autosampler in *Autosampler (for Agilent G1367B WP)* by choosing, from the two methods available, the *standard wash.* This method implies washing the needle before changing the sample.
11. In *Methods* (lower part of the *Acquisition* window) set-up, the method part for the MS in *MS (micrOTOF series)*. The development of this method has to be made with the micrOTOF-Control software. Open micrOTOFControl.
12. Before developing an MS method, the MS instrument should be tuned and calibrated, so that the signal intensity and resolution are optimised. Parameters related to the ion transfer such as the capillary exit, skimmer 1 (note: skimmer 1 = capillary exit/3) and hexapole RF, as well as transfer time and prepulse storage time (parameters on *Source*), are probably the most important for optimisation. According to the user's wishes, these parameters influence the intensity and relative ratio

between lower and higher *m/z* and can be investigated using the calibrant solution. The parameters of the TOF detector, which are related to the maximisation of the resolution, once optimised upon installation of the instrument, should remain unchanged.

13. Calibrate the MS instrument by infusing calibrant solution (sodium formate solution) directly to the source at 180 μL/h (3 μL/min). At this flow rate, the necessary nebuliser pressure should be 0.4 bar, with a dry gas temperature of 180°C at a flow rate of 4.0 L/min. These values can be adjusted in the lower window of the micrOTOF Control, under *Source* (for more MS settings check Table 2). For an *m/z* range suitable for metabolite analysis; 100–1,500 *m/z*, use low values for the capillary exit voltage, hexapole RF and transfer time (see Table 2). Allow the system to obtain a stable signal, in terms of base peak intensity (BPI). On the lowest window of micrOTOF Control, click on *Calibration*. Choose the adequate calibration list, "Na formate" (neg), in the *Reference List*. A customised calibration

Table 2
MS settings in ESI negative mode used for metabolic profiling of metabolites in broccoli by LC-PDA-MS, using a flow rate of 0.2 mL/min. Values in *italic* are used for flow rates of 3 μL/min (e.g. calibration)

Source		Ion optics		TOF	
Source type	ESI	Set capillary exit	-150.0 V	Set corrector fill	47 V
Focus scan	Not active	Set skimmer 1	-50.0 V	Set pulsar pull	820 V
Begin scan	100 *m/z*	Set hexapole 1	-24.0 V	Set pulsar push	820 V
End scan	1,500 *m/z*	Set skimmer 2	-23.0 V	Set reflector	1,700 V
Ion polarity	Negative	Set hexapole 2	-21.0 V	Set flight tube	8,600 V
Set capillary	3,200 V	Set hexapole RF	150.0 V	Set corrector extract	635 V
Set end plate offset	-500 V	Set transfer time	63.0 μs	Set detector TOF	2,010 V
Set nebuliser	1.2 bar (*0.4 bar*)	Pre Puls storage time	1.0 μs		
Set dry heater	200°C (*180°C*)	Set lens 1 storage	-30.0 V	Processing	
Set dry gas	8.0 L/min (*4.0 L/min*)	Set lens 1 extraction	-21.3 V	Summation	15,625×
Set divert valve	Waste	Set lens 2	-9.0 V	Guessed noise	200
		Set lens 3	16.0 V	Peak width	5 pts
Mass calibration	TOF1 calibration mode quadratic	Set lens 4 Set lens 5	0.0 V 26.0 V	Average noise Guessed average	1 100

list can be created for a specific m/z range (and added to the list of possible calibrations). Choose an *Enhanced Quadratic* calibration (as this calibrant solution produces a large number of data points), under *Calibration mode*, and click on *Automatic*. According to the fit, *Score* values in ppm and in percentage (to be seen under *Calibration Status*) are calculated. Accept the calibration if a green colour is displayed, i.e. score > 95%. By clicking on *Properties*, a window displays the coefficients ci, $i = 1, \ldots, 4$ of the enhanced quadratic calibration *fit* and *current* status, as well as the date and time of the last calibration. The TOF-MS should be always externally calibrated before a series of analyses, on a daily basis.

14. Apart from the external calibration, internal calibration can be implemented to make sure that each sample has the best mass accuracy possible. In order to achieve this, the external valve on the TOF-MS instrument is used, Fig. 2. This valve is equipped with a 20 μL loop which is filled with calibrant during the previous run (calculate the flow rate of the syringe pump, according to the length of the LC run; for a 5 mL syringe, the minimum flow rate possible to apply by the pump is 0.03 mL/h) and is injected in the beginning of the analysis. In this way, a calibration plug is produced at the beginning of the LC run, for each sample injected. The MS method should be adapted to include this calibration plug, therefore, three time segments are made with the following time length: first: 0.020–0.120 min (valve to waste), second: 0.120–0.522 min (valve to source), and third: 0.522–60.035 min (valve to waste). The second segment corresponds to the introduction of the calibration plug. Note that the valve positions are referred in micrOTOF control relation to the calibrant ("source" option is on during the calibration plug and "waste" during the analytical run), (see Fig. 2).
15. Connect the effluent tubing from the PDA to position 5 on the MS external valve and the calibrant tubing to position 1

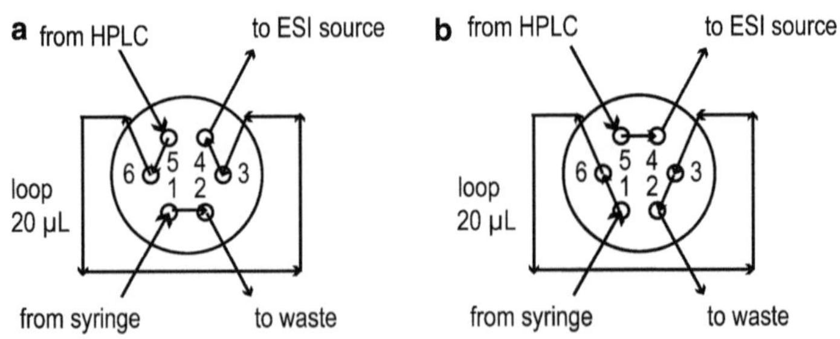

Fig. 2. Valve configuration of TOF-MS, allowing the administration of a calibration plug as internal calibration. Position of the valve when calibrating, (**a**) (in micrOTOF Control "source"), and when analysing the sample, (**b**) (in micrOTOF Control "waste").

(see Fig. 2). Observe the mass spectra obtained from the chromatographic eluents, in order to assess the impact of impurities in the system and their relative intensity.

16. Save the method in micrOTOF Control. In the Sample Table window of HyStar, browse to choose the MS method on the *MS* (*micrOTOF series*) in *Methods* (lower part of the *Acquisition* window).
17. Fill in the sample table with the names of the samples to analyse. Choose the suitable LC method, MS method and autosampler method, the autosampler vial position, volume of injection (in this case 5 μL) and location of the file storage in the computer disk. Perform two to five injections of the same sample for system stabilisation. The list of samples should be randomised to avoid time dependencies. Every ten samples, inject a quality control sample (a 50% water 0.1%FA–50% methanol (v/v) solution of known concentration of, for example, naringenin, chlorogenic acid and rutin) and at the end, include a cleaning gradient (30 min of 100%C).
18. Check all tubing connections (see Figs. 1 and 2, Note 10). Check the sample table, reload the sample table on the *Acquisition* window and make sure all instrument modules are ready (the colour should be green) before pushing the *Start* button (see Notes 11 and 12): *Start sequence.* Click on *Shutdown Settings* of the system (*off* icon): switch off the PDA lamp, the pumps and switch the TOF-MS to standby 5 min after the series has stopped. The analysis in progress mode is indicated by the colour blue.

3.2.3. Data Analysis

1. Open the data files in DataAnalysis.
2. Display the BPI of the mass chromatogram by clicking on *Edit Chromatogram [F7]*, and changing the TIC (total ion chromatogram) into BPI.
3. In *Select range/View spectra [Ctrl]* the mass spectrum can be displayed by clicking on a particular chromatographic signal. The number of digits of the value displayed for the *m/z* signal in the mass spectrum can be changed to four digits in *Parameters, Display, Mass precision.*
4. Perform an automatic internal calibration in the dataset by adjusting the *Internal Calibration* parameters, under *Parameters* (a dedicated window is then displayed) after clicking *Calibrate* in the main toolbar. The *Calibration* group to choose is "ESI" and the *Calibration list* is the one chosen for external calibration, in this case the list in Table 3 (negative mode). The *Mode* used for calibration is "enhanced quadratic" with a *search range* of 0.5 *m/z* and a 0 of *intensity threshold.* The retention time range of the calibrant used for calibration

Table 3
***M/z* values for the obtained clusters Na(NaCOOH)*n*, *n*=2, ..., 21, in the 100–1,500 *m/z* range, for negative and positive mode, present in sodium formate solution used for MS calibration**

n in cluster	*m/z* (negative mode)	*m/z* (positive mode)
2	180.973051	158.964069
3	248.960475	226.951493
4	316.947899	294.938917
5	384.935323	362.926341
6	452.922747	430.913765
7	520.910170	498.901189
8	588.897594	566.888613
9	656.885018	634.876037
10	724.872442	702.863461
11	792.859866	770.850884
12	860.847290	838.838308
13	928.834714	906.825732
14	996.822138	974.813156
15	1,064.809562	1,042.800580
16	1,132.796986	1,110.788004
17	1,200.784410	1,178.775428
18	1,268.771834	1,246.762852
19	1,336.759258	1,314.750276
20	1,404.746682	1,382.737700
21	1,472.734106	1,450.725124

of the dataset can be adjusted under *Automatic Internal Calibration* in the *Parameters* window: *Start*: 0.15 min to *End*: 0.35 min. this window should be adjusted according the length of the calibration plug. The internal calibration of the dataset is then calculated by clicking *Calibrate, Automatic Internal*.

5. Molecular formulae can be calculated for specific mass signals on the spectrum by opening the *SmartFormula manually* [Shift+F8] window under *Chemistry* in the main toolbar. Default atoms taken into account for calculation are C, H, O

are N, but more can be added or removed in the maximum number of atoms (*Max*) and minimum number of atoms (*Min*) as well as stipulate a minimum and maximum number of atoms for certain elements. For example, if there is strong evidence that the ion does not contain N, then "N_0" can be written the *Max* window. For negative ion mode spectra, the *Charge* should be −1 and the *Tolerance* should be set not higher than 10 ppm; otherwise, too many options are calculated and the mass accuracy of the measurements should be under this error, when internal calibration is performed. Furthermore, the *TIP* (*True Isotopic Pattern*) with the *Sigma Fit* algorithm calculates for each putative molecular formula a fit of the theoretical isotopic distribution towards the measured spectrum (see Table 4, Sigma fit). The lowest the fit, the best accordance to reality.

6. For a thorough analysis of as many signals as possible from the LC-PDA-TOF-MS profiles, the datasets can be exported as netCDF files (*.cdf) by clicking on DataAnalyis: *File*, *Export*, *Chromatogram analysis*. In this file format, the datasets can be treated for alignment, baseline correction and data matrix extraction compatible with Microsoft Excel, using metAlign. For more details about using this software for the analysis of LC-MS data consult reference (8).

3.2.4. Putative Metabolite Identification

A typical chromatogram of a 75% methanol (v/v) extract of broccoli is depicted in Fig. 3a. By LC-TOF-MS analysis, putative assignments of metabolites can be made, taking into account the extracted accurate masses and isotopic patterns from which molecular formulae can be computed. For the main metabolites in the broccoli extract, as indicated by the LC-ESI$^-$-TOF-MS chromatogram, several metabolites could be (putatively) assigned (see Table 4).

Metabolites **a**, **b** and **c** are known glucosinolates abundant in *Brassicaceae* previously described in literature (3, 13). Using SciFinder, 22 structures are found for the molecular formula attributed by the accurate mass calculation from the mass spectrum of **a**. One of the structures, glucobrassicin (CAS number: 4356-52-9), has been reported in 641 references (in which 89 contain the search word "broccoli") while for all the others, less than three references were found. Metabolites **b** and **c** have the same accurate mass, and therefore the same molecular formula. In SciFinder, 12 compounds were found for this molecular formula from which one, 1-methoxyglucobrassicin (also named neoglucobrassicin; CAS number: 5187-84-8) has been reported in 363 references (69 in broccoli). Another possibility is the conformational isomer 4-methoxyglucobrassicin (CAS number: 83327-21-3) reported in 280 references (from which 57 in broccoli). According to predicted log *P* values, present in SciFinder for these two metabolites, 4-methoxyglucobrassicin (log $P = 1.853 \pm 1.020$) indicates being

Table 4
Molecular formulae, $[M-H]^-$ theoretical and measured masses, mass error, sigma fit, sigma rank computed by DataAnalysis, and putative identification for the major chromatographic signals present in broccoli (see Fig. 3).

Metabolite	Molecular formula	$[M-H]^-$ theoretical mass	$[M-H]^-$ measured mass	Mass error(ppm)	Sigma fit ($\times 10^{-3}$)	Sigma rank	Puta tiveidentification
a	$C_{16}H_{20}N_2O_9S_2$	447.0537	447.0540	-0.5	9.1	1	Glucobrassicin
b	$C_{17}H_{22}N_2O_{10}S_2$	477.0643	477.0638	1.0	10.3	1	4-methoxyglucobrassicin
c	$C_{17}H_{22}N_2O_{10}S_2$	477.0643	477.0649	-0.9	9.5	1	1-methoxyglucobrassicin
1	$C_{34}H_{42}O_{19}$	753.2248	753.2239	1.1	6.8	1	Disinapoyl-dihexose
2	$C_{33}H_{40}O_{18}$	723.2142	723.2140	0.3	12.2	3	Sinapoyl-feruloyl-dihexose
3	$C_{32}H_{38}O_{17}$	693.2036	693.2037	-0.1	9.4	1	Diferoloyl-dihexose
4	$C_{45}H_{52}O_{23}$	959.2827	959.2823	0.4	13.0	1	Trisinapoyl-dihexose
5	$C_{44}H_{50}O_{22}$	929.2721	929.2714	0.7	9.7	2	Disinapoyl-feruloyl-dihexose

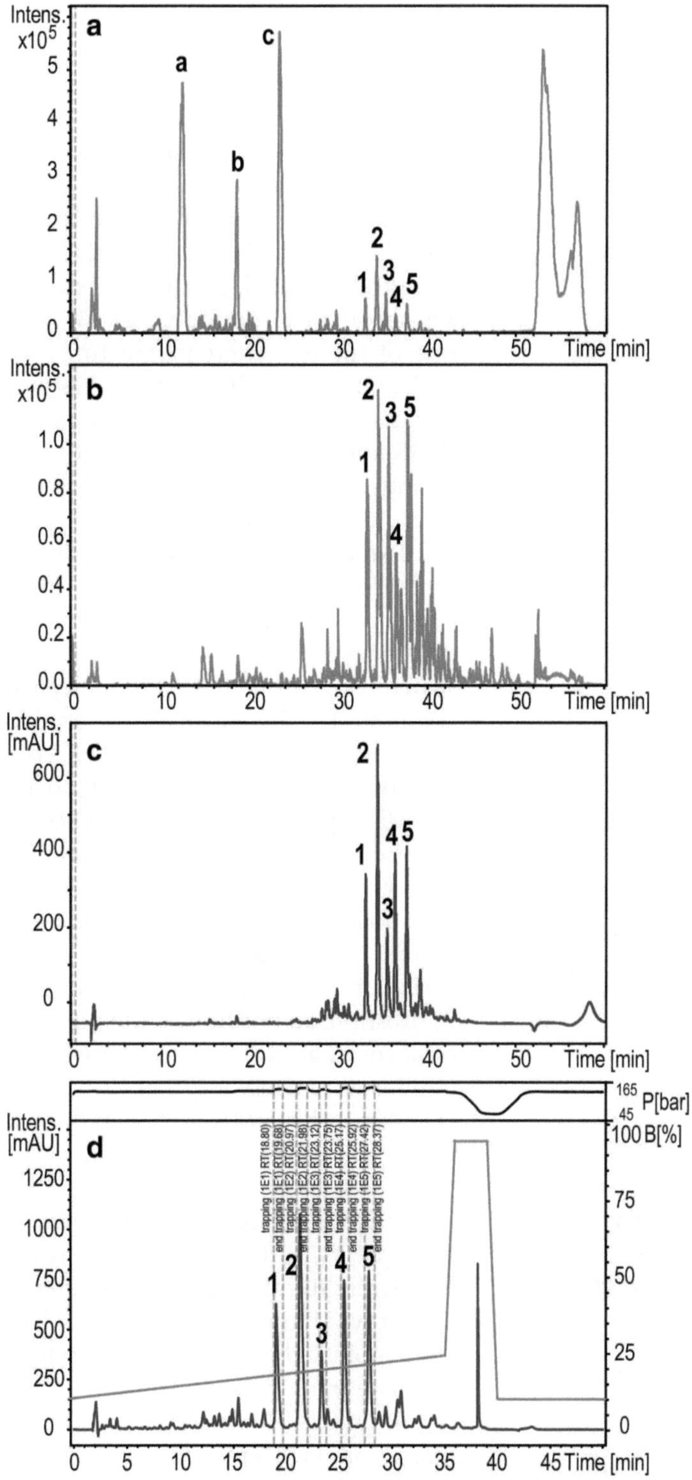

Fig. 3. Chromatograms obtained by negative ion mode LC-ESI-PDA-TOF-MS analysis of broccoli extracts: mass trace of full broccoli extract (**a**); mass trace (**b**), and UV/Vis trace recorded at 329 nm (**c**) of enriched broccoli fraction; UV/Vis trace recorded at 329 nm, percentage of organic mobile phase for the chromatographic separation and pressure profile in a trapping procedure where the metabolites 1–5 where isolated and analysed by LC-PDA-SPE-NMR (**d**). For chemical details of metabolites **a**, **b**, **c**, **1**, **2**, **3**, **4**, and **5** consult Table 4.

more polar than 1-methoxyglucobrassicin (log $P = 2.813 \pm 0.784$). These findings, complemented with information published in literature (3, 13), suggests that **b** is 4-methoxyglucobrassicin and **c** is 1-methoxyglucobrassicin.

Metabolites **1–5** all have different accurate masses and therefore different molecular formulae, even though, chemically, they seem to be related, given their retention time proximity and UV/Vis absorbance (see Table 4). By consulting SciFinder, **1** can be one of the 37 different structures with the same molecular formula. Only one of the options has been found in broccoli (14 references), corresponding to compound 1,2-disinapoylgentiobiose (CAS number: 195006-75-8). For metabolite **2**, 32 possible compounds can be attributed, from which one metabolite, 1-sinapoyl-2-feruloylgentiobiose (CAS number: 195006-74-7) was mentioned in 14 studies of broccoli. Metabolite **3** can be one of the 78 structures documented in SciFinder, suggesting that its identity is a 1,2-diferuloylgentiobiose (CAS number: 553643-73-5) as mentioned in all eight broccoli references. Metabolite **4** has seven structures attributed to his molecular formula from which one compound, 1,2,2′-trisinapoylgentiobiose (CAS number: 155380-01-1) has been reported in all 12 references of broccoli studies. Metabolite **5** is one of eight structures in which compound 1,2′-disinapoyl-2-feruloylgentiobiose (CAS number: 195006-73-6) appears in 12 broccoli references.

Given the connectivity and conformation possibilities of metabolites **1–5**, concerning the phenolic acid moiety and sugar moiety (*E* or *Z* conformation of phenolic acid double bond, glycosilation of the phenolic moiety via the hydroxyl in position 4 or 9, substitution position of the phenolic moieties in the sugar, type of hexoses, glycosidic bond between the two hexoses) these compounds were chosen, as a practical example, to carry out structural elucidation by NMR and therefore, LC-PDA-SPE-NMR experiments were initiated.

3.3. Trapping of Metabolites 1–5 by LC-PDA-SPE-NMR

3.3.1. Sample Preparation and Extraction

An enriched broccoli (*Brassica oleracea* cv Monaco) extract was prepared by extracting 29.2 g (fresh weight) of frozen broccoli powder in 90 mL of methanol. After 15 min sonication, the crude extract was filtrated and the solid debris washed successively with methanol yielding 130 mL of filtrate. Ultrapure water was added to this filtrate, making a final 1:1 proportion of water–methanol (v/v). Of this broccoli extract, 50 mL were used in an offline SPE separation. Five fractions were collected according the applied elution solutions: break through (fraction I); 100% water with 0.1% FA (fraction II); 75% water 0.1% FA–25% methanol (fraction III); 50% water 0.1% FA–50% methanol (fraction IV), and 25% water 0.1% FA–75% methanol (fraction V). This procedure was repeated three times and the samples obtained were combined per fraction. The fractions were evaporated by vacuum at room temperature and freeze-dried. The dried residues were dissolved in 1 mL 50% methanol–50% water 0.1% FA (v/v).

3.3.2. LC-PDA-TOF-MS Applied to Profiling Metabolites in Enriched Broccoli Extracts

Before proceeding with the trapping, the enriched extract should be checked by LC-PDA-TOF-MS to confirm retention times and intensities. Follow the protocol described in Subheading 3.2, as the same set-up was used, for analysing the enriched broccoli fractions after offline SPE concentration. After inspecting the LC-PDA-TOF-MS chromatograms of the five fractions, fraction V (25% water 0.1%FA–75% methanol) contained the highest concentration of metabolites **1**–**5** and therefore was used for trapping experiments, (see Fig. 3b, c).

3.3.3. LC-PDA-SPE-NMR Set-up Applied to the Isolation of Metabolites 1–5 *(Trapping)*

1. Switch on the degasser, pump 1, pump 2, autosampler, PDA and SPE units.
2. Verify that the column installed has 4.6 mm of internal diameter. Make sure the PEEK tubing of the system is in the desired configuration, i.e. effluent from the column is connected to DAD flow cell and from this unit to the liquid waste. Verify that the liquid waste container in not full; otherwise, replace with an empty container.
3. Prepare fresh mobile phase eluents A, B, C and solution for pump 2, D.
4. At the computer, open HyStar. On the main *HyStar/Compass* window, click on *Hardware Setup* and choose the hardware set-up which includes the pumps, the column oven, the autosampler, the PDA, the SPE unit, and the NMR. Load this hardware set-up. Create a sample table and go to the *Acquisition* window.
5. After purging the tubing of pump 1 and 2 and washing the analytical column, check the pressure of the system at 100% C and compare it to previous values obtained at analogous conditions. Extend the washing time if needed. Equilibrate and stabilise the system by setting the eluents composition at the initial HPLC gradient conditions. Verify that the pressure obtained in compare it to previous values obtained at the analogous conditions.
6. Before starting to trap, optimise the chromatographic separation according to the signals of interest, so that these are well separated, facilitating the trapping procedure. Adapt the *LC method part* accordingly (see Note 13).
7. To set up the trapping experiment, go to the *Sample Table* and fill in the series of LC-PDA-SPE-NMR measurements to be performed. In order to trap the signals 1–5 more than once, each in a separate cartridge, several injections of the same sample have to be made. Make a rough calculation from the

UV absorbance signal, how much is needed to load on the column and how many times the trapping procedure should be repeated. In this case, multi-trapping was performed eight times and the *volume* of injection was 50 μL.

8. Adapt the method of analysis for trapping. Start by editing the *LC part*.
   - In LC Parameters, fill in 60 min for Total runtime and Data Acquisition with a 0 min delay.
   - In the LC pump window, fill in the Flow rate of pump 1, 1.0 mL/min, with 400 bar of pressure limit, and the initial conditions of the Solvents: A, 15%, acetonitrile with 0.1% formic acid, B, 85%, ultrapure water with 0.1% formic acid, C, 0%, acetonitrile and D, 0%. For pump 2, the Flow rate is 0 mL/min.
   - In the LC-SPE-NMR window, make sure that in the Peak Trapping options, the Multiple Peak Trapping is ticked. Also, that in the Cartridge washing, there are 0 μL of volume for loaded cartridges after trapping.
   - In LC Timetable, fill in the gradient program, including a washing step and equilibration for the next injection (see Table 1, gradient 2) (this gradient was optimised in step 6). Apply a flow rate gradient for pump 2, by switching on the flow only in the chromatographic region of trapping (see Note 14). In this case, 1.5 mL/min of 100% D were delivered between 15 and 35 min, being the rest of the chromatographic time at 0 mL/min (see Note 15).
   - In Signals1-8, choose the detector signal: signal 1, LC 1 Detector (DAD), 329 nm.
   - In Fraction Treatment, set up the trapping program. A chromatogram can be loaded in Load chromatogram and in this way the treatments in the left Treatment Window can be seen in the chromatogram and adjusted interactively. Once a signal has been added, the Start Time, the End Time and the Detection Mode are listed in the Treatment Window. In this case, trapping was suppressed until 17 min and after 31.53 min (unclick Collect in the Action window). Manual trapping possibilities were allowed from 17 to 31.53 min: add these times in the Treatment Window; click Collect, Prospekt2, auto, in the Action window; and in the Detection Parameters, choose manual as Detection Mode and tick (active) the Signal DAD 329 (see Note 16).
   - Save the method. Browse to choose this method on the LC part of Methods.

9. Select *standard wash* as autosampler method in the *Methods* set-up.
10. Check the sample table. Make sure to perform a dummy run by injecting a 50% water 0.1% FA–50% methanol (v/v) sample, at the same conditions as the real samples, before a series of analysis, and at the end, include a cleaning gradient (30 min of 100%C).
11. Prepare and check all tubing connections (see Figs. 1 and 2). A T-piece should be connected to the outlet of the analytical column together with the tubing from the make-up pump and the tubing to be connected to the SPE unit. Check the valve positions and capillary connections of the SPE unit (consult the Bruker SPE Prospekt2 manual) and the eluent volumes A–E.
12. Load the sample table on the *Acquisition* window and make sure that all instrument modules are ready (colour: green): the acquisition software *HyStar*, pump 1 *Pump*, pump *2K-120*, the autosampler *WPA*, the column oven *Col.*, the SPE unit *Prospekt2*, and the NMR spectrometer *NMR*.
13. Prepare the cartridges for trapping. By right-clicking on Prospekt2 in the main toolbar of the *Acquisition* window of HyStar, click on *Cartridge Control*. This window gives an overview of the history of the two trays of cartridges installed in the system, indicated by different colours (new cartridge, conditioned, equilibrated, loaded, dried, used/empty, in progress) which makes it easier to see which cartridges to use for the next trapping procedure. Right-click again on *Prospekt2*, *Condition and Equilibration*. Choose the cartridges to be washed with 100% E (*Conditioning*) and equilibrated with 100% D (*Equilibration*), by selecting the *First* cartridge and the *Last* cartridge, inclusive. The flow rates and volumes for *Conditioning* and *Equilibrating* can be chosen. In this case, a volume of 500 μL for both events was used, at a flow rate of 6,000 μL/min for *Conditioning* and 1,000 μL/min for *Equilibration* (see Notes 17–19). During this process, the progress is indicated as well as the estimated time to accomplish this procedure. Always prepare more cartridges than the metabolites intended to trap, as a measure of comfort and always just before initiating the trapping procedure. Check the *Service settings*, under *Prospekt2* (right-click), and confirm that there is continuous flushing of nitrogen gas at low flow into the SPE trays and that the flow is not interrupted when switching off the HyStar (see Note 16).
14. Reload the sample table and the method before pushing the *Start* button. After starting the sequence, observe the following sequence of events in which the first unit to start is the autosampler (*WPA*) that is preparing for injection (turns blue).

At this time point, all other modules are ready (green) but not yet on analysis mode (blue). Once the injection is made, the gradient starts (*Pump* turns blue), the PDA (*DAD*) autozeros and starts acquisition (turns blue) and the SPE unit (*Prospekt2*) turns blue (note: if the SPE does not turn blue in the beginning of the run, then it will not allow the trapping).

15. Make sure to click the *manual trapping* in the right side of the *Acquisition* window in order to be able to manually trap (even if previously documented in the LC method). Before pushing the *Start* trapping button, verify that the correct cartridge number on the SPE unit is in bold, as this will be the first cartridge used for trapping. Push *Start* to start trapping, push *End* to stop trapping. Prepare always a dummy cartridge which can be later used to test the transfer procedure (see Note 20).
16. After trapping, the cartridges should be dried with nitrogen gas to prevent the adsorbance of particles and other impurities. The maximum drying time should be applied to minimise the solvent signals on the NMR spectrum, that is, 59 min per cartridge. This procedure can be done overnight. To dry the cartridges, right-click on *Prospekt2*, choose the *First* and the *Last* cartridge to dry, inclusive, input the *drying time*, and click *Start*. While the system is drying, no chromatography can be done (see Notes 21 and 22).
17. To transfer the cartridges, switch to the *Flow Injection* mode (this is another operation window of HyStar), under *Module* in the main toolbar of the *Acquisition* window of HyStar. Check the transfer parameters from the SPE unit to the NMR probe by clicking *Transfer* in the main toolbar of the *Flow Injection* window.

Transfer parameters

- In the field *Wash & Dry NMR probe head*, check the *Settings* and fill in 3 min for *drying*, 300 μL of *volume* of (4) *Deuterated transfer solvent 3* at a *flow* rate of 500 μL/min. Push *Save & Close*.
- In the Transfer box, the *transfer volume* should appear preset, as this value is system dependent and has to be calculated previously and filled in the *Hardware Setup*. In this case, this volume is 227 μL. *Excess volume* is not needed and the transfer is performed at a *flow* of 500 μL/min.
- The system will *finalise* after each transfer, so this statement should be clicked.
- *Dispenser and solvent port* should be chosen: *dispenser right* (2), *solvent port* (4) *deuterated transfer solvent 3*. This solvent corresponds to eluent F.

18. Purge the methanol-d_4 line: in *Prospekt2*, by right-clicking, click on *Direct Control*, choose *ACE* and *syringe 2*. This syringe is dedicated to deuterated solvents (in contrast to syringe 1 which is dedicated to protonated solvents and is used for *conditioning* and *equilibration* of the cartridges). Check the tubing and valve connections. To take up liquid, click on the solvent valve position (e.g. solvent (4)) and push *Aspirate* 500 μL; to discard the solvent, change the valve position to *waste* (5) and push *Dispense* 500 μL. Repeat this procedure until all the air bubbles are expelled from the tubing (see Note 23).
19. Mount the CryoFit. This is delicate operation; please follow rigorously the instructions given by Bruker BioSpin.
20. Select the cartridge to transfer from the *List of sampled cartridges in the Prospekt2 device*. By clicking on a cartridge line, the trapping conditions (% of eluents A and B) are indicated and the chromatographic signal in the chromatogram of the corresponding cartridge.
21. After checking the transfer parameters and the connecting tubing, start transferring the dummy cartridge to test the transfer set-up. Push *Start*. The system will dry the probehead, wash and dry again before eluting (with solvent F) and transferring the contents of the dummy cartridge. Check if the transfer occurred by observing the lock signal at the NMR console. Perform a "bubble test" (pulse sequence imgegp1d2h, 1D gradient echo for gradshim-procedure using lockswitch unit or BSMS 2 H-TX board) to assess whether the probe head is free of air bubbles (see Fig. 4a). Execute a ^{1}H NMR. If everything proceeds well, start transferring a trapped compound, such as one of **1–5**, and proceed with 2D NMR experiments such as COSY or even HMBC, depending on the amount of compound present inside the probe.
22. After NMR analysis of the trapped metabolites, these can be collected via the SPE unit, one by one, and used for further analysis. The confirmation of the mass and putative identification of the isolated metabolites can be done by LC-PDA-TOF-MS (see Note 24).

3.3.4. Data Analysis

The LC-PDA chromatograms of metabolites **1–5** can be seen using Hystar PP, the post-processing software of Hystar. The *unt file created by the acquisition can be directly opened displaying the chromatogram, as well as parameters such as the eluent gradient or system pressure, and indicating the trapping time intervals (see Fig. 3d).

The NMR spectra of the trapped metabolites **1–5** were Fourier-transformed, phased, baseline-corrected, calibrated towards the solvent signals, and visualised in TopSpin (see Fig. 4b). The assignment

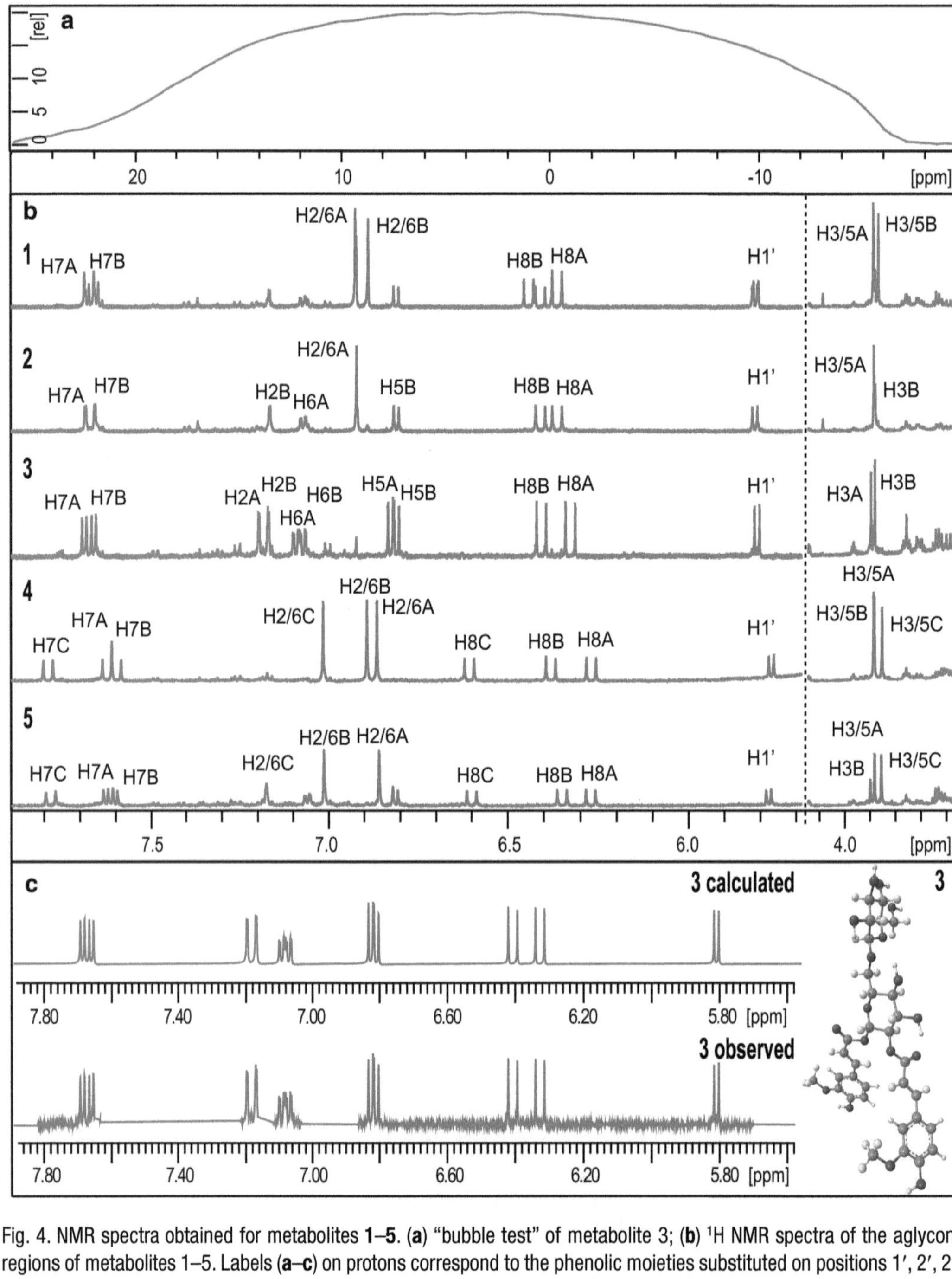

Fig. 4. NMR spectra obtained for metabolites **1–5**. (**a**) "bubble test" of metabolite 3; (**b**) ^{1}H NMR spectra of the aglycone regions of metabolites 1–5. Labels (**a–c**) on protons correspond to the phenolic moieties substituted on positions 1′, 2′, 2″, respectively (see Fig. 5 for complete labelling and metabolite names); (**c**) Experimental and calculated ^{1}H NMR spectra of aglycone region using PERCH for metabolite 3 which 3D structure is depicted.

of the chemical shifts and coupling constants present in the NMR spectra to protons was done using the information generated by the ^{1}H NMR and the ^{1}H–^{1}H COSY (see Table 5). Using PERCH, NMR assignments are checked towards the 3D structure of the

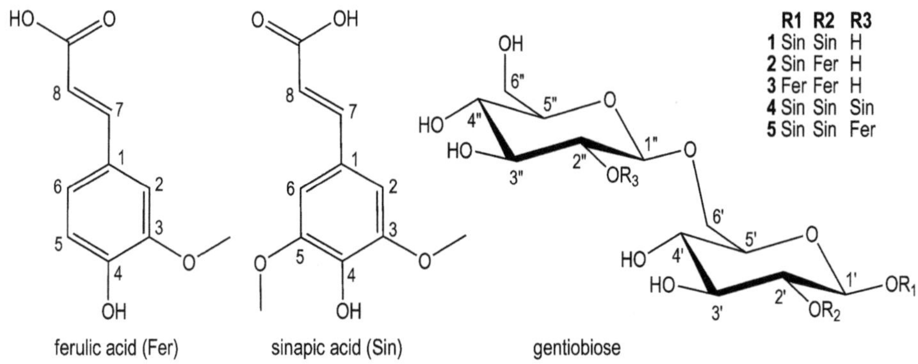

Fig. 5. Two-dimensional chemical structures of metabolites **1–5** with atom labelling.

respective molecule so that chemical shifts and coupling constants are feasible to extract with high precision, even for complex multiplicity patterns (see Fig. 4c).

3.4. Metabolite Identification by LC-PDA-SPE-NMR/MS

The acquisition of NMR spectra (^{1}H and ^{1}H-^{1}H COSY) for metabolites **1–5** enabled the assignments of all the protons present in these molecules. The putative identity and purity of each isolated metabolite was confirmed by LC-PDA-TOF-MS (see Tables 4 and 6) after NMR analysis, which provided basic information about the structure of the molecules: molecular mass, molecular formula, building bocks (ferulic acid, sinapic acid, hexose). By observation of the ^{1}H NMR spectra of **1** and **2**, it can be seen that sample **1** was contaminated with **2**. Nevertheless, in this case, this did not cause major impediments in the elucidation of **1**.

To find out the complete chemical structure of these glycosylated phenolic acids, several chemical items necessary for full identification were addressed by the analysis of the NMR spectra.

First, metabolites **1–5** are chemically related, as the NMR spectra are very similar, in particular the sugar region is analogous for all metabolites. The sugar moiety is constituted by two hexose sugars. By the analysis of the ^{1}H–^{1}H COSY spectra and also in comparison with the NMR properties of other hexoses, such as galactose (14), it can be concluded that these are two glucopyranoses. This is evident from the large coupling constants, ca. 8 Hz, between neighbouring protons in the hemiacetal ring. Because there is an effect on the chemical shifts of H6a/b′ and H1″, the glycosidic bond is established between the two glucoses through a 1→6 bond; therefore, the disaccharide is either an isomaltose or a gentiobiose, depending on the conformation of the anomeric H1″. This proton has a chemical shift of 4.40 ppm and has a large coupling constant, ca. 7.8 Hz, which implies a β configuration; therefore, metabolites **1–5** have a gentiobiose as sugar moiety.

Table 5
^{1}H NMR chemical shifts and coupling constants of metabolites 1–5

		Chemical shifts					Coupling constants (Hz)				
Proton	Multiplicity	1	2	3	4	5	1	2	3	4	5
Gentiobiose											
H1′	d	5.82	5.82	5.81	5.77	5.78	8.3	8.3	8.3	8.2	8.3
H2′	dd	5.12	5.11	5.11	5.03	5.03	8.3; 9.5	8.3; 9.5	8.3; 9.5	8.2; 9.5	8.3; 9.5
H3′	dd	3.77	3.75	3.76	3.60	3.61	9.5; 9.0	9.5; 9.0	9.5; 9.1	9.5; 9.0	9.5; 9.2
H4′	dd	3.64	3.63	3.63	3.41	3.42	9.0; 9.9	9.0; 9.9	9.1; 9.9	[a]	9.2; 9.6
H5′	m	3.72	3.72	3.72	3.61	3.61				9.6; 6.8; 2.0	9.6; 6.0; 1.9
H6′a	dd	4.25	4.25	4.25	4.19	4.18	2.0; −11.7	2.0; −11.7	2.0; −11.8	2.0; −12.0	1.9; −12.1
H6′b	dd	3.88	3.87	3.87	3.84	3.84	5.9; −11.7	5.3; −11.7	5.3; −11.8	6.8; −12.0	6.0; −12.1
H1	d	4.40	4.39	4.39	4.76	4.75	7.8	8.0	8.0	8.0	8.2
H2	dd	3.27	3.26	3.27	4.87	4.81	7.8; 8.8	8.0; 9.0	8.0; 9.4	[b]	[b]
H3	dd	3.39	3.39	3.39	3.61	3.61	[a]	[a]	[a]	9.3; 9.0	9.2; 9.6
H4	dd	3.35	3.35	3.35	3.44	3.44	[a]	[a]	[a]	[a]	9.6; 9.2
H5	m	3.30	3.29	3.30	3.34	3.34	[a]	[a]	[a]	[a]	[a]
H6 a	dd	3.90	3.89	3.90	3.93	3.93	2.3; −11.7	1.6; −11.2	2.1; −11.8	1.9; −12.0	2.0; −11.4
H6 b	dd	3.71	3.71	3.70	3.73	3.73	6.7; −11.7	5.4; −11.2	6.2; −11.8	5.6; −12.0	6.2; −11.4
Phenolic moiety A											
H2	s	6.93	6.92	7.2 (d)	6.86	6.86			1.9		
H5				6.83 (d)					8.2		
H6	s	6.93	6.92	7.09 (dd)	6.86	6.86			8.2; 1.9		
H7	d	7.68	7.67	7.67	7.63	7.62	15.9	15.9	15.9	15.9	15.9
H8	d	6.37	6.37	6.33	6.27	6.27	15.9	15.9	15.9	15.9	15.9
OMe3/5	s	3.89	3.89		3.88	3.88					
OMe5	s			3.90							

Phenolic moiety B											
H2	d	6.89 (s)	7.17	7.17	6.89 (s)	7.18		2.0	1.9		2.0
H5	d		6.81	6.81		6.81		8.3	8.2		8.2
H6	dd	6.89 (s)	7.07	7.07	6.89 (s)	7.06		8.3; 2.0	8.2; 1.9		8.2; 2.0
H7	d	7.65	7.67	7.67	7.60	7.61	15.9	15.9	15.9	15.9	15.9
H8	d	6.44	6.41	6.41	6.38	6.35	15.9	15.9	15.9	15.9	15.9
OMe3/5	s	3.87			3.89						
OMe5	s		3.88	3.88		3.90					
Phenolic moiety C											
H2	s				7.01	7.01					
H6	s				7.01	7.01					
H7	d				7.78	7.79				15.9	15.8
H8	d				6.61	6.6				15.9	15.8
OMe3/5	s				3.85	3.86					

OMe = methyl; other multiplicity than the one on the column is indicated within brackets
[a] Overlaps with MeOD signal
[b] Overlaps with HDO signal

The number of phenolic moieties and their chemical nature (if these correspond to a ferulic acid or to a sinapic acid) was confirmed by the presence and the integral of methyl protons on the 3 and 5 positions from the ^1H NMR spectra (a feruloyl moiety has only a methyl group on the 3 position while a sinapoyl has two methyl groups on the 3 and 5 positions, see Fig. 4b). This means that metabolite **1** has two sinapoyl moieties, **2** has a sinapoyl and a feruloyl, **3** has two feruloyl moieties, **4** has three sinapoyl moieties, and **5** has two sinapoyl moieties and a feruloyl moiety.

The double bond C7 = C8 in all the phenolic moieties is always in the *E* conformer, as it is made obvious from the typical large coupling constants $J_{H7, H8}/J_{H8, H7}$ which are 15.9 Hz.

The sinapoyl and the feruloyl moieties are linked to the sugar moiety through the hydroxyl of the carboxylic acid group and not through the aromatic hydroxyl(s) because shifts on the chemical shifts of the protons H7 and H8 to higher ppm values are observed, compared to the aglycones sinapic acid and ferulic acid (data not shown).

The substitution position of the phenolic moieties in the sugar is made clear by the effects on the chemical shifts of the neighbouring protons of the substitution. In this case, more than 1 ppm shifts are observed for H1′ and H2′ (metabolites **1**–**3**) and H2″ (metabolites **4**–**5**), implying that these are the positions of phenolic substitution. This evidence is further confirmed with literature information where ^{13}C data was also acquired for these metabolites (15–17). Therefore, metabolite **1** is a 1,2-di-*O*-*E*-sinapoyl-β-gentiobiose, **3** a 1,2-di-*O*-*E*-feruloyl-β-gentiobiose, and **4** a 1,2,2'-tri-*O*-*E*-sinapoyl-β-gentiobiose. For metabolites **2** and **5**, the position of substitution of the sinapoyl and feruloyl moieties is made clear through the comparison of chemical shifts in the ^1H NMR with the other related metabolites. For example, the chemical shift of H2/6A in **1** coincides with the H2/6 of **2**, implying that the

Table 6
Mass error (in ppm) computed by DataAnalysis for metabolites 1–5 after collection from the LC-PDA-SPE-NMR and analysed by LC-MS signals present in broccoli (see Table 4)

Metabolite	Mass error (ppm)	Sigma rank
1	-1.8	1
2	-1.9	1
3	-2.7	1
4	2.1	1
5	0.4	2

sinapoyl moiety is in the 1′ position. The feruloyl moiety is in the 2' as also shown by comparison with H2B and H5B of metabolite **3**; therefore, **2** is a 1-*O*-*E*-sinapoyl-2-*O*-*E*-feruloyl-β-gentiobiose. Likewise, by comparison of metabolite **5** to **4**, it can be concluded that the feruloyl moiety in metabolite **5** is in the 2″ position; therefore, **5** is a 1,2'-di-*O*-*E*-sinapoyl-2-*O*-*E*-feruloyl-β-gentiobiose.

The structural identification of metabolites **1**–**5** was achieved by online separation, isolation, concentration, NMR and MS analysis, using a LC-PDA-SPE-NMR/MS set-up. This set-up combines the accurate mass efficiency of a TOF-MS and the sensitivity of a NMR cryogenic 30 μL flow insert. Therefore, the identification of low abundant secondary metabolites present in plants is facilitated, avoiding intensive analytical efforts of scale-up and sample concentration. Combining analytical and database/literature tools, pursuing the full identification of a metabolite is made possible in due time.

4. Notes

1. Organic acids can be flammable, toxic and harmful, therefore should be handled in fume hood with protecting gloves and stored in safety cupboard.
2. Purchase HPLC gradient/MS suitable or ultra-pure reagents, to minimise the occurrence of impurities in the MS spectrometer and NMR.
3. Acids and bases can be harmful and corrosive therefore should be handled in fume hood with protecting gloves and stored in safety cupboard.
4. Deuterated solvents tend to accumulate water after opening. Make sure to seal well the bottles and use immediately after opening. In this way, the water signal in the NMR spectrum is minimised.
5. Samples can be stored at 4°C for months, although chemical stability should be checked, by comparing with analysis performed immediately after extraction.
6. When setting up the sample table in HyStar, as precaution, always open the purge valve of pump 1 before opening the *Acquisition* window so that if the pump switches on, there is not sudden increase of pressure which can damage the analytical column.
7. If performing the chromatography with methanol instead of acetonitrile, as organic modifier, make sure that the system can cope with the overall increase of pressure, as methanol, due to its physical properties offers more pressure than acetonitrile.

8. Keep track of all the events, problems, pressure values, vacuum values (fore and high vacuum pressure values of the MS), alterations of the system, substitution of parts, error messages, etc. by keeping a diary of the system.
9. Regarding disk space, acquire DAD (*.u2) only when absolutely necessary, as these files are very large. Also the MS profile spectra take a large disk space, save only line spectra. However, in this case (line spectra), the Sigma fit algorithm of isotopic distribution works less accurately. In MS and NMR data, deleting the processing files saves disk space and does not interfere with the acquisition files (which should logically be kept).
10. Use red PEEK tubes (outer diameter 1/16 in.) after the HPLC column so that the diffusion of the chromatographic signals is minimised (see Fig. 1).
11. The LC-PDA-MS-SPE-NMR is a versatile and flexible system in terms of connections and set-up; however, this implies also complexity in which a lot of attention needs to be paid in order to synchronise all its modules. Therefore, the most important advice is always to check thoroughly everything (method, connections, volumes, etc.) before starting the analyses.
12. In the LC-MS set-up, when a problem occurs with HyStar and this software needs to be closed, micrOTOF Control should also be closed; otherwise, it will give an error, as it cannot connect with HyStar when re-starting HyStar.
13. Adapt the *LC method part* according to the chromatographic peaks interested to trap in order to optimise chromatography resolution and save time and eluents.
14. When performing trapping experiments, apply a flow on pump 2 only for the chromatographic region needed. This measure will save solvent D.
15. In this method, the ratio of flow rates of pump1 and 2 used for trapping was 2:3. In some cases, depending on the chemistry of the metabolites to be trapped, the stationary phase of the cartridges, etc., this ratio should be adjusted.
16. Automatic settings for trapping (e.g. according to threshold, slope, time slice) are available within HyStar and can be used for the comfort of the user.
17. Equilibrated the cartridges needed for trapping just before trapping and do not leave cartridges equilibrated (100% D) on the system, instead, condition them (100% E). This will prevent possible bacterial growth on the stationary phase of the cartridges, as with the analytical column.
18. Never use cartridges A1 nor B1 for trapping, as these are cartridges very often used by the SPE unit to close the system.
19. The SPE unit does not offer a possibility to choose different solvents to wash the cartridges (*Conditioning*). Nevertheless,

methanol can be used instead of acetonitrile, for instances, by switching solvent bottles and purging the tubing with the new solvent.

20. Do not try to trap two peaks with little time between each other (less than about 10 s), as the SPE unit after trapping one peak needs time to get ready for the next. In the event that this happens, the second peak might be trapped in the same cartridge or the system might crash.
21. Never leave loaded and undried cartridges on the system as these adsorb impurities.
22. To minimise the solvent signals in the NMR spectrum, the loaded cartridges can be dried more than once, e.g. two times the maximum time of drying, i.e. 118 min.
23. Make sure to purge the tubing of syringe 2 of the SPE unit so that these are free of air bubbles before transferring each cartridge.
24. The use of a semi-preparative column can allow the injection of a higher amount of material and therefore a faster isolation of metabolites (18); however, the increase of system pressure should be taken into account.

Acknowledgements

The authors thank Dr. Benoît Biais and the team at INRA Bordeaux Aquitaine for the broccoli samples. The authors acknowledge the financial support from the EU project “META-PHOR”, contract number FOOD-CT-2006-036220.

References

1. FAOSTAT (2009) in “FAOSTAT/Food and Agriculture Organization of the United Nations”.
2. Brennan, P., Hsu, C. C., Moullan, N., Szeszenia-Dabrowska, N., Lissowska, J., Zaridze, D., Rudnai, P., Fabianova, E., Mates, D., Bencko, V., Foretova, L., Janout, V., Gemignani, F., Chabrier, A., Hall, J., Hung, R. J., Boffetta, P., and Canzian, F. (2005) Effect of cruciferous vegetables on lung cancer in patients stratified by genetic status: a mendelian randomisation approach. *Lancet* **366**, 1558–1560.
3. Vallejo, F., Tomás-Barberán, F., and García-Viguera, C. (2003) Health-promoting compounds in broccoli as influenced by refrigerated transport and retail sale period. *Journal of Agricultural and Food Chemistry* **51**, 3029–3034.
4. Vallejo, F., Tomás-Barberán, F. A., and Ferreres, F. (2004) Characterisation of flavonols in broccoli (*Brassica oleracea* L. var. *italica*) by liquid chromatography–UV diode-array detection–electrospray ionisation mass spectrometry. *Journal of Chromatography A* **1054**, 181–193.
5. Bennett, R. N., Mellon, F. A., and Kroon, P. A. (2004) Screening crucifer seeds as sources of specific intact glucosinolates using ion-pair high-performance liquid chromatography negative ion electrospray mass spectrometry. *Journal of Agricultural and Food Chemistry* **52**, 428–438.
6. Cartea, M. E., Velasco, P., Obregon, S., Padilla, G., and de Haro, A. (2008) Seasonal variation in glucosinolate content in *Brassica oleracea* crops grown in northwestern Spain. *Phytochemistry* **69**, 403–410.

7. Moco, S., Forshed, J., De Vos, R. C. H., Bino, R. J., and Vervoort, J. (2008) Intra- and inter-metabolite correlation spectroscopy of tomato metabolomics data obtained by liquid chromatography-mass spectrometry and nuclear magnetic resonance. *Metabolomics* **4**, 202–215.
8. De Vos, R. C. H., Moco, S., Lommen, A., Keurentjes, J. J. B., Bino, R. J., and Hall, R. D. (2007) Untargeted large-scale plant metabolomics using liquid chromatography coupled to mass spectrometry. Nature *Protocols* **2**, 778–791.
9. Moco, S., Bino, R., De Vos, R. C. H., and Vervoort, J. (2007) Metabolomics technologies and metabolite identification. *TrAC Trends in Analytical Chemistry* **26,** 855–866.
10. Moco, S., Bino, R. J., Vorst, O., Verhoeven, H. A., de Groot, J., van Beek, T. A., Vervoort, J., and De Vos, R. C. H. (2006) A liquid chromatography-mass spectrometry-based metabolome database for tomato. *Plant Physiology* **141**, 1205–1218.
11. Exarchou, V., Krucker, M., van Beek, T. A., Vervoort, J., Gerothanassis, I. P., and Albert, K. (2005) LC-NMR coupling technology: recent advancements and applications in natural products analysis. *Magnetic Resonance in Chemistry* **43**, 681–687.
12. Exarchou, V., Godejohann, M., van Beek, T. A., Gerothanassis, I. P., and Vervoort, J. (2003) LC-UV-solid-phase extraction-NMR-MS combined with a cryogenic flow probe and its application to the identification of compounds present in Greek oregano. *Analytical Chemistry* **75**, 6288–6294.
13. Rochfort, S. J., Trenerry, V. C., Imsic, M., Panozzo, J., and Jones, R. (2008) Class targeted metabolomics: ESI ion trap screening methods for glucosinolates based on MS^n fragmentation. *Phytochemistry* **69**, 1671–1679.
14. Moco, S., Tseng, L. H., Spraul, M., Chen, Z., and Vervoort, J. (2006) Building-up a comprehensive database of flavonoids based on nuclear magnetic resonance data. *Chromatographia* **9/10**, 503–508.
15. Baumert, A., Milkowski, C., Schmidt, J., Nimtz, M., Wray, V., and Strack, D. (2005) Formation of a complex pattern of sinapate esters in *Brassica napus* seeds, catalyzed by enzymes of a serine carboxypeptidase-like acyltransferase family? *Phytochemistry* **66**, 1334–1345.
16. Price, K. R., Casuscelli, F., Colquhoun, I. J., and Rhodes, M. J. C. (1997) Hydroxycinnamic acid esters from broccoli florets. *Phytochemistry* **45**, 1683–1687.
17. Rahman, M. A. A., and Moon, S. S. (2007) Antioxidant polyphenol glycosides from the plant *Draba nemorosa*. *Bulletin of the Korean Chemical Society* **28**, 827–831.
18. Miliauskas, G., van Beek, T. A., de Waard, P., Venskutonis, R. P., and Sudholter, E. J. R. (2006) Comparison of analytical and semi-preparative columns for high-performance liquid chromatography–solid-phase extraction–nuclear magnetic resonance. *Journal of Chromatography A* **1112**, 276–284.

Chapter 18

A Strategy for Selecting Data Mining Techniques in Metabolomics

Ahmed Hmaidan BaniMustafa and Nigel W. Hardy

Abstract

There is a general agreement that the development of metabolomics depends not only on advances in chemical analysis techniques but also on advances in computing and data analysis methods. Metabolomics data usually requires intensive pre-processing, analysis, and mining procedures. Selecting and applying such procedures requires attention to issues including justification, traceability, and reproducibility. We describe a strategy for selecting data mining techniques which takes into consideration the goals of data mining techniques on the one hand, and the goals of metabolomics investigations and the nature of the data on the other. The strategy aims to ensure the validity and soundness of results and promote the achievement of the investigation goals.

Key words: Data mining process, Metabolomics, Scientific data mining, Data mining technique selection

1. Introduction

Data mining uses a wide range of modelling techniques involving machine learning, pattern recognition, statistics, and clustering algorithms (1–3). In metabolomics, data mining is performed either in a hypothesis-driven fashion where it seeks an answer to a preset research question or in a data-driven fashion where it seeks to discover patterns, trends, or associations which might be completely different from those intended when the data were originally acquired. However, hypothesis-driven and data-driven investigations can both be seen as part of the *knowledge cycle*, (2) where each might lead to the other. The first is used for deducing knowledge through testing a preset hypothesis, while the second might be used for inducing knowledge from data and generating new hypotheses for further investigations (2, 3).

Nigel W. Hardy and Robert D. Hall (eds.), *Plant Metabolomics: Methods and Protocols*, Methods in Molecular Biology, vol. 860, DOI 10.1007/978-1-61779-594-7_18, © Springer Science+Business Media, LLC 2012

Formalizing a framework strategy for conducting data mining, which focuses on providing a mechanism for the selection of data mining techniques, provides several benefits. It encourages the achievement of the aims of a metabolomics study as well as ensuring justifiability of technique choice throughout the analysis. It also provides traceability of the procedures applied and ultimately, supports the reproducibility of the investigation outcomes.

In this chapter, we describe a strategy for selecting data mining modelling techniques. In Subheading 2, we provide an overview of the inputs required for the selection, while in Subheading 3 we describe the methods to be used for performing the steps of the strategy. Notes are provided to define concepts, suggest alternatives, or to expand the discussion.

2. Materials (Inputs for the Selection)

Here, we describe the important inputs to the selection of techniques. The first focuses on understanding the aims of the metabolomics *study* and their relation to the research *investigation* and the data acquisition *assays* (see Note 1). The second input is related to the understanding of the general goals of data mining, the tasks which are performed and the techniques used to achieve these goals. The third concerns the nature and quality of metabolomics data.

In addition to the inputs discussed in this section, it is also important to consider other factors concerning the application of the techniques in practice. These include data pre-processing and data acclimatization in addition to management and technical issues such as planning, project management, feasibility, and the availability of software tools and expertise (4–6).

2.1. The Aims of a Metabolomics Study

Data mining modelling techniques are used in metabolomics, either in an hypothesis-driven or in a data-driven fashion, to fulfil the aims of a study and consequently answer the question of the research investigation. Accordingly, the aims of a metabolomics study are derived from the goals of the research investigation. The study might then require one or more assays to acquire the required data. Furthermore, and in order to perform a successful, *justifiable*, *traceable* and *reproducible* analysis of metabolomics data (see Note 2) the *aims* of the study must be narrowed, and afterwards expressed in terms of data mining *objectives* which must be specific, measurable, realistic, and achievable, while still corresponding to the original investigation goals (see Note 3).

2.2. Data Mining Goals, Tasks and Techniques

When selecting data mining techniques, it is crucial to understand data mining approaches, goals and tasks (see Fig. 1) as well as the techniques they use to achieve their modelling objectives. The hypothesis-driven data mining approach tests a pre-existing

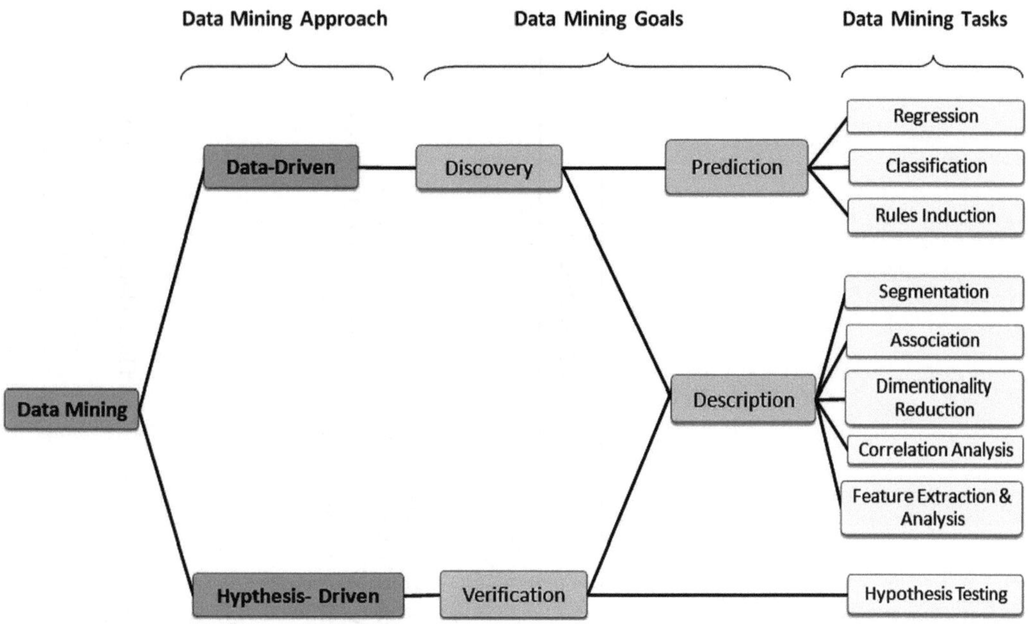

Fig. 1. Data mining approaches, goals and tasks.

hypothesis regarding the relationships among data and is achieved either through description or verification. By contrast, the data-driven approach aims to uncover novel knowledge in the data regardless of the original purpose of their acquisition. This is usually performed either through prediction or description (7–9), e.g. predicting biomarkers for a disease or classifying samples into healthy and diseased.

Data-driven mining is used for the purpose of knowledge discovery. In this case, the objectives of data mining focus on finding interesting and novel patterns, trends or associations in the data, even if the data were originally acquired for a different purpose (7, 10, 11). Hypothesis-driven objectives are generally motivated by the goals of the research investigation and the aims of its subsequent studies (11).

In order to achieve its goals, data mining employs a wide spectrum of machine learning, statistical and pattern recognition techniques which perform a narrow set of tasks, e.g. segmentation, classification. Figure 1 illustrates data mining approaches, goals, and tasks, while Table 1 describes those tasks and provides examples of their modelling techniques showing whether these are supervised or unsupervised (see Note 4).

2.3. Metabolomics Data

Both the quality and nature of metabolomics data influence the selection of data mining techniques as well as their relation with the research investigation, study and assay. Metabolomics data consist of both the data set as acquired by the instruments and its associated meta-data. The data set is acquired by *chemical analysis instruments*, e.g. NMR, LC/GC-MS, HPLC, FT-IR, etc. (40–47)

Table 1
Data mining tasks

Data mining task	Description	Data mining example techniques: Supervised	Data mining example techniques: Unsupervised
Regression	Build a model that uses data to predict new continuous numerical data.	Multiple Linear Regression (MLR), Partial Least Squares (PLS) (12, 13), Support Vector Machine (SVM) (14), Linear Regression (LR) (15), Regression Trees (16).	
Classification	Build a model that is capable of classifying data in order to predict new discreet or categorical data.	Artificial Neural Networks(ANN) (1), Decision Trees, Random Forest (17), Linear Discriminant Analysis (LDA), Discriminant Function Analysis (DFA) (17–19), Support Vector Machine (SVM), Soft Independent Modelling of Class Analogy (SIMCA) (19), Genetic Programming (20), Genetic Algorithm (21).	Kohonen Neural Networks Self-Organizing Map (SOM) Cluster Analysis Techniques (22, 23).
Rules inductive	Extract useful rules from the data set based on significance.	Genetic Programming, Genetic Algorithm Classification and Regression Trees (CART), Inductive Logic Programming (1, 24, 25).	
Segmentation	Identify the natural grouping among the data set and classify the data accordingly.	Discriminant Function Analysis (DFA) (17–19) Genetic Programming (2, 20), Genetic Algorithm (21).	Hierarchical Clustering Analysis (HCA) (19, 23), K-Means (22, 26), fuzzy c-means (27) Self-Organizing Map (SOM) (22).

Association	Identify the relationships within the data set and the probability of their occurrence		Association Rules (28–31), Apriory (32).
Dimensionality reduction	Create an optimized data set on which to base a model and eliminating non-informative features	Linear Discriminant Analysis (LDA) (12), Partial Least Squares (PLS) (33), Discriminant Analysis (PLS-DA) (12, 34) Orthonormalized Partial Least Squares (OPLS) (33).	Independent Component Analysis (ICA) (35, 36) Principle Component Analysis (PCA)(33) Factor Analysis (FA) (22).
Feature extraction and analysis	Gain insight into the rationale underlying class divisions (12).	Partial least squares discriminant analysis (PLS-DA), Random Forest feature selection (12).	
Correlation analysis	Determine the association between the changes in the value of one variable with the changes in another variable.	Covariance analysis (37, 38).	
Hypothesis testing	Test assertion about the data set based on the concept of proof by contradiction	Chi-test, *z*-test, *f*-test, Goodness of fit, Analysis of Variance (ANOVA) (22), Multivariate analysis of variance (MANOVA) (39).	

in assays. The choice of the instrument depends on the goals of the investigation and their relation with the aims of the study and the design of the assay on the one hand, and with the metabolic approaches (see Note 5) on the other (1).

The assay data set is usually generated in the form of spectra which vary in their detailed structure depending on the data acquisition instrument and on the transformation used to convert the spectra from one format into another, e.g. Fourier transformation for NMR, peak lists, spectra bins, or concentration profiles (48). Metabolomics meta-data concerns the recorded information in the study regarding the factors which might influence the data set, e.g. bio-source, sample preparation, metabolic approach, data acquisition instruments, administration, chemical and other study related factors (38, 49–51).

2.3.1. The Nature of the Data

Factors related to the nature of metabolomics data including size, data types, data structures, and format must be considered in the selection of the modelling technique.

Different techniques may vary in their ability to handle large volumes of data whether in terms of number of attributes, number of examples (52), or their ratio. Some techniques require reducing the dimensionality of data (33), e.g. regression (12, 13, 15) or DFA (17–19), while others are able to handle a larger number of variables, e.g. decision trees (7, 53). On the other hand, some techniques are able to handle some types of data better than others, e.g. *classification techniques* handle discrete data better than continuous data, *regression techniques* are more efficient in handling continuous data, *neural networks* are able to handle numerical data only (52). Decision trees are able to handle both nominal and numerical data (54). Furthermore, conversion of data structures and formats might also be required during data acclimatization (see Subheading 2.3.2). The level and intensity of the conversion depends on the requirements of the modelling technique implementation and indirectly affect the selection when considering management and other technical factors.

2.3.2. Quality of Data

Careful examination of the quality of data may be vital for the selection of modelling techniques and eventually the success and soundness of data mining results. Some techniques are more tolerant to issues such as missing values (55, 56), outliers, and unusual distributions of data (57). Several procedures might be required to improve the quality of the data and make it more suitable for modelling; this can be done either through data pre-processing or acclimatization.

Data Pre-processing: Data pre-processing is usually performed either at the level of the instrument or externally as a precursor to model building. The extent of pre-processing which the data may require affects the choice of data mining technique and covers issues such as the aims of the study, quality of data, project management and other practical trade-offs. Pre-processing activities cover a wide range of operations including the handling of outliers and missing values, normalization, phasing, peak picking, alignment, baseline correction, bucketing, data reduction, extraction, etc. (38, 42).

Data Acclimatization: The level and intensity of data acclimatization depends on the objectives of modelling as well as on the selected technique. Different techniques may require different levels of acclimatization depending on the type, quality, format, and the structure of the data. The aim of data acclimatization is to make the data suit the modelling technique. Examples of acclimatization activities include the following: (1) *Conversions:* transforming data from one type into another might be required. (2) *Merging:* combining attributes that imply redundant information. (3) *Splitting*: separating attributes that imply more than one piece of information. (4) *Formatting*: configuring input files to suit the requirements of the modelling tools, e.g. tabular, textual, xml, etc. (58–60). Other more sophisticated procedures might also be required, particularly when combining more than one modelling technique, e.g. reducing the dimensionality of data before building the model (60).

3. Methods

The strategy defines a framework for selecting data mining techniques and providing the appropriate justification. Figure 2 illustrates the framework of the strategy, while a demonstration of its applicability, based on examples from metabolomics literature, is provided later (see Note 6).

The strategy consists of three major steps: Setting Objectives; Data Exploration; and Matching Objectives to Data Mining Technique(s). The strategy defines the flow of these steps and shows their relationships with other data mining phases. It also defines the inputs and deliverables of each step.

3.1. Setting Objectives

The modelling objectives can be expressed either in an hypothesis-driven fashion or in a data-driven fashion depending on the aims of the study (see Subheading 2). Modelling objectives should be in line with the goals of the original investigation, consistent with the aims of its subsequent studies, measurable, feasible and should be achievable generally through data mining and knowledge discovery.

The Activities:

1. Decide the type of objectives to be set either as hypothesis-driven or as data-driven objectives based on the general understanding of data mining approaches as discussed in Subheading 2.2.

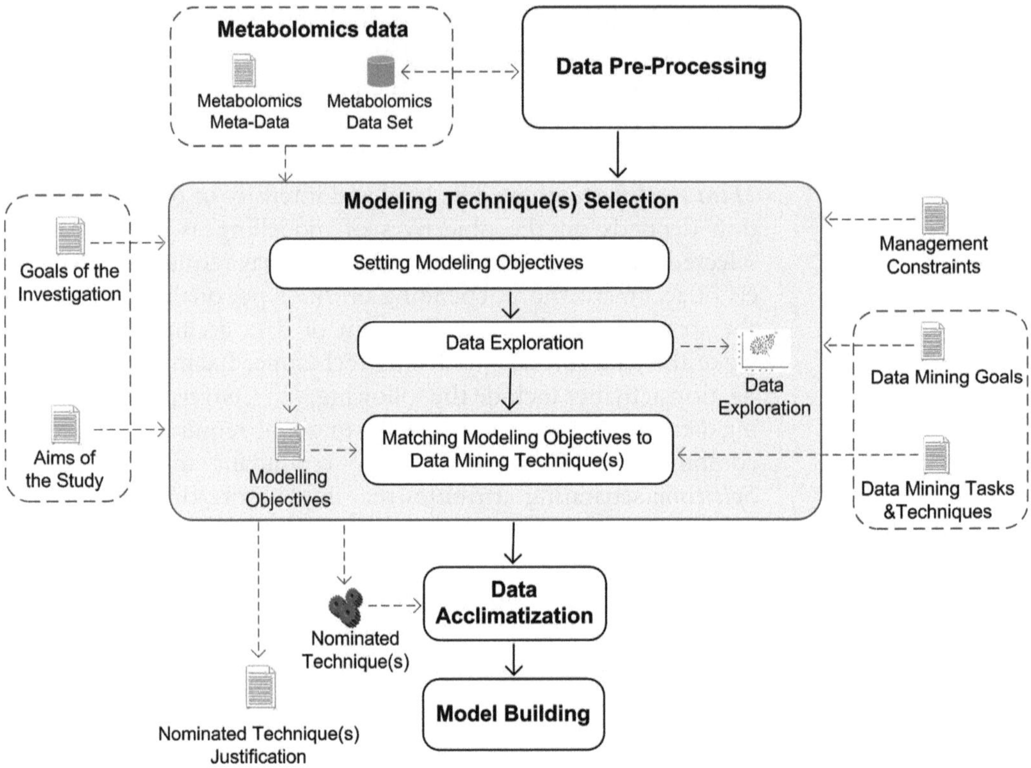

Fig. 2. The framework of the strategy.

2. Examine the goals of the research investigation and the aims of the metabolomics study which the assay has been designed to achieve.
3. Translate the goals of the research investigation and the aims of the study into definable draft modelling objectives based on the general understanding of data mining goals, and tasks as discussed in Subheading 2.2.
4. Assess the achievability of the draft objectives in terms of the availability, relevance, and adequateness of appropriate data.
5. Assess the feasibility of fulfilling the draft objectives in light of management and technical constraints.
6. Depending on the results of the assessment in steps 4 and 5, retain the objectives which passed the assessment criteria and discard the ones which failed.
7. Define success criteria and measurements to be applied to evaluate the results and assess the fulfilment of defined modelling objectives.

3.2. Data Exploration

Data exploration gives insight into the data to which the technique will be applied. It must be comprehensive and thorough, covering all aspects which may contribute towards the selection of the technique including (1) *Data Investigation*, which examines the nature and quality of the data as discussed in Subheading 2.3. (2) *Data prospecting*, which concerns seeking interesting distributions and trends (61) and (3) *Data explanation*, which describes the meaning of data items and their scope (i.e. the acceptable range of possible values) and describes relationships among the variables. The output of this step takes the form of a report containing details regarding the activities performed and their outcomes.

The Activities:

1. Examine the nature of data, e.g. data types, structure, size, and format (see Subheading 2.3).
2. Investigate the quality of the data, e.g. missing values, statistical outliers, and distribution.
3. Verify data understandability by explaining the meaning and the scope (possible values) of each attribute and its relation with other variables, e.g. dependent versus independent variables.
4. Prospect the data for interesting trends and distributions using basic statistical measures, e.g. variance, mean, deviation, etc., or using more complex statistical techniques, e.g. PCA, regression, or correlation, to gain more insight in the data.
5. Confirm the relevance, sufficiency, and adequacy of data to fulfil the defined objectives.

3.3. Matching Objectives to Data Mining Techniques

In this step, the objectives defined in step 1 are matched to the goals, tasks and possible data mining techniques. The final selection of the techniques must consider the practical achievability of the defined objectives through the chosen technique, its applicability to the targeted data, its technical and management feasibility, as well as both the level and degree of data pre-processing and acclimatization procedures that it may require.

The outputs of this step include both the selection and a justification report including results of assessment and showing all the factors which have been considered.

The Activities:

1. Using data mining goals (see Fig. 2) and for each objective defined in step 1:
   (a) Depending on the modelling objective and its relation with the aims of the study as discussed in Subheadings 2.1 and 2.2, determine which data mining approach is more appropriate to use (data-driven or hypothesis driven).
   (b) Depending on the data mining goals (see Fig. 2), match the modelling objective to the data mining goals.
   (c) Match the objectives to the appropriate data mining subgoals, e.g. prediction, description.

(d) Match the modelling objective to the objectives of the data mining tasks as demonstrated in Fig. 1 and Table 1, taking into consideration the results of data exploration on the one hand and the tasks inputs and results on the other.

(e) Select the data mining technique that would fulfil these objectives. The selection should be based on the results of data exploration in step 2 and the background knowledge regarding each technique, its modelling objectives, the inputs it takes, and the output it produces.

2. Based on the data investigation, validate the tolerance of the candidate technique to the nature, quality, and distribution of the data, as well as its applicability to the types of data to be mined.
3. Assess the expected fulfilment of the defined objectives by the candidate technique.
4. Assess the level of additional pre-processing procedures required to improve the quality of data if required by the candidate technique.
5. Assess the expected level of acclimatization required to adapt the data to the candidate data mining modelling technique, e.g. dimensionality reduction.
6. Assess the technical and management constraints including cost and time feasibility, and the availability of the software tools and modelling expertise.
7. Consider alternatives and combinations of the candidate techniques then re-evaluate each through the steps 1–7 (see Note 7).

4. Notes

1. The terminologies used here are based on those proposed by *RSBI* (62) and used in *ISA-TAB* (11), where the word experiment is deliberately avoided and replaced by more precise terminologies. "*Investigation*" refers to the highest level concept of scientific enquiry that can be seen as a multi-faceted research activity. "*Study*" refers to the experimental design and its related variables. Subsequently one or more studies are designed to carry out an investigation where each examines one side of the overall investigation. Finally, "*Assay*" refers to smallest level of experimentation, where the data acquisition instrument's run is used to generate the data (11, 62–64).
2. The scientific nature of biological data requires attention to explanatory issues when performing data mining (65). Justifiability refers to the availability of evidence for the applicability of a particular data mining technique based on the

desired objectives which data mining hopes to achieve and the nature of the data to be mined. Traceability implies recording both the decision to choose a data mining technique and the factors which contributed to that decision which permits change of the decision if the parameters which led to it change. Finally, reproducibility, which is a desirable attribute of scientific work, refers to the ability to repeat scientific procedures (in this case the technique choice) and always come to the same result provided that the experimental conditions (in this case, the selection parameters) remain the same. The reproducibility of the final results is supported by the traceability of steps and their intermediate results, while traceability is enabled by the justifiability of all decision procedures.

3. Despite the similar definitions of *goal*, *aim*, and *objective* in an English dictionary (66), these words are frequently used in academic literature to describe different levels of abstraction and generality. Goal refers to the highest level of generality and abstraction, while aim is used to imply a narrower and less abstract meaning. Objective is used to describe a much narrower, more specific and measurable meaning. In this chapter, we use these words to imply the differences described above, in the way they are used in research methodology and project management contexts, e.g. SMART (4–6).
4. *Supervised methods learn* through finding a model that represents association between inputs (*X* variables or predictors) which are typically the meta-data of the study with the outcomes (*Y* variables or responses) which are typically the assay results, e.g. classification, regression, etc. *Unsupervised methods* learn from data through finding patterns or groups within the inputs (*X* variables) and are performed with no such guidance, e.g. segmentation or data reduction. In metabolomics, the inputs represent the data set, while outcomes represent the traits or classes (1).
5. Metabolic approaches include the following: *True metabolomics:* an unbiased (43) and comprehensive analysis of the overall metabolome in a particular condition (1, 42); *Metabolite profiling:* a quantitative analysis which is conducted over a set of predefined metabolites in a particular biochemical pathway, or on profiled subgroups of chemical classes (42, 43, 67); *Targeted metabolite analysis:* a form of metabolite profiling that targets particular metabolites of a specific biological system or biochemical pathway such as enzymes which are directly influenced by a specific type of environmental or genetic perturbations (1, 42); *Metabolite fingerprinting:* a rapid, global, high-throughput analysis which aims to discover patterns and classify samples without the need to identify or quantify the metabolites involved (43).
6. Table 2 demonstrates the applicability of the strategy based on examples from metabolomics literature. The table illustrates

Table 2
Matching data mining goals, tasks, and modelling objectives to the goals of metabolomics investigations and studies

Data mining goals		Data mining tasks	e.g. Goals of investigation	e.g. Aims of study	e.g. Modelling objectives
Discovery	Prediction	Regression	Toxic effects, Gene functional classes and annotation (68).	Identify the potential bio-markers, identify the significant features which causes the classification (69).	Analyse the relationship between independent and dependent variables and predict the response based on predictors (70).
		Classification	Classification of mutant genes with unknown function by comparison of their co-response pattern to the set of known genes (23, 37).	Classify samples (finger printing) (71). Gene function analysis (37). Identify biomarkers that classify samples into diseased or healthy controls (20).	Predict a class for new unknown data Using the classifier model (19). Understand the difference between groups or classes (72). Mapping unknown samples to preset classes (73, 74).
		Rule induction	Investigating complex biological systems at the whole-tissue level (75)	Identify metabolites involvement in bio processes (20).	Inference rules from data based, generate optimized mapping between inputs and outputs (1).
	Description	Segmentation	Classifying samples according to their origin (76).	Classify unknown sample by their closeness to known gene knockouts (guilt by association) (1).	Classify samples into its natural classes (38). Comparison and Visualization of similarities and differences between data (46).
		Association	Find biomarker that assist Early diagnosis of disease (32).	Characterized metabolic changes through metabolite concentration profiling (32)	Generate a set of association rules that uncover relationships among the data(31) and satisfying certain support and confidence constraints (32).

	Dimensionality reduction	Investigating the role of metabolites in genotype discrimination (77). Studying Genetically Modified food (78).	Distinguish between genotypes (78).Evaluate the contribution of each metabolite towards the total information of metabolome (71, 79).	Transform large related data set into a smaller uncorrelated set ignoring irrelevant data (73, 77, 80), Visualizing data in a reduced dimensionality (23, 34, 38).
	Features extraction and analysis	Study disease mechanism (34). Metabolic networks, diet studies (12).	Finding genetic markers relevant in interactions with other markers or environmental variables (12), Find metabolites associated with researches (e.g. diseases, biomarkers) (34)	Gain insight into the rationale underlying class divisions, discovery significant features represent class discriminating metabolites and eliminating non-informative features (12, 34).
	Correlation	Systems biology, metabolic network and pathways studies (37, 81, 82).	Investigate metabolites dependency and identify correlated metabolites (12, 20). Uncover silent mutation (82). Comparing different genotypes (81).	Visualize the relation between data and allow identifying the pattern of the correlation (37, 38).
Verification	Hypothesis testing	Drugs discovery and development, diseases biomarkers (83, 84).	Test biological relevance of hypothesis obtained from metabolomics data (76, 85). Test the individual metabolites that increase or decrease significantly between classes and groups (38).	Verify truth or falsity of a proposition, on the basis of empirical evidence (86). Assess the significance of the ratio of the variation within and between classes (85).

matching data mining goals, tasks and modelling objectives to the goals of metabolomics investigations and studies.

7. Alternative techniques might be useful to see results from different perspectives or to propagate new questions to be answered or even to seek explanations for results. On the other hand, combining more than one technique might be useful to tackle the weakness or to enhance the selected technique.

References

1. Goodacre, R., Vaidyanathan, S., Dunn, W. B., Harrigan, G. G. and Kell, D. B. (2004) Metabolomics By Numbers: Acquiring Understanding Global Metabolite Data. *Trends Biotech* **22**, 245–252.
2. Kell, D. B. (2002) Genotype-phenotype mapping: genes as computer programs. *Trends Genetics* **18**, 555–559.
3. Kell, D. B. and Oliver, S. G. (2004) Here is the evidence, now what is the hypothesis? The complementary roles of inductive and hypothesis-driven science in the post-genomic era. *BioEssays* **26**, 99–105.
4. Heldman, K. (2005) *Project Management Jumpstart.* 2nd ed. SYBEX Inc., San Francisco, CA.
5. Heldman, K. (2007) *PMP: Project Management Professional Exam Study Guide.* 5th ed. Wiley Publishing Inc., Indianapolis, IN.
6. Lewis, J. P. (2007) *Fundamentals of Project Management.* 3rd ed. American Management Association, New York, NY.
7. Maimon, O. and Rokach, L. (2005) *Data Mining and Knowledge Discovery Handbook.* Springer, New York, NY.
8. Maimon, O. and Rokach, L. (2005) *Decomposition methodology for knowledge discovery and data mining: theory and applications.* Series in machine perception and artificial intelligence Vol. 61. World Scientific, Singapore.
9. Sumathi, S. and Sivanandam, S. N. (2006) Data Mining Tasks, Techniques, and Applications, in *Introduction to Data Mining and its Applications* (S. Sumathi, ed.), Springer, New York, NY/Berlin. pp. 195–216.
10. Fayyad, U., Piatetsky-Shapiro, G. and Smyth, P. (1996) Knowledge Discovery and Data Mining: Toward a Unifying Framework. in *The Second Int Conf on Knowledge Discovery and Data Mining (KDD96).* Portland, OR, AAAI Press. Menlo Park, CA.
11. Taylor, C. F., Field, D., Sansone, S., Aerts, J., Apweiler, R., Ashburner, M., et al. (2008) Promoting coherent minimum reporting guidelines for biological and biomedical investigations: the MIBBI project. *Nat Biotech* **26**, 889–896.
12. Bryan, K., Brennan, L. and Cunningham, P. (2008) MetaFIND: A feature analysis tool for metabolomics data. *BMC Bioinformatics* **9**, 470.
13. Hayashi, S., Akiyama, S., Tamaru, Y., Takeda, Y., Fujiwara, T., Inoue, K., et al. (2009) A novel application of metabolomics in vertebrate development. *Biochem & Biophys Res Comm* **386**, 268–272.
14. Truong, Y., Lin, X. and Beecher, C. (2004) Learning a complex metabolomic dataset using random forests and support vector machines. in *Proc Tenth ACM SIGKDD Int Conf Knowledge Discovery and Data Mining.* Seattle, WA, ACM Press, Menlo Park, CA.
15. Sanchez, D. H., Redestig, H., Kramer, U., Udvardi, M. K. and Kopka, J. (2008) Metabolome-ionome-biomass interactions: What can we learn about salt stress by multiparallel phenotyping? *Plant Signal Behav* **3**, 598–600.
16. Hollywood, K., Brison, D. R. and Goodacre, R. (2006) Metabolomics: Current technologies and future trends. *Proteomics* **6**, 4716–4723.
17. Enot, D. P., Lin, W., Beckmann, M., Parker, D., Overy, D. P. and Draper, J. (2008) Preprocessing, classification modeling and feature selection using flow injection electrospray mass spectrometry metabolite fingerprint data. *Nat Protocols* **3**, 446–470.
18. Ye, J., Janardan, R., Li, Q. and Park, H. (2004) Feature extraction via generalized uncorrelated linear discriminant analysis. in *The Twenty-First Int Conf Machine Learning.* Banff, Alberta, ACM, New York, NY.
19. Lindon, J. C., Holmes, E. and Nicholson, J. K. (2001) Pattern recognition methods and applications in biomedical magnetic resonance. *Progress in Nuclear Magnetic Resonance Spectroscopy* **39**, 1–40.
20. Brown, M., Dunn, W. B., Ellis, D. I., Goodacre, R., Handl, J., Knowles, J. D., et al. (2005) A

metabolome pipeline: from concept to data to knowledge. *Metabolomics* **1**, 39–51.

21. Johnson, H. E., Broadhurst, D., Goodacre, R. and Smith, A. R. (2003) Metabolic fingerprinting of salt-stressed tomatoes. *Phytochem* **62**, 919–928.
22. Steuer, R., Morgenthal, K., Weckwerth, W. and Selbig, J. (2007) A Gentle Guide to the Analysis of Metabolomic Data, in *Metabolomics: Methods and Protocols* (W. Weckwerth, ed.), Humana Press, Totowa, NJ. pp. 105–126.
23. Sumner, L. W., Mendes, P. and Dixon, R. A. (2003) Plant metabolomics: large-scale phytochemistry in the functional genomics era. *Phytochem* **62**, 817–836.
24. Goodacre, R. (2007) Metabolomics of a Superorganism. *J Nutrition* **137**, 259–266.
25. Goodacre, R. (2005) Making sense of the metabolome using evolutionary computation: seeing the wood with the trees. *J. Exp Bot* **56**, 245–254.
26. Cuperlović-Culf M, Belacel N et al. (2009) NMR metabolic analysis of samples using fuzzy K-means clustering. *Magnetic Resonance in Chem* **47**, S96–S104.
27. Li, X., Lu, X., Tian, J., Gao, P., Kong, H. and Xu, G. (2009) Application of Fuzzy c-Means Clustering in Data Analysis of Metabolomics. *Anal Chem* **81**, 4468–4475.
28. Thakkar, D., Ruiz, C. and Ryder, E. F. (2007) Hypothesis-Driven Specialization of Gene Expression Association Rules. in *Proc 2007 IEEE Int Conf Bioinformatics and Biomedicine*. Fremont, CA, IEEE Computer Society.
29. Hipp, J., Güntzer, U. and Nakhaeizadeh, G. (2002) Data Mining of Association Rules and the Process of Knowledge Discovery in Databases, in *Advances in Data Mining* (P. Perner, ed.), Springer, Berlin/Heidelberg. pp. 207–226.
30. Agrawal, R., Imieliski, T. and Swami, A. (1993) Mining association rules between sets of items in large databases. in Proc 1993 ACM SIGMOD Int Conf on Management of Data. Washington, DC, ACM, New York, NY.
31. Gupta, R. K. and Agrawal, D. P. (2009) Improving the Performance of Association Rule Mining Algorithms by Filtering Insignificant Transactions Dynamically. *Asian J Information Management* **3**, 7–17.
32. Osl, M., Dreiseitl, S., Pfeifer, B., Weinberger, K., Klocker, H., Bartsch, G., et al. (2008) A new rule-based algorithm for identifying metabolic markers in prostate cancer using tandem mass spectrometry. *Bioinformatics* **24**, 2908–2914.
33. Yamamoto, H., Yamaji, H., Abe, Y., Harada, K., Waluyo, D., Fukusaki, E., et al. (2009) Dimensionality reduction for metabolome data using PCA, PLS, OPLS, and RFDA with differential penalties to latent variables. *Chemometrics & Intelligent Lab Sys* **98**, 136–142.
34. Kim, Y., Park, I. and Lee, D. (2007) Integrated Data Mining Strategy for Effective Metabolomic Data Analysis. in *Optimization and Systems Biology, The First Int Symp, OSB'07*. Beijing, China, ORSC & APORC.
35. Scholz, M., Gatzek, S., Sterling, A., Fiehn, O. and Selbig, J. (2004) Metabolite fingerprinting: detecting biological features by independent component analysis. *Bioinformatics* **20**, 2447–2454.
36. Scholz, M. and Selbig, J. (2006) Visualization and Analysis of Molecular Data, in *Metabolomics* (W. Weckwerth, ed.), Humana Press, NJ. pp. 87–104.
37. Mendes, P. (2002) Emerging bioinformatics for the metabolome. *Briefings Bioinformatics* **3**, 134–145.
38. Goodacre, R., Broadhurst, D., Smilde, A., Kristal, B., Baker, J., Beger, R., et al. (2007) Proposed minimum reporting standards for data analysis in metabolomics. *Metabolomics* **3**, 231–241.
39. Johnson, H., Lloyd, A., Mur, L., Smith, A. and Causton, D. (2007) The application of MANOVA to analyse Arabidopsis thaliana metabolomic data from factorially designed experiments. *Metabolomics* **3**, 517–530.
40. McGregor, M. (1997) Nuclear Magnetic Resonance Spectroscopy in *Handbook of instrumental techniques for analytical chemistry* (F.A. Settle, ed.), Prentice Hall, Upper Saddle River, NJ/London. pp. 309–337.
41. Brown, P. and DeAntonis, K. (1997) High-performance Liquid Chromotography, in *Handbook of instrumental techniques for analytical chemistry* (F.A. Settle, ed.), Prentice Hall, Upper Saddle River, NJ/ London. pp. 309–337.
42. Dettmer, K., Aronov, P. A. and Hammock, B. D. (2007) Mass spectrometry-based metabolomics. *Mass Spectrometry Rev* **26**, 51–78.
43. Dunn, W. B. and Ellis, D. I. (2005) Metabolomics: Current analytical platforms and methodologies. *Trends Anal Chem* **24**, 285–294.
44. Hites, R. A. (1997) Gas Chromotography Mass Spectrometry, in *Handbook of instrumental techniques for analytical chemistry* (F.A. Settle, ed.), Prentice Hall, Upper Saddle River, NJ/London. pp. 609–626.
45. Krishna, C., Sockalingum, G., Bhat, R., Venteo, L., Kushtagi, P., Pluot, M., et al. (2007) FTIR and Raman microspectroscopy of normal,

benign, and malignant formalin-fixed ovarian tissues. *Analytical & Bioanalytical Chem* **387**, 1649–1656.

46. Jain, A. K., Murty, M. N., et al. (1999). Data clustering: A review. *ACM Comput Surv* **31**(3), 264–323.
47. Sherman Hsu, C. P. (1997) Infrared Spectroscopy in *Handbook of instrumental techniques for analytical chemistry* (F.A. Settle, ed.), Prentice Hall, Upper Saddle River, NJ/London. pp. 309–337.
48. Xia, J., Psychogios, N., Young, N. and Wishart, D. S. (2009) MetaboAnalyst: a web server for metabolomic data analysis and interpretation. *Nucleic Acids Res* **37**, W652–660.
49. Spasic, I., Dunn, W., Velarde, G., Tseng, A., Jenkins, H., Hardy, N., et al. (2006) MeMo: a hybrid SQL/XML approach to metabolomic data management for functional genomics. *BMC Bioinformatics* 7, 281.
50. Sumner, L. W., Amberg, A., Barrett, D., Beale, M. H., Beger, R., Daykin, C. A., et al. (2007) Proposed minimum reporting standards for chemical analysis. *Metabolomics* **3**, 211–221.
51. Jenkins, H., Johnson, H., Kular, B., Wang, T. and Hardy, N. (2005) Toward supportive data collection tools for plant metabolomics. *Plant Physiol* **138**, 67–77.
52. Goebel, M. and Gruenwald, L. (1999) A survey of data mining and knowledge discovery software tools. *SIGKDD Explorations Newsletter.* **1**, 20–33.
53. Rokach, L. and Maimon, O. Z. (2008) *Data mining with decision trees: theory and applications.* Series in machine perception and artificial intelligence. Vol. 69. World Scientific, Singapore.
54. Clare, A. (2003) *Machine Learning and Data Mining for Yeast Functional Genomics* PhD. University of Wales, Aberystwyth
55. Michalski, R. S., Bratko, I. and Kubat, M. (1998) *Machine Learning and Data Mining: Methods and Applications.* John Wiley & Sons, Chichester, UK.
56. Pelckmans, K., De Brabanter, J., Suykens, J. A. K. and De Moor, B. (2005) Handling missing values in support vector machine classifiers. *Neural Networks* **18**, 684–692.
57. Jingke, X. (2008) Outlier Detection Algorithms in Data Mining. in *Intelligent Information Technology Application, 2008. IITA '08. Second International Symposium on.* Shanghai, IEEE Computer Society.
58. Chapman, P., Clinton, J., Kerber, R., Khabaza, T., Reinartz, T., Shearer, C., et al., *CRISP-DM 1.0 Step-by-step data mining guide.* 2000, SPSS Inc.
59. Wirth, R. and Hipp, J. (2000) CRISP-DM: Towards a Standard Process Model for Data Mining. in *Proc 4th Int Conf Practical Application of Knowledge Discovery and Data Mining.* Manchester, UK
60. Xia, J.m., Wu, X.j., and Yuan, Y.j. (2007) Integration of wavelet transform with PCA and ANN for metabolomics data-mining. *Metabolomics* **3**, 531–537.
61. Trochim, W. and Donnelly, J. (2007) *The Research Methods Knowledge Base.* 3rd ed. Atomic Dog Publishing.
62. Sansone, S., Rocca-Serra, P., Tong, W., Fostel, J., Morrison, N. and Jones, A. R. (2006) A Strategy Capitalizing on Synergies: The Reporting Structure for Biological Investigation (RSBI) Working Group. *OMICS: A J of Integrative Biology* **10**, 164–171.
63. Sansone, S., Rocca-Serra, P., Brandizi, M., Brazma, A., Field, D., Fostel, J., et al. (2008) The First RSBI (ISA-TAB) Workshop: Can a Simple Format Work for Complex Studies? *OMICS: A J of Integrative Biology* **12**, 143–149.
64. Smith, B., Ashburner, M., Rosse, C., Bard, J., Bug, W., Ceusters, W., et al. (2007) The OBO Foundry: coordinated evolution of ontologies to support biomedical data integration. *Nat Biotech* **25**, 1251–1255.
65. Langley, P., Shiran, O., Shrager, J., Todorovski, L. and Pohorille, A. (2006) Constructing explanatory process models from biological data and knowledge. *Artificial Intelligence in Medicine* **37**, 191–201.
66. Merriam-Webster Inc. (2005) *The Merriam-Webster dictionary.* Merriam-Webster, Springfield, MA.
67. Kell, D. B. (2004) Metabolomics and system Biology, making the Sense of the Soup. *Curr Opin Biotech* 7, 296–307.
68. Barrett, S. J. and Langdon, W. B. (2006) Advances in the Application of Machine Learning Techniques in Drug Discovery Design and Development. in *Applications of Soft Computing: Recent Trends.* Springer, Berlin/Heidleberg/New York, NY
69. Mahadevan, S., Shah, S. L., Marrie, T. J. and Slupsky, C. M. (2008) Analysis of metabolomic data using support vector machines. *Anal Chem* **80**, 7562–7570.
70. Chatterjee, S. and Hadi, A. S. (2006) *Regression analysis by example.* 4th ed. Wiley series in probability and statistics. Wiley-Interscience, Hoboken, N.J.

71. Fukusaki, E. and Kobayashi, A. (2005) Plant metabolomics: potential for practical operation. *J Bioscience and Bioengineering* **100**, 347–354.
72. Enot, D. P., Beckmann, M., Overy, D. and Draper, J. (2006) Predicting interpretability of metabolome models based on behavior, putative identity, and biological relevance of explanatory signals. *PNAS* **103**, 14865–14870.
73. Kotsiantis, S., Zaharakis, I. and Pintelas, P. (2006) Machine learning: a review of classification and combining techniques. *Artificial Intelligence Rev* **26**, 159–190.
74. Kotsiantis, S. B. (2007) Supervised Machine Learning a Review of Classification techniques. *Informatica* **31**, 249–268
75. Johnson, H. E., Gilbert, R. J., Winson, M. K., Goodacre, R., Smith, A. R., Rowland, J. J., et al. (2000) Explanatory Analysis of the Metabolome Using Genetic Programming of Simple, Interpretable Rules. *Genetic Programming & Evolvable Machines* **1**, 243–258.
76. Fiehn, O. (2001) Combining Genomics, Metabolome Analysis, and Biochemical Modelling to Understand Metabolic Networks. *Comparative & Functional Genomics* **2**, 155–168.
77. Taylor, J., King, R., Altmann, T. and Fiehn, O. (2002) Application of Metabolomics to Plant Genotype Discrimination Using Statistics and Machine Learning *BioInformatics* **18**, 241–248.
78. Catchpole, G. S., Beckmann, M., Enot, D. P., Mondhe, M., Zywicki, B., Taylor, J., et al. (2005) Hierarchical metabolomics demonstrates substantial compositional similarity between genetically modified and conventional potato crops. *PNAS* **102**, 14458–14462.
79. Wishart, D. S. (2008) Metabolomics: applications to food science and nutrition research. *Trends in Food Sci & Tech* **19**, 482–493.
80. Badjio, E. F. and Poulet, F. (2005) User Guidance: From Theory to Practice, the Case of Visual Data Mining. in *Proceedings of the 17th IEEE International Conference on Tools with Artificial Intelligence*. Hong Kong, IEEE Computer Society.
81. Camacho, D., de la Fuente, A. and Mendes, P. (2005) The origin of correlations in metabolomics data. *Metabolomics* **1**, 53–63.
82. Roessner-Tunali, U. (2007) uncovering the plant metabolome: current and future challenges, in *Concepts in Plant Metabolomics* (B.J. Nikolau and E.S. Wurtele, eds.), Springer, Dordrecht. pp. 71–85.
83. Xu, E., Schaefer, W. and Xu, Q. (2009) Metabolomics in pharmaceutical research and development: Metabolites, mechanisms and pathways. *Current Opinion in Drug Discovery & Development* **12**, 40–52.
84. Rozen, S., Cudkowicz, M. E., Bogdanov, M., Matson, W. R., Kristal, B. S., Beecher, C., et al. (2005) Metabolomic analysis and signatures in motor neuron disease. *Metabolomics* **1**, 101–108.
85. Broadhurst, D. and Kell, D. (2006) Statistical strategies for avoiding false discoveries in metabolomics and related experiments. *Metabolomics* **2**, 171–196.
86. Smelser, N. J. and Baltes, P. B. (2001) *International encyclopedia of the social & behavioral sciences*. 1st ed. Elsevier, Amsterdam/ New York, NY.

INDEX

Nigel W. Hardy and Robert D. Hall (eds.), *Plant Metabolomics: Methods and Protocols*, Methods in Molecular Biology, vol. 860, DOI 10.1007/978-1-61779-594-7, © Springer Science+Business Media, LLC 2012

MIX
Papier aus verantwortungsvollen Quellen
Paper from responsible sources
FSC® C105338

If you have any concerns about our products,
you can contact us on
ProductSafety@springernature.com

In case Publisher is established outside the EU,
the EU authorized representative is:
Springer Nature Customer Service Center GmbH
Europaplatz 3, 69115 Heidelberg, Germany

Printed by Libri Plureos GmbH
in Hamburg, Germany